高 等 学 校 教 材

环境科学与工程专业英语

第四版

华南理工大学　钟　理　主编

化学工业出版社

·北京·

全书共分为八个部分(PART),每个部分含若干个单元(Unit),共42个单元。每个单元由一篇课文和一篇阅读材料组成,共计84篇。其中第一部分(1~4单元)介绍环境、环境科学与工程概述和历史;第二部分(5~11单元)介绍大气化学和空气污染及其防治;第三部分(12~19单元)介绍水污染来源种类、水化学、各种污水处理技术、处理单元设备等;第四部分(20~23单元)为固体废物及处理方法;第五部分(24~27单元)介绍噪声污染与控制、全球气候变化、土壤污染与腐蚀、热污染等其他污染及控制技术;第六部分(28~35单元)为环境管理与政策,包括介绍环境影响及其评价、环境监控和分析及采样程序、环境政策与策略、环境风险评价、毒素和食品安全;第七部分(36~38单元)为生态系统与生态群落,介绍环境生物与生态圈,环境生态学,生态与生命系统与生物多样性等;第八部分(39~42单元)介绍环境可持续发展,绿色化学与技术,清洁生产,清洁生产工艺及其绿色化学过程。每篇课文均配有与课文相对应的练习,主要以主观练习题为主,包括阅读与词汇练习,英译汉与汉译英,用英语回答问题及写出课文或某一段落的小结(summary)等。为便于学生自学,本书每课配有单词和词组表,并对课文的难点作必要的注释,全书最后附有总词汇表。

本书根据《大学英语教学大纲》(理工科本科用)专业阅读部分的要求编写,是高等学校环境科学、环境工程或相关专业的教材,也可供同等英语程度环境科学工作者及环境工程师或相关领域的科技人员使用。

图书在版编目(CIP)数据

环境科学与工程专业英语/钟理主编. —4版. —北京:
化学工业出版社,2020.2(2025.5重印)
高等学校教材
ISBN 978-7-122-35635-2

Ⅰ.①环⋯ Ⅱ.①钟⋯ Ⅲ.①环境科学-英语-高等学校-教材②环境工程-英语-高等学校-教材 Ⅳ.①X

中国版本图书馆CIP数据核字(2019)第252606号

责任编辑:王文峡　　　　　　　　　装帧设计:张　辉
责任校对:王素芹

出版发行:化学工业出版社(北京市东城区青年湖南街13号　邮政编码100011)
印　　装:北京天宇星印刷厂
787mm×1092mm　1/16　印张18½　字数449千字　2025年5月北京第4版第8次印刷

购书咨询:010-64518888　　售后服务:010-64518899
网　　址:http://www.cip.com.cn
凡购买本书,如有缺损质量问题,本社销售中心负责调换。

定　价:49.00元　　　　　　　　　　　　　　　　　　　版权所有　违者必究

前　言

专业英语是高等院校理工科《英语教学大纲》所要求的内容，目的是使本科生在专业内容方面进行英语阅读的系统训练。对非英语专业的理工科学生而言，英语水平和英语使用的培养与提高不仅是综合素质的重要部分之一，而且也是能力的补充与文化素质的提高延伸。在该阶段的英语学习中，主要训练能正确、快速地阅读英语科技文献能力，初步学会专业英语写作与概括，掌握一定数量科技词汇及习惯用法，了解专业英语特点等，把学到的基础英语进行专业化训练，同时保持大学英语学习不断线。

本教材是在《环境科学与工程专业英语》（第三版）基础上，结合环境工程学科与其他学科的发展，根据目前高等院校教学改革与质量工程建设、环境科学与工程类专业英语学习及要求，对教材进行了修订，在保证篇幅不增加的前提下，对第二版教材内容进行了适当完善，增加环境科学与工程方面的一些新内容，包括什么是环境、环境历史、酸雨、废水高级氧化技术、全球气候变暖、毒素、食品安全、环境政策等。教材涉及环境科学与工程专业英语的词汇和常用的化学、化工等领域的科技词汇，词汇出现率较高。

为了使学生从基础英语学习过程的"单纯学习"转变到"使用"英语解决实际问题上来，教材的练习主要是以词汇、短语或句子的英译汉或汉译英，用英语回答问题或概括论文的主要观点，阅读文章后给出某一章节或段落或文章的摘要或小结（summary）等主观题型为主，而有别于基础英语学习采用客观选择题为主的练习，本书旨在为环境科学与工程类专业的学生提供一本比较系统的专业英语教学用书，达到强化该专业学生用书面英语表达科技信息和使用英语的能力。

本教材分为八个部分（PART）。每个部分含若干个单元（Unit），全书共42个单元。每个单元由一篇课文和一篇阅读材料组成。阅读材料提供与课文相应的背景知识或课文的续篇，以进一步拓宽课文的内容，根据课文内容，配有相应的练习题、注释和词汇表。课文和阅读材料均选自原版英文教科书、科技报告、著作、专业期刊、国际会议论文集等，题材较广，从纵横两个方面覆盖环境科学与工程专业的相关内容。其中：

PART1介绍环境、环境科学与工程概述和历史，环境的研究，城市环境，环境分析与环境经济学概括等；

PART2介绍大气化学和空气污染及其防治，包括大气化学，环境大气化学，酸雨，大气中的颗粒，大气中的氮化物及硫化物等的氧化过程，空气污染类型及来源，室内空气质量与污染，空气污染的一般性治理技术等；

PART3介绍水污染来源种类，水化学，各种污水处理技术，包括高级氧化技术，废水处理过程的单元设备及优化等；

PART4为固体废物及处理方法，包括固体废物种类及来源，有害物质及处理方法，固体废物处理过程与技术；

PART5介绍其他污染及控制技术，包括噪声及噪声控制，全球变化，土壤污染与腐蚀，热污染及控制等；

  PART6 为环境管理与政策，包括介绍环境影响及评价，环境数据库统计分析的设计，空气质量环境评价，环境监控在污染科学中的作用，环境监控分析及采样程序，空气质量监控，毒素，食品安全，环境政策与策略等；

  PART7 为环境生态系统与生态群落，介绍环境生物与生态圈，环境生态学，生态与生命系统及多样性，工业生态学等；

  PART8 为环境的可持续发展与绿色化学，包括可持续发展由来与困惑，绿色化学与技术，通过绿色化学与技术的可持续发展，清洁生产，替代溶剂，利用藻类获取经济效益和可持续发展的清洁生产技术等。

  书后附有总词汇表。为了便于学习，可参考《双语学习　英汉环境科学与工程专业词典》。词典给出词或词组的中文和英文，同时对重要和常用的词或词组还给出它们英文的定义、含义及释义。

  本教材由华南理工大学钟理主编。其中 PART1、PART3、PART7 和 PART8 由华南理工大学钟理编写，PART2 和 PART6 由浙江大学金一中、史惠祥及华南理工大学钟理编写，PART4 由郑州大学刘宏、魏新利及华南理工大学钟理编写，PART5 由四川大学王跃川及华南理工大学钟理编写，全书最后由钟理统稿，华南理工大学研究生张腾云、黄君涛和熊帆参加了文字编排、部分习题及注释。在编写过程中得到化工类及相关专业大学专业英语阅读教材编审委员会、华南理工大学教务处及化学工业出版社的大力支持。谨在此一并表示衷心感谢。

  本教材涉及内容较广，虽经多次补充完善，限于编者水平，书中不足之处在所难免，望读者不吝指正，使本书在使用过程中不断得到改进。

编　者

2019 年 6 月

第一版前言

组织编审出版系列的专业英语教材，是许多院校多年来共同的愿望。在高等教育面向 21 世纪的改革中，学生基本素质和实际工作能力的培养受到了空前重视。对非英语专业的学生而言，英语水平和能力的培养不仅是文化素质的重要部分，在很大程度上也是能力的补充和延伸。在此背景下，教育部（原国家教委）几次组织会议研究加强外语教学问题，制订有关规范，使外语教学更加受到重视。教材是教学的基本要素之一，与基础英语相比，专业英语教学的教材问题此时显得尤为突出。

国家主管部门的重视与广大院校的呼吁引起了化学工业出版社的关注，他们及时地与原化工部教育主管部门和全国化工类专业教学指导委员会请示协商后，组织全国十余所院校成立了本套专业英语教材编委会。在经过必要的调查和研究后，根据学校需求，编委会优先从各校教学（交流）讲义中确定选题，同时组织力量开展编审工作。本套教材涉及的专业主要包括化学工程与工艺、石油化工、机械工程、信息工程、工业过程自动化、应用化学、生物工程、环境工程、精细化工及制药工程、材料科学与工程、化工商贸等。

根据"全国部分高校化工类及相关专业大学英语专业阅读教材编审委员会"的要求和安排编写的《环境工程专业英语》教材，可供环境工程及相关专业本科生使用，也可作为同等程度（通过大学英语四级）的专业技术人员自学教材。

内容与结构　教材分为七部分（PART），每个部分含 4~5 个单元（Unit），共 29 个单元，每个单元由一篇课文和一篇阅读材料组成（第七部分除外）。阅读材料提供与课文相应的背景知识或是课文的续篇，以进一步拓宽课文内容。根据课文与阅读材料的内容，配有相应的练习题、注释和词汇表。课文与阅读材料共计 54 篇，均选自原版英文教科书、科技报告、著作、专业期刊、产品说明书、专利及文摘等。体裁较广，从纵横两个方面覆盖环境工程专业的相关内容。其中：

PART 1 为环境工程概述，包括环境工程的历史，环境的研究，城市环境，能源开采与环境，环境分析概况等；

PART 2 为空气污染及控制，包括空气污染物类型及来源，空气污染对气候及生态的影响，空气污染治理的一般性技术和新的处理方法；

PART 3 为水污染及废水处理，包括水污染来源及类型，废水处理技术及方法，废水处理装置控制及优化；

PART 4 为固体废物及处理，包括固体废物种类及来源，有害物质及处理方法，固体废物及能量回收；

PART 5 为其他污染及控制技术，包括声音与噪声，噪声控制，能耗与噪声，热污染及控制；

PART 6 为环境影响评价，包括环境影响评价概况，废水对河水影响评价，空气质量环境评价，噪声影响评价等；

PART 7 为专利、广告、说明书，包括专利文摘，化学文摘，销售广告，招聘广告，CD-ROM 的使用说明。

书后附有总词汇表。

词汇与练习 专业英语练习是高等院校理工科《英语教学大纲》所要求的内容，目的是使本科生在专业内容方面进行英语阅读的系统训练。在这阶段英语学习中，主要是提高学生正确、快速地阅读英语科技文献的能力，初步学会专业英语的写作方法，掌握一定数量的科技词汇及其习惯用法，了解专业英语的特点等，把学生学到的基础英语进行专业化训练。本教材包括环境工程专业英语词汇和相当数量的常用科技词汇，词汇复现率较高。习题设有词汇练习，以利学生掌握基本词汇。为使学生学习英语从"形式"用法提高到"实际"运用上来，练习主要以英译汉、汉译英、用英语回答问题及写出课文或某一段落的摘要等主观题型为主，而不是基础英语中的客观题（选择题）为主，从而强化学生用英语书面表达科技信息的能力。

致谢 本教材在编写过程中得到了化工类及相关专业大学英语专业阅读教材编审委员会、华南理工大学教务处、各编写单位以及化学工业出版社的大力支持。教材是四所院校的六位教师共同劳动的结晶。其中第一和第三部分由华南理工大学钟理编写，第二和第六部分由浙江大学金一中和史惠祥编写，第四和第七部分由郑州工业大学的刘宏和魏新利编写，第五部分由四川联合大学王跃川编写，全书最后由钟理统稿，大连理工大学周集体审阅了全书，并提出了许多宝贵意见，清华大学环境科学与工程系张晓健教授提供了详尽的改进意见，谨在此一并表示衷心感谢。由于时间所限，对张晓健教授提出需作较大调整的内容暂未能改动，在此向张教授致歉。本教材涉及内容较广，可能出现错漏，希望读者不吝指正，使本书在使用过程中不断得到改进。

编 者

1999 年 1 月

第二版前言

专业英语练习是高等院校理工科《英语教学大纲》所要求的内容，目的是使本科生在专业内容方面进行英语阅读的系统训练。在这阶段英语学习中，主要是提高学生正确、快速地阅读英语科技文献的能力，初步学会专业英语的写作方法，掌握一定数量的科技词汇及其习惯用法，了解专业英语的特点等，把学生学到的基础英语进行专业化训练。本教材包括环境工程专业英语词汇和相当数量的常用科技词汇，词汇复现率较高。为使学生学习英语从"形式"用法提高到"实际"运用上来，练习主要以句子、段落或词汇的英译汉或汉译英，用英语回答问题及写出课文或某一段落的摘要等主观题型为主，而不是基础英语中的客观题（选择题）为主，从而强化学生用英语书面表达科技信息的能力。

教材第二版分为九个部分（PART），每个部分含若干个单元（Unit），共37个单元，每个单元由一篇课文和一篇阅读材料组成。阅读材料提供与课文相应的背景知识或是课文的续篇，以进一步拓宽课文内容。根据课文与阅读材料的内容，配有相应的练习题、注释和词汇表。课文与阅读材料均选自原版英文教科书、科技广告、著作、专业期刊及国际论文集等，体裁较广，从纵横两个方面覆盖环境工程专业的相关内容。其中：

PART 1　为环境工程概述，包括环境工程的历史，环境的研究，城市环境，废物减量，环境分析概况等；

PART 2　为空气污染及控制，包括空气污染类型及来源，空气污染对气候及生态的影响，空气污染的一般性治理技术和新的处理方法等；

PART 3　为水污染及废水处理，包括水污染来源及类型，废水处理的各种技术及方法，废水处理单元设备及优化等；

PART 4　为固体废物处理，包括固体废物种类及来源，有害物质及处理技术与方法，固体废物处理过程中的能量与材料回收等；

PART 5　为其他污染及控制技术，包括噪声，噪声控制，能耗与噪声，热污染及控制等；

PART 6　为环境影响评价，包括环境影响评价概况，废水对河水影响评价，空气质量环境评价等；

PART 7　为环境监控，包括山区、陆地覆盖环境的监控，环境数据库统计分析的设计，以及沉积物和土壤的采样程序介绍等；

PART 8　为环境政策与管理，包括当前新的环境政策和污染管理策略，还包括固体废物以及核污染废物的管理等；

PART 9　为环境与可持续发展，包括微生物清洁生产技术、绿色化学合成工艺技术以及藻类产氢消除温室气体技术等。

书后附有总词汇表。为了便于学习，我们编写了配套的《双语教学　英汉环境科学与工程专业词典》。该词典收集了本专业常用的词汇、词组，并且运用简明的英文对概念进行解释。

本教材在编写过程中得到了化工类及相关专业大学英语专业阅读教材编审委员会、华南理工大学教务处、各编写单位以及化学工业出版社的大力支持。本教材第二版是在第一版基础上并结合当前形势发展，进行了较大的修改和补充。其中PART1、PART3、PART7、PART8和PART9由华南理工大学钟理编写，PART2和PART6由浙江大学金一中、史惠祥及华南理工大学钟理编写，PART4由郑州工业大学的刘宏和魏新利以及华南理工大学的钟理编写，PART5由四川联合大学王跃川及华南理工大学钟理编写，全书最后由钟理统稿，华南理工大学研究生张腾云、黄君涛和熊帆参加了文字编排、部分习题及注释，谨在此一并表示衷心感谢。

由于时间所限，本教材涉及内容较广，可能出现错漏，希望读者不吝指正，使本书在使用过程中不断得到改进。

编 者
2005年2月

第三版前言

专业英语学习是高等院校理工科《英语教学大纲》所要求的内容，目的是使本科生在专业内容方面进行英语阅读的系统训练。对非英语专业的理工科学生而言，英语水平和英语使用的培养与提高不仅是综合素质的重要部分之一，而且也是能力的补充与文化素质提高延伸。在该阶段英语学习中，主要是提高学生正确、快速地阅读英语科技文献能力，初步学会专业英语写作与概括，掌握一定数量科技词汇及习惯用法，了解专业英语特点等，把学生学到的基础英语进行专业化训练，同时保持大学英语学习不断线。本教材是在《环境工程专业英语》（第一版）和（第二版）基础上，结合21世纪环境工程学科与其他学科的发展，根据目前高等院校环境类专业英语学习及要求，对原来教材进行了改编，增加了环境科学方面的内容。教材涉及环境科学与环境工程专业英语的词汇和常用的化学、化工等领域的科技词汇，词汇出现率较高。为了使学生从基础英语学习过程"单纯学习"转变到"使用"英语解决实际问题上来，教材的练习主要是以词汇、短语或句子的英译汉或汉译英，用英语回答问题或概括论文的主要观点，阅读文章后给出某一章节或段落或文章的摘要或小结（summary）等主观题型为主，而有别于基础英语学习采用客观选择题为主的练习，本书旨在为环境类专业的学生提供一本比较系统的专业英语教学用书，达到强化该专业学生用书面英语表达科技信息和使用英语的能力。

本书分为八个部分（PART）。每个部分含若干个单元（Unit），共42个单元。每个单元由一篇课文和一篇阅读材料组成。阅读材料提供与课文相应的背景知识或课文的续篇，以进一步拓宽课文的内容，根据课文内容，配有相应的练习题、注释和词汇表。课文和阅读材料均选自原版英文教科书、科技报告、著作、专业期刊、国际会议论文集等，题材较广，从纵横两个方面覆盖环境科学与工程专业的相关内容。其中

PART1　介绍环境科学与工程概述和历史，环境的研究，废物减量，城市环境，环境分析概括等；

PART2　介绍大气化学和空气污染及其防治，包括大气化学，环境大气化学，大气中的颗粒，大气中的氮化物及硫化物等的氧化过程，空气污染类型及来源，室内空气，空气污染的一般性治理技术和新技术等；

PART3　介绍水污染来源种类，水化学，各种污水处理技术，废水处理过程的单元设备及优化等；

PART4　为固体废物及处理方法，包括固体废物种类及来源，有害物质及处理方法，固体废物处理过程的能量与材料回收；

PART5　介绍其他污染及控制技术，包括噪声及噪声控制，全球变化，土壤污染，热污染及控制等；

PART6　为环境管理，包括介绍环境影响及评价，环境数据库统计分析的设计，空气质量环境评价，环境监控在污染科学中的作用，环境监控分析及采样程序，空气质

量监控，环境政策与策略；

PART7 为环境生态系统与生态群落，介绍环境生物与生态圈，环境生态学，生态与生命系统，可持续资源利用的工业生态学等；

PART8 为绿色化学与环境的可持续发展，包括绿色化学与技术，通过绿色化学与技术的可持续发展，清洁生产，替代溶剂，藻类产氢消除温室气体技术等。

书后附有总词汇表。为了便于学习，我们编写了配套的《双语学习 英汉环境科学与工程专业词典》。词典给出词或词组的中文和英文，同时对重要和常用的词或词组还给出它们英文的定义、含义及释义。

本教材在编写过程中得到化工类及相关专业大学专业英语阅读教材编审委员会、华南理工大学及化学工业出版社的大力支持。本教材对《环境工程专业英语》第一版和第二版作了较大修改补充。其中PART1、PART3、PART7和PART8由华南理工大学钟理编写，PART2和PART6由浙江大学金一中、史惠祥及钟理编写，PART4由郑州工业大学的刘宏、魏新利及钟理编写，PART5由四川大学王跃川及钟理编写，全书最后由钟理统稿，华南理工大学的研究生张腾云、黄君涛和熊帆参加了文字编排、部分习题及注释，谨在此一并表示衷心感谢。

本教材涉及内容较广，虽经多次补充完善，限于编者水平，书中不足之处在所难免，希望读者不吝指正，使本书在使用过程中不断得到改进。

编 者
2011 年 8 月

Contents

PART 1 INTRODUCTION TO ENVIRONMENT AND ENVIRONMENTAL SCINECE AS WELL AS ENGINEERING ··· 1
  Unit 1 Text: What is Environment? ··· 1
          Reading Material: A Little Environmental History ················· 5
  Unit 2 Text: What are Environmental Science and Engineering? ········· 7
          Reading Material: Environmental Science ··························· 12
  Unit 3 Text: Environmental Engineering ··· 13
          Reading Material: Studying the Environment ······················ 17
  Unit 4 Text: Environmental Analysis ·· 20
          Reading Material: Overview of Environmental Economics ······ 24
Part 2 ATMOSPHERIC CHEMISTRY AND AIR POLLUTION AS WELL AS CONTROL ········ 27
  Unit 5 Text: Chemistry of the Atmosphere ···································· 27
          Reading Material: Reactions of Atmospheric Nitrogen and its Oxides ········ 30
  Unit 6 Text: Acid Rain ·· 32
          Reading Material: Reactions of Atmospheric Sulfur Compounds ············· 35
  Unit 7 Text: Introduction to Environmental Chemistry of the Atmosphere ············· 37
          Reading Material: Oxidation Process in the Atmosphere ······ 43
  Unit 8 Text: Type and Sources of Air Pollutants [I] ····················· 45
          Reading Material: Type and Sources of Air Pollutants [II] ····················· 47
  Unit 9 Text: Indoor Air Quality ··· 50
          Reading Material: Indoor Air Pollution ······························· 54
  Unit 10 Text: New Technologies of Air Pollution Control [I] ········· 56
          Reading Material: New Technologies of Air Pollution Control [II] ········· 60
  Unit 11 Text: Effects of Air Pollution ·· 63
          Reading material: Control of Air Pollution by Oxidation ······ 66
PART 3 WATER AND WASTE-WATER TREATMENT ···································· 69
  Unit 12 Text: Water Pollution and Pollutants ································· 69
          Reading Material: Wastewater ·· 72
  Unit 13 Text: Pollution of Inland Waters and Oceans ····················· 74
          Reading Material: Water Supply ··· 79
  Unit 14 Text: Water Purification ··· 83
          Reading Material: Principles of Wastewater Treatment ········ 87
  Unit 15 Text: Water Treatment Processes ······································· 89
          Reading Material: Freshwater Systems ································ 94
  Unit 16 Text: Biological Wastewater Treatment [I] ························· 96

Reading Material: Biological Wastewater Treatment [Ⅱ] ……………… 100
Unit 17　Text: Advanced Oxidation Processes (AOPs)—Wastewater
　　　　　Decontamination by Solar Photo-catalysis ……………………………… 104
　　　　　Reading Material: Precipitation ……………………………………………… 107
Unit 18　Text: Oxidation of Wastewater [Ⅰ] ……………………………………… 110
　　　　　Reading Material: Oxidation of Wastewater [Ⅱ] ………………………… 114
Unit 19　Text: Unit Operations of Pretreatment ………………………………… 118
　　　　　Reading Material: Wastewater Treatment ………………………………… 122

PART 4　SOLID WASTES AND DISPOSAL …………………………………… 125
Unit 20　Text: Sources and Types of Solid Wastes …………………………… 125
　　　　　Reading Material: Quantities of Wastes …………………………………… 129
Unit 21　Text: Everybody's Problems—Hazardous Waste …………………… 131
　　　　　Reading Material: Municipal Solid—Waste Management …………… 135
Unit 22　Text: Methods of Waste Disposal ……………………………………… 138
　　　　　Reading Material: Incineration of Hazardous Waste in the U. S. A. ……… 141
Unit 23　Text: Disposal of Solid Wastes ………………………………………… 144
　　　　　Reading Material: Emerging Technologies in Hazardous Waste
　　　　　Management: An Overview ………………………………………………… 148

PART 5　OTHER POLLUTION AND CONTROL TECHNOLOGIES …………… 151
Unit 24　Text: Noise Control [Ⅰ] ………………………………………………… 151
　　　　　Reading Material: Noise Control [Ⅱ] …………………………………… 152
Unit 25　Text: Global Change ……………………………………………………… 157
　　　　　Reading Material: Global Warming ………………………………………… 161
Unit 26　Text: Soil …………………………………………………………………… 163
　　　　　Reading Material: Soil Erosion ……………………………………………… 166
Unit 27　Text: Thermal Pollution ………………………………………………… 168
　　　　　Reading Material: The Major Greenhouse Gases ……………………… 169

PART 6　ENVIRONMENTAL MANAGEMENT AND POLICY ………………… 172
Unit 28　Text: Summary of Environmental Impact Assessment (EIA) …………… 172
　　　　　Reading Material: Introduction to Methods for Environmental
　　　　　Impact Assessment …………………………………………………………… 175
Unit 29　Text: Impact of Wastewater Effluents on Water Quality of River ……… 179
　　　　　Reading Material: The Aims and Objectives of Environmental
　　　　　Impact Assessment …………………………………………………………… 181
Unit 30　Text: Environmental Impact Assessment of Air Quality ……………… 186
　　　　　Reading Material: Risk Assessment ………………………………………… 189
Unit 31　Text: The Role of Environmental Monitoring in Pollution Science ……… 192
　　　　　Reading Material: Environmental Chemical Processes and Chemicals …… 195
Unit 32　Text: Toxins ………………………………………………………………… 197
　　　　　Reading Material: Toxins Basics …………………………………………… 200

Unit 33	Text: Sampling Sediment and Soil	203
	Reading Material: EMAP Overview—Objectives, Approaches, and Achievements	208
Unit 34	Text: Pollution Control Strategies [I]	211
	Reading Material: Pollution Control Strategies [II]	214
Unit 35	Text: Food Security	217
	Reading Material: Challenging Food Security	220

PART 7 THE BIOSPHERE: ECOSYSTEMS AND BIOLOGICAL COMMUNITIES 223

Unit 36	Text: Life and the Biosphere	223
	Reading Material: Nutrient Cycles for Ecosystem	226
Unit 37	Text: Ecology	230
	Reading Material: Benefits of Biodiversity	234
Unit 38	Text: Ecology and Life Systems	236
	Reading Material: The Five Major Components of an Industrial Ecosystem	240

PART 8 ENVIRONMENTAL SUSTAINABLE DEVELOPMENT AND GREEN SCIENCE AND TECHNOLOGY 243

Unit 39	Text: The Dilemma of Sustainability	243
	Reading Material: The Origins of Sustainable Development	247
Unit 40	Text: Sustainability	250
	Reading Material: Newer Synthetic Methods	255
Unit 41	Text: Green Science and Technology and Development	258
	Reading Material: Green Chemistry	262
Unit 42	Text: Clean Technologies through Microbial Processes for Economic Benefits and Sustainability	265
	Reading Material: Green Development: Reformism or Radicalism?	269

GLOSSARY 272

PART 1　INTRODUCTION TO ENVIRONMENT AND ENVIRONMENTAL SCINECE AS WELL AS ENGINEERING

Unit 1

Text：What is Environment?

The word "environment" is used in so many contexts that its intended meaning is not immediately clear. We hear about businesses that care about the "environment", we know people who are "environmentalists", we hear about "environmental change" in the news. We can think of the environment as the physical realm from which we draw resources to sustain our lives and our societies, but we have also created understandings of what the environment is and what our relationship to it is. Where is the environment, for example, in an aquarium? People visit an aquarium to see exotic fish and, presumably, to learn about the places from which those fist come. Aquarium developers have learned that people do not come to see "brown" fish. They want to see colorful fish from distant parts of the world that they may not actually be able to travel to themselves. What many aquarium visitors may not know is that demand for exotic, colorful fish has fostered illegal trade in fish which often depends on a practice known as cyanide fishing[①]. Cyanide fishing involves stunning fish with a blast of cyanide, usually issued with a plastic squeeze bottle by fishers who have few other options for employment. Stunned fish are easy to catch and sell to middlemen who then make these exotic fish available on a global black market. There is more than a single "environment" in this scenario. There is the environment of the aquarium, which is a constructed place where people expect to go and learn about "nature". Aquaria tend to foster the idea that there is a pristine environment "out there" that we can appreciate by looking at members of its ecosystem. Yet there is also the environment where fish are being stunned and sold and where people are engaging in this activity out of economic desperation. How did that environment come to be? The point is that there is no single "environment", but instead there are many environments demonstrating a complex interplay of "nature" and society.

　　The environment is considered the surrounding in which an organism operates, including air, water, land natural resources, flora, fauna, humans and their interaction. It is this environment which is both so valuable, on the one hand, and so endangered on the other. And it is people which are by and large ruining the environment both for themselves and for all other organisms.

What difference does it make if we thinks of ourselves—as people and as societies—as distinct or separate from our physical, "natural" environment?② It is an interesting question because we might imagine that if we are separate from the environment, there is some distance between our activities and their affects on other places. If we think of ourselves as distinct from our environment, we might also think that the distance between us and the environment allows us a degree of control or an ability to manipulate ecosystems without harming our own potential for well-being. This view of humans as separate from nature has roots in Judeo-Christian thinking and is evident in many ways in Western societies today. Even before industrialization took off in Britain, for example, Francis Bacon (1561—1626) promoted his view that "science would restore [man's] dominion over nature". This perspective encouraged the view that the physical, natural environment was something to be tamed and controlled for the benefit of society. We can see even in more recent times the idea that humans are separate and "in control" of the environment.

Gifford Pinchot, grandfather of the National Forest system in the US, is famous for coining the phrase "Conservation means the wise use of the earth and its resource… for the greatest good of the greatest number for the longest time"③. His view of forestry was that "Forestry is Tree Farming". Forestry is handling trees so that one crop follows another. To grow trees as a crop is Forestry. Nature, clearly, was something to be managed. When Pinchot looked upon the status of forests in the USA in late 1800s, he saw that there were no guidelines to how people were utilizing the natural resources of the country. He made it his mission to establish a theory and practice of forestry that would enable the continued production and use of natural resources for national economic benefit. After devoting over half of a century of his life to forestry, Pinchot shared his observation that "The earth and its resources belong of right to its people" and that "The first duty of the human race… is to control the use of the earth and all that therein is". ④ Clearly this view distinguishes the environment as distinct from humans and as something that should be managed and controlled by humans for the benefit of humans. This perspective persists to this day and is practiced by groups such as the US Forestry Service, which is based on a multiple use approach allowing hiking, timber harvesting, and sometimes mining in federally owned forests. Ducks Unlimited exemplifies a category of organizations that draw upon this philosophy too, as they seek to maintain habitat for ducks and other waterfowl for the benefit of recreational hunters. The very principle of conservation is that environmental resources can and should be managed for humans' economic benefit as well as for the maintenance of ecosystem services. This view necessarily distinguishes humans from the environment.

Geographers have long recognized humans' capacity to alter the environment. For example, George Perkins Marsh's book, The Earth as Modified by Human Action, published in 1874, is still cited today as a classic text of the discipline. Recently, we have become increasingly aware of the reach and depth of changes made to the physical environment by humans: by altering entire watersheds and ecosystems with chemical fertilizers; through repeated nuclear weapons testing, use, and nuclear accidents such as Chernobyl; by releasing industrial volumes of chemicals such as chlorofluorocarbons and sulfur dioxide into the air sufficient to alter the atmosphere at regional and global scales; by generating and emitting long-lasting, cross-media persistent organic pollutants that have negative health effects on people and animals from the tropics to the poles; by removing a significant percentage of fish from the oceans; by increasing energy consumption; by building dams and river diversion projects—just to name a few examples. Scholars have made the case that humans have essentially brought about a new geologic era known as the Anthropocene which is characterized by irreversible, human-induced change to the globe. Simon Dalby has observed that recognizing the Anthropocene opens up the opportunity to reconsider not only human-environment relationships but also the very basis of our understanding of spatial arrangements of power:

Security threats to modernity, long the preoccupation of the discipline of international relations, have usually assumed that threats are external to states, a matter of manipulation of external environments. But in the case of environment it is clear that such formulations are seriously misleading because it is the consequences of industrial production, and the appropriation of resources and displacement of populations as a result of these appropriations, which are causing the environmental changes that are supposedly a threat in the first place.

Taking the idea of the Anthropocene seriously, Dalby argues, gives us reason to examine how our economic and political systems have contributed to the irreversible manipulation of the air, water, and ecosystems on which we depend. He challenges us to reconsider ways in which ecological issues tend to enter into political dialogue and to pay attention to spatial patterns associated with those ecological issues. Where are the "haves" and the "have nots" when it comes to resource issues, and how does our society identify and prioritize environmental "problems" and the solutions? Who benefits from "our" environmental priorities, and which places or groups of people are left with few alternatives or limited options for improving their own chances for survival? How might we conceive of rearranging our systems of governance and ways in which political power and decision-making are distributed in recognition of humans' integral relationship with the environment?

Selected from "*Shannon O'Lear*, Environmental Politics: Scale and Power, *Cambridge University Press, New York, US 2010*"

Words and Expressions

exotic　*n.* 恶习（性）
cyanide fishing　氰化物捕集
stunning fish　令人眩晕的鱼
a blast of cyanide　氰化物爆炸
out there　向那边，到战场，现在的情况下
coining the phrase　解释词组，编造词组
all that therein　里面的一切都是

spatial patterns　空间模式，空间格局
the haves and the have nots　富人与穷人
prioritize　*v*. 优先处理

Notes

① What many aquarium visitors may not know is that demand for exotic, colorful fish has fostered illegal trade in fish which often depends on a practice known as cyanide fishing. 可译为：许多水族馆的参观者不知道的是，人们的恶习以及对多姿多彩鱼类的需求培育了非法鱼类贸易和交易，该贸易通常依赖于称为氰化物捕集的操作。

② What difference does it make if we thinks of ourselves—as people and as societies—as distinct or separate from our physical, "natural" environment? 可译为：如果我们把自己（作为人类和社会）看成是从我们的自然环境分离出来的话，会有什么区别？

③ Gifford Pinchot, grandfather of the National Forest system in the US, is famous for coining the phrase "Conservation means the wise use of the earth and its resource… for the greatest good of the greatest number for the longest time". 可翻译为：美国吉福德·平肖——国家森林系统之父，以提倡："保护（环境）意味着聪明地利用地球和它的资源，以便最长时间地获得最多和最大的好处利益"而著名。

④ After devoting over half of a century of his life to forestry, Pinchot shared his observation that "The earth and its resources belong of right to its people" and that "The first duty of the human race… is to control the use of the earth and all that therein is". 可译为：平肖将他的大半生奉献给了森林业，他分享了他的看法，"地球和地球资源理所当然属于人类，不同人种的首要任务是控制开发利用地球，所有活动都是如此。"

Exercises

1. Put the following into Chinese.

（1）Cyanide fishing involves stunning fish with a blast of cyanide, usually issued with a plastic squeeze bottle by fishers who have few other options for employment.

（2）Aquaria tend to foster the idea that there is a pristine environment "out there" that we can appreciate by looking at members of its ecosystem.

（3）He challenges us to reconsider ways in which ecological issues tend to enter into political dialogue and to pay attention to spatial patterns associated with those ecological issues.

（4）Who benefits from "our" environmental priorities, and which places or groups of people are left with few alternatives or limited options for improving their own chances for survival?

2. Put the following into English.

恶习　令人眩晕的鱼　氰化物捕集　向那边　富人与穷人

3. Based on the first paragraph

（1）what is the environment?

（2）where is the environment?

Reading Material: A Little Environmental History

A brief historical explanation will help clarify what we seek to accomplish. Before 1960, few people had ever heard the word ecology, and the word environment meant little as a political or social issue. Then came the publication of Rachel Carson's landmark book, Silent Spring. At about the same time, several major environmental events occurred, such as oil spills along the coasts of Massachusetts and southern California, and highly publicized threats of extinction of many species, including whales, elephants, and songbirds. The environment became a popular issue.

As with any new social or political issue, at first relatively few people recognized its importance. Those who did found it necessary to stress the problems—to emphasize the negative—in order to bring public attention to environmental concerns. Adding to the limitations of the early approach to environmental issues was a lack of scientific knowledge and practical know—how. Environmental sciences were in their infancy. Some people even saw science as part of the problem.

The early days of modern environmentalism were dominated by confrontations between those labeled "environmentalists" and those labeled "anti-environmentalists". Stated in the simplest terms, environmentalists believed that the world was in peril. To them, economic and social development meant destruction of the environment and ultimately the end of civilization, the extinction of many species, and perhaps the extinction of human beings. Their solution was a new worldview that depended only secondarily on facts, understanding, and science. In contrast, again in simplest terms, the anti-environmentalists believed that whatever the environmental effects, social and economic health and progress were necessary for people and civilization to prosper. From their perspective, environmentalists represented a dangerous and extreme view with a focus on the environment to the detriment of people, a focus they thought would destroy the very basis of civilization and lead to the ruin of our modern way of life.

Today, the situation has changed. Public-opinion polls now show that people around the world rank the environment among the most important social and political issues. There is no longer a need to prove that environmental problems are serious.

We have made significant progress in many areas of environmental science (although our scientific understanding of the environment still lags behind our need to know). We have also begun to create legal frameworks for managing the environment, thus providing a new basis for addressing environmental issues. The time is now ripe to seek truly lasting, more rational solutions to environmental problems.

The solution to specific environmental problems requires specific knowledge. The six themes listed above help us see the big picture and provide a valuable background. The opening case study illustrates linkages among the themes, as well as the importance of details.

Our Rapid Population Growth

The most dramatic increase in the history of the human population occurred in the last part of the 20th century and continues today into the early 21st century. As mentioned, in merely the past 40 years the human population of the world more than doubled, from 2.5 billion to about 6.8 billion. This shows the population explosion, sometimes referred to as the "population bomb".

Human population growth is, in some important ways, the underlying issue of the environment. Much current environmental damage is directly or indirectly the result of the very large number of people on Earth and our rate of increase. As you will see in chapter 4, where we consider the human population in more detail, for most of human history the total population was small and the average long-term rate of increase was low relative to today's growth rate.

Although it is customary to think of the population as increasing continuously without declines or fluctuations, the growth of the human population has not been a steady march. For example, great declines occurred during the time of the Black Death in the 14th century. At that time, entire towns were abandoned, food production declined, and in England one-third of the population died within a single decade.

Famine and Food Crisis

Famine is one of things that happen when a human population exceeds its environmental resources. Famines have occurred in recent decades in Africa. In the mid-1970s, following a drought in the Sahel region, 500,000 Africans starved to death and several million more were permanently affected by malnutrition. Starvation in African nations gained worldwide attention some 10 years later, in the 1980s.

Famine in Africa has had multiple interrelated causes. One, as suggested, is drought. Although drought is not new to Africa, the size of the population affected by drought is new. In addition, deserts in Africa appear to be spreading, in part because of changing climate but also because of human activities. Poor farming practices have increased erosion, and deforestation may be helping to make the environment drier. In addition, the control and destruction of food have sometimes been used as a weapon in political disruptions. Today, malnutrition contributes to the death of about 6 million children per year. Low- and middle-income countries suffer the most from malnutrition, as measured by low weight for age.

The emerging global food crisis in the first decade of the 21st century has not been caused by war or drought but by rising food costs. The cost of basic items, such as rice, corn, and wheat, has risen to the point where low- and moderate-income countries are experiencing a serious crisis. In 2007 and 2008, food riots occurred in many locations, including Mexico, Haiti, Egypt, Yemen, Bangladesh, India, and Sudan. The rising cost of oil used to produce food (in fertilizer, transportation, working fields, etc.) and the conversion of some corn production to biofuels have been blamed. This situation involves yet another key theme:

science and value. Scientific knowledge has led to increased agricultural production and to a better understanding of population growth and what is required to conserve natural resources. With this knowledge, we are forced to confront a choice: Which is more important, the survival of people alive today or conservation of the environment on which future food production and human life depend?

Answering this question demands value judgments and the information and knowledge with which to make such judgments. For example, we must determine whether we can continue to increase agricultural production without destroying the very environment on which agriculture and, indeed, the persistence of life on Earth depend. Put another way, a technical, scientific investigation provides a basis for a value judgment.

The human population continues to grow, but humans' effects on the environment are growing even faster. People cannot escape the laws of population growth. The broad science-and-values question is: What will we do about the increase in our own species and its impact on our planet and on our future?

Selected from " *Daniel B. Botkin, Edward A. Keller. Environmental Science, Eight Edition, John Wiley & Sons, Inc. Printed in Asia, 2012*"

Unit 2

Text: What are Environmental Science and Engineering?

Natural Science

In the broadest sense, science is systematized knowledge derived from and tested by recognition and formulation of a problem, collection of data through observation, and experimentation. We differentiate between social science and natural science in that the former deals with the study of people and how they live together as families, tribes, communities, races, and nations, and the latter deals with the study of nature and the physical world. Natural science includes such diverse disciplines as biology, chemistry, geology, physics, and environmental science.

Environmental Science

Environmental science is a group of sciences that attempt to explain how life on the Earth is sustained, what leads to environmental problems, and how these problems can be solved. **Environmental science** is the study of how the natural world works, how our environment affects us, and how we affect our environment. We need to understand how we interact with our environment in order to devise solutions to our most pressing challenges. It can be daunting to reflect on the sheer magnitude of environmental dilemmas that confront us today, but these problems also bring countless opportunities for creative solutions.

Whereas the disciplines of biology, chemistry, and physics (and their subdisciplines of microbiology, organic chemistry, nuclear physics, etc.) are focused on a particular aspect of natural science, environmental science in its broadest sense encompasses all the fields of natural science. The historical focus of study for environmental scientists has been, of course, the natural environment. By this, we mean the atmosphere, the land, the water and their inhabitants as differentiated from the built environment. Modern environmental science has also found applications to the built environment or, perhaps more correctly, to the effusions from the built environment.

What is the "Science" in Environmental Science?

Many sciences are important to environmental science. These include biology (especially ecology, that part of biology that deals with the relationships among living things and their environment), geology, hydrology, climatology, meteorology, oceanography, and soil science.

Quantitative Environmental Science

Science or, perhaps more correctly, the scientific method, deals with data, that is, with recorded observations. The data are, of course, a sample of the universe of possibilities. They may be representative or they may be skewed. Even if they are representative, they will contain some random variation that cannot be explained with current knowledge. Care and impartiality in gathering and recording data, as well as independent verification, are the cornerstones of science[1].

When the collection and organization of data reveal certain regularities, it may be possible to formulate a generalization or hypothesis[2]. This is merely a statement that under certain circumstances certain phenomena can generally be observed. Many generalizations are statistical in that they apply accurately to large assemblages but are no more than probabilities when applied to smaller sets or individuals.

In a scientific approach, the hypothesis is tested, revised, and tested again until it is proven acceptable.

If we can use certain assumptions or tie together a set of generalizations, we formulate a theory. For example, theories that have gained acceptance over a long time are known as laws. Some examples are the laws of motion, which describe the behavior of moving bodies, and the gas laws, which describe the behavior of gases. The development of a theory is an important accomplishment because it yields a tremendous consolidation of knowledge. Furthermore, a theory gives us a powerful new tool in the acquisition of knowledge for it shows us where to look for new generalizations. Thus, the accumulation of data becomes less of a magpie collection of facts and more of a systematized hunt for needed information. It is the existence of classification and generalization, and above all theory that makes science an organized body of knowledge.

Logic is a part of all theories. The two types of logic are qualitative and quantitative logic[3]. Qualitative logic is descriptive. For example we can qualitatively state that when the amount of wastewater entering a certain river is too high, the fish die. With qualitative logic we cannot identify what "too high" means—we need quantitative logic to do that.

When the data and generalizations are quantitative, we need mathematics to provide a theory that shows the quantitative relationships. For example, a quantitative statement about the river might state that "when the mass of organic matter entering a certain river equals x kilograms per day, the amount of oxygen in the stream is y."

Perhaps more importantly, quantitative logic enables us to explore "What if?" questions about relationships. For example, "if we reduce the amount of organic matter entering the stream, how much will the amount of oxygen in the stream increase?" Furthermore, theories, and in particular, mathematical theories, often enable us to bridge the gap between experimentally controlled observations and observations made in the field[4]. For example, if we control the amount of oxygen in a fish tank in the laboratory, we can determine the minimum amount required for the fish to be healthy. We can then use this number to determine the acceptable mass of organic matter placed in the stream.

Given that environmental science is an organized body of knowledge about environmental relationships, then quantitative environmental science is an organized collection of mathematical theories that may be used to describe and explore environmental relationships.

How Is Environmental Science Different from other Sciences?

It involves many sciences.

It includes sciences, but also involves related nonscientific fields that have to do with how we value the environment, from environmental philosophy to environmental economics.

It deals with many topics that have great emotional effects on people, and there are subject to political debate and to strong feelings that often ignore scientific information.

Engineering

Engineering is a profession that applies science and mathematics to make the properties of matter and sources of energy useful in structures, machines, products, systems, and processes.

Environmental Engineering

The Environmental Engineering Division of the American Society of Civil Engineers (ASCE) has published the following statement of purpose that may be used to show the relationship between environmental science and environmental engineering:

Environmental engineering is manifest by sound engineering thought and practice in the solution of problems of environmental sanitation, notably in the provision of safe, palatable, and ample public water supplies[5]; the proper disposal of or recycle of wastewater and solid wastes; the adequate drainage of urban and rural areas for proper sanitation; and the control of water, soil, and atmospheric pollution, and the social and environmental impact of these solutions. Furthermore it is concerned with engineering problems on the field of public health, such as control of arthropod-borne diseases, the elimination of industrial health hazards, and the provision of adequate sanitation in urban, rural, and recreational areas, and the effect of technological advances on the environment.

Neither environmental science nor environmental engineering should be confused with heating, ventilating, or air conditioning (HVAC), nor with landscape architecture. Neither should they be confused with the architectural and structural engineering functions associated with built environments, such as homes, offices, and other workplaces.

Selected from *"Mackenzie L Davis, Susan J Masten. Principles of Environmental Engineering and Science, The McGraw-Hill Companies, Inc. USA 2004"*

Words and Expressions

tribes *n.* 部落，群落
communities *n.* 社区
subdisciplines *n.* 分支学科
effusions *n.* 流出物，出口流体
quantitative *a.* 定量的
qualitative logic 定性逻辑学
quantitative logic 定量逻辑学
skewed *n./adj.* 偏离，曲解
cornerstones *n.* 基石
hypothesis *n.* 假想，假设
consolidation *n.* 加强，协同，合并，凝固
magpie collection 胡乱收集，混杂收集
logic *n.* 逻辑学
bridge the gap 填补差距
manifest *v./a.* 表明，显示
palatable *a./ad.* 可口的，受欢迎的
hazards *n.* 有害物

Notes

① Care and impartiality in gathering and recording data, as well as independent verification, are the cornerstones of science. 可译为：在数据收集与记录过程中仔细认真、无偏见和独立核实是科学的基石。

② When the collection and organization of data reveal certain regularities, it may be possible to formulate a generalization or hypothesis. 可译为：当对数据收集和整理披露了某些规律时，可能归纳出概况或假设。

③ The two types of logic are qualitative and quantitative logic. 可译为：两类逻辑学分别为

定性与定量逻辑学。

④ Theories, and in particular, mathematical theories, often enable us to bridge the gap between experimentally controlled observations and observations made in the field. 可译为：理论，尤其是数学理论，通常使我们可以弥补（缩小）实验过程观察的结果与该领域理论推导的结果间的差距。

⑤ Environmental engineering is manifest by sound engineering thought and practice in the solution of problems of environmental sanitation, notably in the provision of safe, palatable, and ample public water supplies; 可译为：环境工程是在解决环境卫生问题过程中，具有代表性的，提供安全、可口与充足的公共供水过程，用正确的工程思想与合理的实施显现出来。

Exercises

1. Put the following words or phrases into Chinese.

Biology, races, disciplines, encompass, inhabitant, built environment, sound engineering, statistical, probability, organic matter, ample public water supply, disposal of wastewater

2. Put the following into English.

| 协同 | 胡乱收集 | 曲解 | 基石 |
| 流出物 | 假想 | 可口的 | 生物学 |

3. Translate the following passage into Chinese.

　　The environmental physical sciences have traditionally been concerned with individual environmental compartments. Thus, geology is centred primarily on the solid earth, meteorology on the atmosphere, oceanography upon the salt-water basins, and hydrology upon the behaviour of freshwaters. In general (but not exclusively) it has been the physical behaviour of these media which has been traditionally perceived as important. Accordingly, dynamic meteorology is concerned primarily with the physical processes responsible for atmospheric motion, and climatology with temporal and spatial patterns in physical properties of the atmosphere (temperature, rainfall, etc.). It is only more recently that chemical behaviour has been perceived as being important in many of these areas. Thus, while atmospheric chemical processes are at least as important as physical processes in many environmental problems such as stratospheric ozone depletion, the lack of chemical knowledge has been extremely acute as atmospheric chemistry (beyond major component ratios) only became a matter of serious scientific study in the 1950s.

4. What is the difference between social science and natural science based on the text? What is the difference between qualitative logic and quantitative logic according to the text?

5. Give a brief summary of Natural and Environmental Science.

6. What is the difference between Environmental Science and Environmental Engineering based on the text?

Reading Material: Environmental Science

It may surprise the student of today to learn that "the environment" has not always been topical and indeed that environmental issues have becomes a matter of widespread public concern only over the past 20 years or so. Nonetheless, basic environmental science has existed as a facet of human scientific endeavour since the earliest days of scientific investigation. In the physical sciences, disciplines such as geology, geophysics, meteorology, oceanography, and hydrology, and in the life sciences, ecology, have a long and proud scientific tradition. These fundamental environmental sciences underpin our understanding of the natural world and its current-day counterpart perturbed by human activity, in which we all live.

The environmental physical sciences have traditionally been concerned with individual environmental compartments. Thus, geology is centred primarily on the solid earth, meteorology on the atmosphere, oceanography upon the salt-water basins, and hydrology upon the behaviour of freshwaters. In general (but not exclusively) it has been the physical behaviour of these media which has been traditionally perceived as important. Accordingly, dynamic meteorology is concerned primarily with the physical processes responsible for atmospheric motion, and climatology with temporal and spatial patterns in physical properties of the atmosphere (temperature, rainfall, etc.). It is only more recently that chemical behaviour has been perceived as being important in many of these areas. Thus, while atmospheric chemical processes are at least as important as physical processes in many environmental problems such as stratospheric ozone depletion, the lack of chemical knowledge has been extremely acute as atmospheric chemistry (beyond major component ratios) only became a matter of serious scientific study in the 1950s.

There are two major reasons why environmental chemistry has flourished as a discipline only rather recently. Firstly, it was not previously perceived as important. If environmental chemical composition is relatively invariant in time, as it was believed to be, there is little obvious relevance to continuing research. Once, however, it is perceived that composition is changing (e. g. CO_2 in the atmosphere; ^{137}Cs in the Irish Sea) and that such changes may have consequences for humankind, the relevance becomes obvious. The idea that using an aerosol spray in your home might damage the stratosphere, although obvious to us today, would stretch the credulity of someone unaccustomed to the concept. Secondly, the rate of advance has in many instances been limited by the available technology. Thus, for example, it was only in the 1960s that sensitive reliable instrumentation became widely available for measurement of trace concentrations of metals in the environment. This led to a massive expansion in research in this field and a substantial downward revision of agreed typical concentration levels due to improved methodology in analysis. It was only as a result of James Lovelock's invention of the electron capture detector that CFCs were recognized as minor atmospheric constituents and it became possible to monitor increases in their concentrations, as shown in Table 1. The table exemplifies the sensitivity of analysis required since concentrations are at the ppt level (1ppt is one part in 10^{12} by volume in the atmosphere) as well as the substantial increasing trends in atmospheric halocarbon concentrations, as measured up to 1990. The implementation of the Montreal Protocol, which requires controls on production of CFCs and some other halocarbons, has led to a slowing and even a reversal of annual concentration trends since 1992.

Table 1 Atmospheric halocarbon concentrations and treads

Halocarbon	Concentration(ppt)		Annual change(ppt)		Lifetime(years)
	Pre-industrial	2000	To 1999	1999~2000	
CCl_3F	0	261	+9.5	−1.1	50
CCl_2F_2	0	543	+16.5	+2.3	102
$CClF_3$ (CFC-13)	0	3.5			400
$C_2Cl_3F_3$ (CFC-113)	0	82	+4~5	−0.35	85
$C_2Cl_2F_4$ (CFC-114)	0	16.5			300
C_2ClF_5 (CFC-115)	0	8.1		+1.6	1700
CCl_4	0	96.1	+2.0	−0.94	42
CH_3CCl_3	0	45.4	+6.0	−8.7	4.9

Selected from *"Roy M Harrison. Principles of Environmental Chemistry, The Royal Society of Chemistry. UK, 2007"*

Unit 3

Text: Environmental Engineering

What Is This Book About?

The objective of this book is to introduce engineering and science students to the interdisciplinary study of environmental problems: their causes, why they are of concern, and how we can control them. The book includes:
- Description of what is meant by environment and by environmental systems
- Information on the basis causes of environmental disturbances
- Basis scientific knowledge necessary to understand the nature of environmental problems and to be able to quantify them
- Current state of the technology environmental control in its application to water, air and pollution problems
- Considerable gaps in our current scientific knowledge of understanding and controlling many of the complex interactions between human activities and nature
- Many environmental problems which could be eliminated or reduced by the application of current technology, but which are not dealt with because of society's lack of will to do so, or in many instances because of a lack of resources to do so.

Some Important Definitions

Where they are first used in this book, definitions are introduced in block form, as shown here, or printed in bold type.

> **Environment** is the physical and biotic habitat which surrounds us; that which we can see, hear, touch, smell, and taste.
>
> **System**, according to Wehster's dictionary[①], is defined as "a set or arrangement of things so related or connected as to form a unit or organic whole; as, a solar system, irrigation system, supply system, the world or universe".
>
> **Pollution** can be defined as an undesirable change in the physical, chemical, or biological characteristics of the air, water, or land that can harmfully affect the health, survival, or activities of humans or other living organisms.

When the goal of improving environmental quality is taken to be improving human well-being, the word "environment" broadens to include all kinds of social, economic, and cultural aspects. Such broadness is unworkable in many real situations and impractical in a textbook designed for a one-semester course. Our examination of environmental problems is therefore limited by our definition of "environment".

Interaction of Systems

A number of different environmental problems are associated with water, air, or land systems. Many of these problems will apply only within one of these systems, justifying the breakdown into these categories. Such a classification is also useful for easier comprehension of related problems within one system. Moreover, it is sensible because, for managerial and administrative reasons[②], such subfields as air pollution, water supply, wastewater disposal, and solid waste disposal are often dealt with separately by governmental agencies.

Unfortunately, many important environmental problems are not confined to an air, water, or land system, but involve interactions between systems. A current example is the acid rain problem stemming from the emission of sulfur dioxide and nitrogen oxide gases into the atmosphere from the stacks of generating stations[③], smelters, and automobile exhausts. These gases are then transported by air currents over wide regions. Rainfall "washes them out", creating acid rain which is harmful to aquatic life, forests, and agricultural crops. Two examples of interaction between systems that cause major environmental disturbances are presented-the buildup of atmospheric carbon dioxide, a global problem, and the acid rain problem, normally of regional nature.

Environmental Disturbances

Many major improvements to our standard of living can be attributed to the application of science and technology. A few examples are noted here. Can you think of others?

- The production of more and better quality food.
- The creation of housing as protection from extremes from climates and as living

space.
- The building of fast and reliable means of transportation.
- The invention of various systems of communication.
- The invention of machines to replace human or animal power.
- The supply of safe water and the disposal of wastes.
- The elimination of many infectious diseases.
- The elimination of most water-borne diseases in the developed world through improved water technology.
- The availability of leisure time through greater productivity, providing the opportunity for cultural and recreational activities.
- The protection from the worst effects of natural disasters such as floods, droughts, earthquakes, and volcanic eruptions.

With these improvements, however, have come disturbing side effects, such as lost arable land, disappearing forests, environmental pollution, and new organisms resistant to controls. Many effects originally considered to be just nuisances are now recognized as potential threats to nature and to humans. In an agrarian society, people lived essentially in harmony with mature, raising food, gathering firewood, and making clothing and tools from the land. The wastes from animals and humans were returned to the soil as fertilizer. Few, if any, problems of water, land, or air pollution occurred.

The cities of ancient times, particularly those of the Roman Empire[4], had systems to supply water and to dispose of wastes. The aqueducts supplying the ancient city of Rome (population about 1 million) with safe water from the Cloaca Maxima[5], the best known and one of the earliest sewers to be built, are examples of such systems. The municipal technology of ancient cities seems to have been forgotten for many centuries by those who built cities throughout Europe. Water supply and waste disposal were neglected, resulting in many outbreaks of dysentery, cholera, typhoid, and other waterborne diseases. Until the middle of the nineteenth century, it was not realized that improper wastes disposal polluted water supplies with disease-carrying organisms. The industrial revolution in nineteenth-century Britain, Europe, and North America aggravated the environmental problems since it brought increased urbanization with the industrialization. Both phenomena, urbanization and industrialization, were and are fundamental causes of water and air pollution which the cities of that time were unable to handle.

Rapid advances in technology for the treatment of water and the partial treatment of wastewater took place in the developed countries over the next few decades. This led to a dramatic decrease in the incidence of waterborne diseases[6]. Note that all wastes discharge into the environment, and thus pollute our water, air, and land systems.

Selected from "*Henry, Gray W. Heinke. Environmental Science and Engineering, Prentice-Hall International Editions, Pretice. Hall Englewood Cliffs, NJ, USA, 1989*"

Words and Expressions

smelter　*n*. 熔炉；冶金厂，冶炼者
aqueduct　*n*. 渠；水管
dysentery　*n*. 痢疾
cholera　*n*. 霍乱
typhoid　*n*. 伤寒，似班伤疹伤寒；*a*. 伤寒的
sulfur dioxide　二氧化硫
nitrogen oxide　氧化氮
carbon dioxide　二氧化碳
arable　*a*. 可耕的
agrarian　*a*. 土地的；农民的；农业的
urbanization　*n*. 城市化
harmony　*n*. 协调，一致
environmental disturbance　环境破坏
expose the considerable gaps　揭示大的差别
aquatic life　水生物
discharge　*v*. 排出；*n*. 排出物

Notes

① Webster's dictionary　韦氏词典，由美国 Merrian-Webster 公司出版。该公司出版的《韦氏国际英语大词典》和《韦氏大学词典》等是世界公认的权威性工具书。
② for managerial and administrative reasons　行政管理上的原因。
③ stacks of generating stations　许多发电站，这里 stacks of 相当于 lots of。
④ Roman Empire　罗马皇帝。
⑤ Cloaca Maxima　古罗马的大排泄沟，污秽物的普通储藏库（the main drain of ancient Rome）。
⑥ waterborne diseases　水传染的疾病，如痢疾、伤寒等。

Exercises

1. Put the following into Chinese.
   life expectancy, poverty-stricken, smog-laden air, global conditions, haves and have-nots, underprivileged, savanna, predator, environmental disruptions

2. Put the following into English.

农药	化肥	有机废物	微生物	衰减
阻滞的	稀释	添加剂	合成塑料	再生

3. Translate the part of "The Human Condition" into Chinese.
4. How many kinds of environmental problems are there according to Reading Material?

Reading Material: Studying the Environment

The Human Condition

Consider the state of humankind on Earth, the planet we call home. In the past, life was hard and short for most people. Then, at the beginning of the twentieth century, rapid progress in medicine, agriculture, and industrial techniques seemed to promise that everyone might soon be able to enjoy long life, decent food, satisfying employment, and adequate housing. This promise has not been realized. In 1982 there were about 5 billion people on Earth. One quarter of them had a greater life expectancy and lived in greater luxury than anyone a hundred years ago would have believed possible. But at the same time three quarters of all people had inadequate or unsatisfactory water and shelter, and more than one third suffered from malnutrition and hunger. More people starved to death in 1982 than in any year since the beginning of time.

Most of the people who starve to death live in poverty-stricken developing nations. But now, at the end of the twentieth century, the wealthy nations are in trouble too. According to most estimates, the average standard of living in North America and Western Europe peaked in about 1967. Even the wealthiest nations are running out of fuel, hardwoods, and some minerals. As a result, necessities such as housing, food, and fuel are demanding more and more of the family budget, leaving less available for luxuries. Pollutants contaminate cities, towns, and even rural environments. Sewage or poisonous pesticides in waterways, smog-laden air, and garbage in streets or parks lower the standard of living of everyone, no matter how wealthy.

Our vast and swiftly growing population consumes the Earth's resources of agricultural land, minerals, water, and fuel faster than natural processes can replace them. This is a serious problem in itself. But global conditions are further endangered because these resources are not distributed evenly. One quarter of the population, those that live in the developed nations, use nearly 80 percent of the resources consumed by humankind in any one year. The other three quarters of the population consume only about 20 percent of the resources used in a year. The gap between the haves and the have-nots is growing wider. Today, when transistor radios are to be found even in the most remote African villages, the world's poor know how underprivileged they are. This knowledge produces political instability. Years ago, political upheaval in one nation meant little to the rest of the world. But times have changed, because modern technology ensures that nearly all nations possess weapons that can wreak havoc far beyond their own borders. As a result, no nation can afford to ignore the problems of another.

Classifying Environmental Problems

Environmental problems are always interrelated. Sometimes a solution to one problem actually creates another problem. For example, when people are sick and dying from disease, it is natural to want to improve human health. When health is improved and infant mortality is reduced, a population explosion may result. To feed this growing pollution, natural habitats are often destroyed by turning them into farmland. As natural habitats are destroyed, the wild plants, predatory animals, and parasites that once lived there are killed as well. Because of the lack of predators and parasites, outbreaks of insect pests become more common. Farmers use pesticides to control the pests and protect the crops, but in the process the environment becomes polluted. The development of this entire cycle in itself consumes fossil fuel supplies that are becoming scarce. In addition, when fuels are burned, air pollutants are generated.

How does a person begin to study such a network of interlocking problems? To make the task a bit more manageable, we will divide environmental disruptions into five main types.

(a) Overpopulation. Overpopulation may be defined as the presence in a given area of more people than can be supported adequately by the resources available in that area. Many people argue that the population explosion that has taken place in the twentieth century is now the most important problem we face. It is important first because overpopulation is a major cause of all other environmental problems: Fewer people would use less oil, chop down fewer trees, and pour less sewage into rivers. Second, overpopulation and the starvation that accompanies it are generally higher on our list of priorities than other environmental concerns. It is hard to argue that an area should be set aside as parkland to preserve a vanishing forest or savanna when that land might be used to raise crops that would prevent fellow human beings from starving to death.

(b) Pollution. Pollution is a reduction in the quality of the environment by the introduction of impurities. Smoke pollutes the air; sewage pollutes the waters; junk cars pollute the land. We know that such contamination exists; it can be seen, smelled, or even tasted. The effects of pollution on human welfare or on the economy, however, may be matters of considerable disagreement.

There are two distinctly different types of pollution.

1. *Concentration of Organic Wastes.* All living organisms produce waste products; wastes are associated with the act of living. Upon death, the entire organism becomes a waste product. Before modern civilization, most organic wastes did not accumulate in the environment because they were consumed by other organisms and thereby recycled. In modern times, the natural decomposition of organic wastes does not always operate efficiently.

2. *Introduction of Synthetic Chemicals into the Environment.* Everything is make of chemicals-people, eagles, trees, lakes, plastic-everything. Although many natural chemical compounds have existed for billions of years, people have recently learned to make new chemical compounds, called **synthetic chemicals**. The quantity and variety of new synthetic chemicals are staggering. They are present in paints, dyes, food additives, drugs, pesticides, fertilizers, fire retardants, building materials, clothes, cleaning supplies, cosmetics, plastics, and so on.

Synthetic chemicals are noted for the variety of their properties. Some of them are drugs that save millions of lives every year, and others are poisons. But because most of them are new to the environment, the traditional patterns of decay and recycling do not necessarily apply. Some synthetic chemicals break down rapidly in the environment by the action of sunlight, air, water, or soil, and some are eaten by living organisms. Such processes may take place over a span of minutes, hours, or days. A material that decomposes in the environment as a result of biological action is said to be **biodegradable.**

Many compounds, however, do not disappear so readily. Synthetic plastics, for example, remain in the environment for a long time because organisms that feed on them and break them down are rare. Plastic shampoo bottles may produce unsightly litter, but they are not biologically active. Many other synthetic chemicals, however, are harmful. For example, DDT was developed as an insecticide, but experience with it has shown that it has undesirable environmental effects as well.

(c) **Depletion of Resources.** A resource is any source of raw materials. Fuels, minerals, water, soil, and timber are all resources. A material is depleted, or used up, as it becomes less available for its intended function. Material resources can become depleted in three different ways. First, a substance can be *destroyed*, that is, converted into something else. Fuels are destroyed when they are used: Coal is converted to ashes and gas; uranium is converted to radioactive waste products. The ashes or waste products are no long fuels.

Second, a substance can be lost by being *diluted*, or by being *displaced* to some location from which it cannot easily be recovered. If you open a helium-filled balloon, the gas escapes to the atmosphere. Not one atom of helium is destroyed, but nevertheless the gas is lost because it would be impossible, as a practical matter, to recover it. The same concept of loss by dilution applies to minerals.

Third, a substance can be rendered unfit for use by being *polluted*. In this way pollution and depletion are related to each other. If industrial or agricultural wastes are discharged into a stream, or if they percolate down through soil and porous rock to reach a supply of groundwater, then these water resources become less fit for drinking or, in the case of the stream, for recreation or for the support of aquatic life.

Finally, conservation is often seen as a measure whose benefits will be realized later, perhaps only by our children or grandchildren, and not all makers of policy are equally concerned about future generations.

(d) **Changes in the Global Condition.** Scientists have begun only recently to wonder whether human activities might affect the global environment. Aerosol sprays and aircraft exhaust may be destroying the ozone layer in the atmosphere that filters out ultraviolet radiation. Burning fossil fuels releases carbon dioxide that could affect planetary weather patterns. Pollution of the oceans destroys plant life that produces oxygen, and such pollution might eventually reduce the oxygen content of the air we breathe. Throughout much of the world, forests, jungles, shrublands, and other natural systems are being converted to farmland. In many areas, this process is depleting the fertility of the soil, altering the climate, and causing the extinction of literally thousands of species of plants and animals. Except in emotional terms, people often do not know precisely what has been lost when a species becomes extinct. Scientists are convinced that many endangered species of plants or animals should be saved because they may be essential in breeding valuable crops or domestic animals.

(e) War. In many ways, war is a combination of all environmental problems rather than a separated category. From time immemorial, overpopulation and want have led human groups into wars over food, land, or some other coveted resource. In modern times war and the preparation for war have led to pollution and depletion of resources that are far more extreme than any single peacetime activity. War reduces population, although the effect is trivial: more people were born in 6 months in 1982 than were killed throughout the first and second world wars. Finally, a nuclear war places the global systems of the Earth, human civilization, and even the human species itself at risk.

Selected from "*Jonathan Turk, Amos Turk. Environmental Science, Third Edition, Saunders College Publishing, The Drylen Press, USA, 1984*"

Unit 4

Text: Environmental Analysis

Environmental analysis involves the performance of chemical, physical, and biological measurements in an environmental system. This system may involve either the natural or the polluted environment, although the term "environmental analysis" is increasingly used to refer only to situations in which measurements are made of pollutants.

Normally, four types of measurements are made: qualitative analysis to identify the species present; quantitative analysis to determine how much of the species is present; speciation or characterization to establish details of chemical form and the manner in which the pollutant is actually present (such as being adsorbed onto the surface of a particle); and impact analysis in which measurements are made for the specific purpose of determining the extent to which an environmental impact is produced by the pollutants in question.

The overall objective of environmental analysis is to obtain information about both natural and pollutant species present in the environment so as to make a realistic assessment of their probable behavior. In the case of pollutants, this involves assessment of their actual or potential environmental impact, which may be manifest in several ways. Thus, a pollutant species may present a toxicological hazard to plants or animals. It may also cause contamination of resources (such as air, water, and soil) so that they cannot be utilized for other purposes. The effects of pollutants on materials, especially building materials, may be another area of concern and one which is often very visible and displeasing (for example, the defacing of ancient statues by sulfur dioxide in the atmosphere). A further area of environmental impact involves esthetic depreciation such as reduced visibility, dirty skies, and unpleasant odors. Finally, it is important to recognize that an environmental impact may not always be discernible by normal human perception so that detection may require sophisticated chemical or physical analyses[①].

In order to provide a meaningful description of the general field of environmental analysis, this field may be considered from three points of view: the basic concepts underlying the reasons for and choice of the analyses which are normally performed; the available techniques and methodology commonly used; and the current status of capabilities in environmental analysis.

Some of the philosophical concepts which form the bases for environmental analyses are as follows.

Purpose. Collection and analysis of an environmental sample may be undertaken for the purpose of research or monitoring, or as a spot check. A spot check analysis is used to obtain rapid information about the approximate extent or nature of an environmental problem.

Authenticity. Whatever the information being sought, it is vitally important to obtain an authentic sample which represents the particular system being investigated. In fact, the ability to obtain an authentic sample is probably one of the most difficult aspects of any environmental analysis due to the considerable complexity and heterogeneity of most environmental systems.

Detection limits. A statement of the detection limits which can be attained by the analytical method being employed must always be included in providing the results of any environmental analysis. This is because considerable confusion exists about the meaning of a "zero" level of concentration. In fact, it is probably never possible to state that none of the atoms or molecules of the species in question are present so that no true zero exists.

Precision and accuracy. In reporting an environmental analysis it is also necessary to specify the precision and accuracy associated with the measurements. Thus, many environmental measurements involve comparison of results obtained in different systems or under different conditions (of temperature, time, pollutant concentration, and so on) in the same system so that it is necessary to establish whether two numbers which are different are in fact indicative of different conditions.

State of matter. In making an environmental analysis it is necessary to designate the physical form of the species being analyzed. Most simply, this involves the actual state of matter in which the species exists (whether it is solid, liquid, or gas) since many species (both inorganic and organic) may exist concurrently in different states. For example, certain organic gases can exist either as gases or adsorbed onto the surface of solid particles, and the analytical procedures employed for determination of each form are quite different.

Element/compound distinction. One of the most strongly emphasized aspects of environmental analysis involves the distinction between a chemical element and the chemical compound in which that element exists. It is appropriate to establish the chemical form in which a given element exists in an environmental sample rather than simply to specify the fact that the element is present at a given concentration. While such a concept is philosophically acceptable, analytical methodology has not reached the stage where specification of inorganic compounds present at trace levels can readily be achieved.

Particle surfaces. Where pollutant species are present in or associated with a condensed phase, it is sometimes necessary to establish whether the pollutant is part of the bulk system or present on its surface. Such a consideration is particularly meaningful since material present on the surface of an airborne particle, for example, comes into immediate contact with the external environment, whereas that which is distributed uniformly throughout the particle is effectively present at a much lower concentration and can exert a much lower chemical intensity at the particle surface. Since airborne particles can be inhaled, surface predominance can result in high localized concentrations of chemical species at the points of particle deposition in the lung[②].

Availability. While not one of the analyses normally performed by analytical chemists, determination of the availability of a chemical species is often necessary[③]. To exert a meaningful environmental effect, a pollutant species must almost always enter solution. For instance, toxic compounds associated with airborne particles must be dissolved by lung fluids before either a local or systemic effect can be produced. Thus, laboratory-scale simulation[④] of the availability characteristics of a chemical species can provide a necessary link between its presence and its eventual impact.

Environmental effect. The final link in the analytical/environmental effects chain involves determination of the actual environmental effect produced. In most cases, this involves some form of biological measurement or bioassay which determines the toxicological effect upon a biological organism. Due to the expense, complexity, and time-consuming nature of bioassays, it is usual to substitute a chemical analysis for purposes of monitoring toxicological effects, and in setting standards for compliance. In doing so, however, it is necessary to establish a so-called dose-response relationship between the level of one or more pollutant species and the degree of toxic impact.

Selected from "*McGraw-Hill, Encyclopedia of Environmental Science, 2nd Edition, McGraw-Hill Boo. Company, NJ. USA, 1980*"

Words and Expressions

qualitative *a.* 定性的
quantitative *a.* 定量的
speciation *n.* 物种形式
characterization *n.* 表征，性能描写
toxicological *a.* 毒物学的，毒理学的
defacing *n.* 磨损，损伤外观
esthetic depreciation 感觉下降

unpleasant odors 难闻气味
in question 上述的，所讨论的
methodology *n*. 方法论
impact analysis 作用分析
philosophical *a*. 哲学的；理性的；自然科学研究的
discernible *a*. 鉴别的，识别的
authenticity *n*. 可靠性
heterogeneity *n*. 不均匀性，多相性
trace levels 微量，痕量
airborne *a*. 空气中的；空降的，空运的
time-consuming 耗时的
bioassay *n*. 生物鉴定，活体鉴定
compliance *n*. 符合，一致
dose-response 用量（剂量）响应

Notes

① sophisticated chemical or physical analysis 高级（先进）的化学或物理分析，如气相色谱、液相色谱、红外、质谱分析等。

② Since airborne particles can be inhaled, surface predominance can result in high localized concentrations of chemical species at the points of particle deposition in the lung. 参考译文：由于可以吸入空气中的颗粒，表面控制可以导致肺部颗粒沉积点化学物质局部浓度升高。

③ While not one of the analysis normally performed by analytical chemists, determination of the availability of a chemical species is often necessary. 可译为：虽然分析化学家并不可能正常地做出每一种分析，但是确定化学物质的种类通常是必要的。

④ laboratory-scale simulation 实验室范围（规模）的模拟。在生产过程中，通常先在实验室模拟，再经过逐级放大后最后才进入工业过程。

Exercises

1. Put the following into Chinese.
   precision and accuracy, bulk collection, matrix material, analytical sequence, multivariate statistics, interactive effect, insofar, overall analytical scheme
2. Put the following into English.

灵敏度	采样	真实时间	样品预处理	稳定性	曲线拟和
吸附	累积	分析评价	物理分离	因次图	标准方差

3. Give the four steps for chemical analysis.
4. Give the basic concepts which form the bases for environmental analyses.
5. Translate paragraphs 1~4 into Chinese.

Reading Material: Overview of Environmental Economics

The history modern environmental law can be traced back to the 1960s and the beginnings of the modern social and political movement we know as environmentalism. Its foundation is the "three E's": ecology, engineering, and economics. Although in this environmental science textbook we will devote most of our time to the scientific basis—ecology, geology, climatology, and all the other sciences involved in environmental analysis—economics underlies much of the discussion. It is always a factor in finding solutions that work, are efficient, and are fair. This is why we are devoting one of our early overview chapters to environmental economics.

Environmental economics is not simply about money; it is about how to persuade people, organizations, and society at large to act in a way that benefits the environment, keeping it as free as possible of pollution and other damage, keeping our resources sustainable, and accomplishing these goals within a democratic framework. Put most simply, environmental economics focused on two broad areas: controlling pollution and environment damage in general, and sustaining renewable resources—forests, fisheries, recreational lands, and so forth. Environmental economists also explore the reasons why people don't act in their own best interests when it comes to the environment. Are there rational explanations for what seem to be irrational choice? If so, and if we can understand them, perhaps we do something about them. What we do, what we can do, and how we do it are known collectively as **policy instruments.**

Environmental decision-making often, perhaps even usually, involves analysis of tangible and intangible factors. In the language of economics, a **tangible factor** is one you can touch, buy, and sell. A house lost in a mudslide due to altering the slope of the land is an example of a tangible factor. For economists, **an intangible factor** is one you can't touch directly, but you value it, as with the beauty of the slope before the mudslide. Of the two, the intangibles are obviously more difficult to deal with because they are harder to measure and to value economically. Nonetheless, evaluation of intangibles is becoming more important. As you will see in later chapters, huge amounts of money and resources are involved in economic decisions about both tangible and intangible aspects of the environment. There are the costs of pollution and the loss of renewable resources, and there are the costs of doing something about these problems.

In every environmental matter, there is a desire on the one hand to maintain individual freedom of choice, and on the other to achieve a specific social goal. In ocean fishing, for example, we want to allow every individual to choose whether or not to fish, but we want to prevent everyone from fishing at the same time and bringing fish species to extinction. This interplay between private good and public good is at the heart of environmental issues.

In this chapter we will examine some of the basic issues in environmental economics: the environment as a commons; risk-benefit analysis; valuing the future; and why people often do not act in their own best interest.

The Environment as a Commons

One reason has to do with what the ecologist Garrett Hardin called "tragedy of the commons". When a resource is shared, an individual's personal share of profit from its exploitation is usually greater than his or her share of resulting loss. A second reason has to do with the low growth rate, and therefore, low productivity, of a resource.

A **commons** is land (or another resource) owned publicly, with public access for private uses. The term **commons** originated from land owned publicly in English and New England towns and set aside so that all the farmers of the town could graze their cattle. Sharing the grazing area worked as long as the number of cattle was low enough to prevent overgrazing. It would seem that people of goodwill would understand the limits of a commons. But take a dispassionate view and think about the benefits and costs to each farmer as if it were a game. Phrased simply, each farmer tries to maximize personal gain and must periodically consider whether to add more cattle to the herd on the commons. The addition of one cow has both a positive and a negative value. The positive value is the benefit when the farmer sells that cow. The negative value is the additional grazing by the cow. The personal profit from selling a cow is greater than the farmer's share of the loss caused by the degradation of the commons. Therefore, the short-term successful game plan is always to add another cow.

Risk-Benefit Analysis

Death is the fate of all individuals, and almost every activity in life involves some risk of death or injury. How, then, do we place a value on saving a life by reducing the level of a pollutant? This question raises another important area of environmental economics: **risk-benefit analysis,** in which the riskiness of a present action in terms of its possible outcomes is weighed against the benefit, or value, of the action. Here, too, difficulties arise.

With some activities, the relative risk is clear. It is much more dangerous to stand in the middle of a busy highway than to stand on the sidewalk, and hang gliding has a much higher mortality rate than hiking. The effects of pollutants are often more subtle, so the risks are harder to pinpoint and quantify. Table 1 gives the lifetime risk of death associated with a variety of activities and some forms of pollution. In looking at the table, remember that since the ultimate fate of everyone is death, the total lifetime risk of death from all causes must be 100%. So if you are going to die of something and you smoke a pack of cigarettes a day, you have 8 chances in 100 that your death will be a result of smoking. At the same time, your risk of death from driving an automobile is 1 in 100. Risk tells you the chance of an event but not its timing. So you might smoke all you want and die from the automobile risk first.

Table 1　Risk of Death from Various Causes

Cause	Result	Risk of Death (Per Lifetime)	Lifetime of Death	Comment
Cigarette Smoking(pack a day)	Cancer, effect on heart, lung, etc.	8 in 100	8%	
Breathing radon-containing Air in the home	Cancer	1 in 100	1%	naturally occurring
Automobile driving		1 in 100	1%	
Death from a fall		4 in 1000	0.4%	
Droning		3 in 1000	0.3%	
Fire		3 in 1000	0.3%	
Artificial chemicals in the home	Cancer	2 in 1000	0.2%	paints, cleaning agents
Sunlight exposure	melanoma	2 in 1000	0.2%	of those exposed to sunlight
Electrocution		4 in 10000	0.04%	
Air outdoors in industrial area		1 in 10000	0.01%	
Artificial chemicals in water		1 in 100000	0.001%	
Artificial chemicals in foods		<1 in 10000	0.001%	
Airplane passenger		<1 in 10000	0.001%	

source: from *Guide to Environmental Risk* (1991), USA EPA Region 5 Publication Number 905/91/017

Valuing the Beauty of Nature

The beauty of nature—often termed landscape aesthetics—is an environmental intangible that has probably been important to people as long as our species has existed. We know it has been important since people have written, because the beauty of nature is a continuous theme in literature and art. Once again, as with forests cleaning the air, we face the difficult question: How do we arrive at a price for the beauty of nature? The problem is even more complicated because among the kinds of scenery we enjoy are many modified by people. For example, the open farm fields in Vermont improved the view of the mountains and forests in the distance, so when farming declined in the 1960s, the state began to provide tax incentives for farmers to keep their fields open and thereby help the tourism economy.

One of the perplexing problems of aesthetic evaluation is personal preference. One person may appreciate a high mountain meadow far removed from civilization; a second person may prefer visiting with others on a patio at a trailhead lodge; a third may prefer to visit a city park; and a fourth may prefer the austere beauty of a desert. If we are going to consider aesthetic factors in environmental analysis, we must develop a method of aesthetic evaluation that allows for individual differences—another yet unsolved topic.

Selected from " *Daniel B. Botkin, Edward A. Keller. Environmental Science, Eight Edition, John Wiley & Sons, Inc. Printed in Asia, 2012*"

Part 2 ATMOSPHERIC CHEMISTRY AND AIR POLLUTION AS WELL AS CONTROL

Unit 5

Text: Chemistry of the Atmosphere

The thin gaseous envelope that surrounds our planet is integral to the maintenance of life on earth. The composition of the atmosphere is predominately determined by biological processes acting in concert with physical and chemical change. Though the concentrations of the major atmospheric constituents oxygen and nitrogen remain the same, the concentration of trace species, which are key to many atmospheric processes are changing. It is becoming apparent that man's activities are beginning to change the composition of the atmosphere over a range of scales, leading to, for example, increased acid deposition, local and regional ozone episodes, stratospheric ozone loss and potentially climate change[①]. In this part, we will look at the fundamental chemistry of the atmosphere derived from observations and their rationalization.

In order to understand the chemistry of the atmosphere we need to be able to map the different regions of the atmosphere. The atmosphere can be conveniently classified into a number of different regions which are distinguished by different characteristics of the dynamical motions of the air. The lowest region, from the earth's surface to the tropopause at a height of 10-15 km, is termed the troposphere. The troposphere is the region of active weather systems which determine the climate at the surface of the earth. The part of the troposphere at the earth's surface, the planetary boundary layer, is that which is influenced on a daily basis by the underlying surface.

Above the troposphere lies the stratosphere, a quiescent region of the atmosphere where vertical transport of material is slow and radiative transfer of energy dominates. In this region lies the ozone layer which has an important property of absorbing ultraviolet (UV) radiation from the sun, which would otherwise be harmful to life on earth. The stratopause at approximately 50 km altitude marks the boundary between the stratosphere and the mesosphere, which extends upwards to the mesopause at approximately 90 km altitude. The mesosphere is region of large temperature extremes and strong turbulent motion in the atmosphere over large spatial scales.

Above the mesopause is a region characterized by a rapid rise in temperature, known as the thermosphere. In the thermosphere, the atmospheric gases, N_2 and O_2, are dissociated to a significant extent into atoms so the mean molecular mass of the atmospheric species falls. The pressure is low and thermal energies are significantly departed from the Boltzmann equilibrium. Above 160 km gravitational separation of the constituents becomes significant and atomic hydrogen atoms, the lightest natural species, moves to the top of the atmosphere. The other characteristic of the atmosphere from mesosphere upwards is that above 60km, ionisation is important. This region is called the ionosphere. It is subdivided into three regimes, the D, E and F region, characterized by the types of dominant photo ionisation.

With respect to atmospheric chemistry, though there is a great deal of interesting chemistry taking place higher up in the atmosphere, we shall focus in the main on the chemistry of the troposphere and stratosphere.

Sources of trace gases in the atmosphere

As previously described, the troposphere is the lowest region of the atmosphere extending from the earth's surface to the tropopause at 10-18 km. About 90% of the total atmospheric mass resides in the troposphere and the greater part of the trace gas burden is found there. The troposphere is well mixed and its bulk composition is 78% N_2, 21% O_2, 1% Ar and 0.036% CO_2 with varying amounts of water vapor depending on temperature and altitude. The majority of the trace species found in the atmosphere is emitted into the troposphere from the surface and are subject to a complex series of chemical and physical transformations. Trace species emitted directly into the atmosphere are termed to have primary sources, e.g. trace gases such as SO_2, NO and CO. Those trace species formed as a product of chemical and/or physical transformation of primary pollutants in the atmosphere, e.g. ozone, are referred to as having secondary sources or being secondary species.

Emissions into the atmosphere are often broken down into broad categories of anthropogenic or "mass-made sources" and biogenic or natural sources with some gases also having geogenic sources. For the individual emission of a primary pollutant there are a number of factors that need to be taken into account in order to estimate the emission strength, these include the range and type of sources and the spatio-and temporal-distribution of the sources. Often these factors are compiled into the so-called emission inventories that combine the rate of emission of various sources with the number and type of each source and the time over which the emissions occur[2]. It is clear from the data obtained, for example, SO_2 has strong sources from public power generation whereas ammonia has strong sources from agriculture. In essence, it has been apportioned spatially according to magnitude of each source category (e.g. road transport, combustion in energy production and transformation, solvent use). For example, the major road routes are clearly visible, showing NO_2 has a major automotive source. It is possible to scale the budgets of many trace gases to a global scale.

It is worth noting that there are a number of sources that do not occur within the boundary layer (the decoupled lowest layer of the troposphere), such as lightning production of nitrogen oxides and a range of pollutants emitted from the combustion-taking place in aircraft engines[3]. The non-surface sources often have a different chemical impact owing to their direct injection into the free troposphere (the part of the troposphere that overlays the boundary layer).

In summary, there are a range of trace species present in the atmosphere with a myriad of sources varying both spatially and temporally. It is the chemistry of the atmosphere that acts to transform the primary pollutants into simpler chemical species.

Initiation of photochemistry by light

Photodissociation of atmospheric molecules by solar radiation plays a fundamental role in the chemistry of the atmosphere. The photodissociation of trace species such as ozone and formaldehyde contributes to their removal from the atmosphere, but probably the most important role played by these photoprocesses is the generation of highly reactive atoms and radicals. Photodissociation of trace species and the subsequent reaction of the photoproducts with other molecules is the prime initiator and driver for the bulk of atmospheric chemistry[④].

Selected from "*Roy M Harrison. Principles of Environmental Chemistry,*
The Royal Society of Chemistry. UK, 2007"

Words and Expressions

rationalization　*n.* 合理，有理化
tropopause　*n.* 对流层顶，休止层
quiescent　*a.* 宁静的，平静的
stratopause　*n.* 同温层顶
mesosphere　*n.* 中间层，散逸层（同温层上部最低温度区，高度在 50～90m 处）
mesopause　*n.* 中间层顶
turbulent　*a.* 湍流的，湍动的
Boltzmann equilibrium　波耳兹曼平衡
troposphere　*n.* 对流层
stratosphere　*n.* 平流层
biogenic　*a.* 生物的，生命所需的
geogenic　*a.* 断裂地貌的，破裂带地貌的
spatio-and temporal-distribution　空间与时间分布
decoupled　*a.* 去偶的，分离的

Notes

① It is becoming apparent that man's activities are beginning to change the composition of the atmosphere over a range of scales, leading to, for example, increased acid deposition, local and regional ozone episodes, stratospheric ozone loss and potentially climate change. 可译为：很明显，人类的活动在一定的尺度范围正改变大气组成，例如，导致越来越多酸的积累、局部和地区性臭氧变化异常、平流层臭氧的损失以及潜在的气候变化。

② Often these factors are compiled into the so-called emission inventories that combine the rate of emission of various sources with the number and type of each source and the time over which the emissions occur. 可译为：通常，这些因素整理成所谓的释放清单，该清单将不同污染物源的释放速度与每一种污染物源的数目和种类以及释放时间组合起来。

③ It is worth noting that there are a number of sources that do not occur within the boundary layer (the decoupled lowest layer of the troposphere), such as lightning production of nitrogen oxides and a range of pollutants emitted from the combustion-taking place in aircraft engines. 可译为：值得注意的是，在边界层内（对流层分离的最底层）有许多不会发生的污染源，如闪电产生的氮氧化物，飞机发动机燃烧释放的污染物。

④ Photodissociation of trace species and the subsequent reaction of the photoproducts with other molecules is the prime initiator and driver for the bulk of atmospheric chemistry. 可译为：痕量物质以及光化学产物与其他分子后续反应的光分解是大气化学的初始诱发剂和发动剂。

Exercises

1. Put the following words or phrases into Chinese.

   Turbulent, biogenic, geogenic, chemical and physical transformation, prime initiator

2. Put the following into English according to the text.

   合理　去偶的　空间与时间分布　平流层　对流层　光化学激发　痕量物

3. Give a brief description of Sources of trace gases in the atmosphere.
4. Translate "the first paragraph" of the text into Chinese.
5. Give a brief summary of atmosphere classification.

Reading Material: Reactions of Atmospheric Nitrogen and its Oxides

Reactions of Atmospheric Nitrogen and its Oxides

The 78% by volume of nitrogen contained in the atmosphere constitutes an inexhaustible reservoir of that essential element. The nitrogen cycle and nitrogen fixation by microorganisms and several other microbially mediated reactions important in the nitrogen cycle were discussed before. A small amount of nitrogen is thought to be fixed (chemically bound to other elements) in the atmosphere by lightning, and some is also fixed by combustion processes, as in the internal combustion engine.

Before the use of synthetic fertilizers reached its current high level, chemists were concerned that denitrification processes in the soil would lead to nitrogen depletion on the Earth. Now, with millions of tons of synthetically fixed nitrogen being added to the soil each year, concern has shifted to possible excess accumulation of nitrogen in soil, fresh water, and the oceans.

Unlike oxygen, which is almost completely dissociated to the monatomic form in higher regions of the thermosphere, molecular nitrogen is not readily dissociated by ultraviolet radiation. However, at altitudes exceeding approximately 100 km, atomic nitrogen is pro-

duced by photochemical reactions:
$$N_2 + h\nu \longrightarrow N + N \tag{1}$$

Most stratospheric ozone is probably removed by the action of nitric oxide, which reacts with ozone as follows:
$$O_3 + NO \longrightarrow NO_2 + O_2 \tag{2}$$
$$NO_2 + O \longrightarrow NO + O_2 \text{ (regeneration of NO from NO}_2) \tag{3}$$

Pollutants oxides of nitrogen, particularly NO_2, are key species involved in air pollution and the formation of photochemical smog. For example, NO_2 is readily dissociated photochemically to NO and reactive atomic oxygen by electromagnetic radiation of less than 398 nm wavelength:
$$NO_2 + h\nu \longrightarrow NO + O \tag{4}$$

This is the most important primary photochemical process involved in smog formation.

Air Pollutant Nitrogen Compounds

Nitrogen compounds, especially nitrogen oxides, are among the most significant air pollutants. The three oxides of nitrogen normally encountered in the atmosphere are nitrous oxide (N_2O), nitric oxide (NO), and nitrogen dioxide (NO_2). Microbially generated nitrous oxide is relatively unreactive and probably does not significantly influence important chemical reactions in the lower atmosphere. However, colorless, odorless nitric oxide and pungent red-brown nitrogen dioxide, collectively designated NO_x, are very important in polluted air. Regionally high pollutant NO_2 concentrations can result in severe air quality deterioration. Practically are anthropogenic NO_x enters the atmosphere as NO from the combustion of fossil fuels in both stationary and mobile sources. The contribution of automobiles to nitric oxide production has become significantly lower as newer automobiles with nitrogen oxide pollution controls have become more common.

Atmospheric Reactions of NO_x

Nitrogen dioxide is a very reactive and significant species in the atmosphere. It absorbs light throughout the ultraviolet and visible spectrum penetrating the troposphere. The photochemical dissociation of NO_2 (reaction 4) and the highly reactive O atom product set off several significant inorganic reactions and many atmospheric reactions involving organic species. Atmospheric chemical reactions convert NO_x to nitric acid, inorganic nitrate salts, organic nitrates, and oxidant peroxyacetyl nitrate as discussed with respect to photochemical smog later. Although NO is the primary form in which NO_x is released to the atmosphere, the conversion of NO to NO_2 is relatively rapid in the troposphere.

Harmful Effects of Nitrogen Oxides

Nitric oxide is less toxic than NO_2. (In recent years the role of NO as an important intermediate in biochemical processes has gained recognition.). Acute exposure to NO_2 can be quite harmful to human health and sufficiently high exposures to this gas can be fatal. For exposures ranging from several minutes to one hour, a level 50 to 100 ppm of NO_2 causes inflammation of lung tissue for a period of 6 to 8 weeks, after which time the subject normally recovers. Exposure of the subject to 150 to 200 ppm of NO_2 causes brochiolitis fibrosa obliterans, a condition fatal within 3 to 5 weeks after exposure. Death generally results within 2 to 10 days after inhalation of air containing 500 ppm or more of NO_2. Although extensive damage to plants is observed in areas receiving heavy exposure to NO_2, most of this damage probably comes from secondary products of nitrogen oxides, such as PAN formed in smog.

Ammonia as an Atmospheric Pollutant

Ammonia is the only nonoxygenated gaseous inorganic nitrogen compound that is likely to be a significant atmospheric pollutant. The only water-soluble base is present at significant levels in the atmosphere. It is toxic and can be a significant localized pollutant in specific cases. Its most important effect in the atmosphere is the formation of corrosive pollutant salts including NH_4NO_3, NH_4HSO_4, and $(NH_4)_2SO_4$.

Selected from "*Roy M Harrison. Principles of Environmental Chemistry, The Royal Society of Chemistry. UK, 2007*"

Unit 6

Text: Acid Rain

Acid rain is precipitation in which the pH is below 5.6. The pH, a measure of acidity and alkalinity, is the negative logarithm of the concentration of the hydrogen ion (H^+). Because the pH scale is logarithmic, a pH value of 3 is 10 times more acidic than a pH value of 4 and 100 times more acidic than a pH value 5. Automobile battery acid has a pH value of 1. Many people are surprised to learn that all rainfall is slightly acidic; water reacts with atmospheric carbon dioxide to produce weak carbonic acid. Thus, pure rainfall has a pH of about 5.6, where 1 is highly acidic and 7 is neutral. Natural rainfall in tropical rain forests has been observed in some instances to have a pH of less than 5.6; this is probably related to acid precursors emitted by the trees.

Acid rain includes both wet (rain, snow, fog) and dry (particulate) acidic depositions. The depositions occur near and downwind of areas where the burning of fossil fuels produces major emissions of sulfur dioxide (SO_2) and nitrogen oxides (NO_x). Although these oxides are the primary contributors to acid rain, other acids are also involved. An example is hydrochloric acid emitted from coal-fired power plants.

Acid rain has likely been a problem at least since the beginning of the Industrial Revolution. In recent decades, however, it has gained more and more attention, and today it is a major, global environmental problem affecting all industrial countries. In the United States, nearly all of the eastern states are affected, as well as West Coast unban centers, such as Seattle, San Francisco, and Los Angeles. The problem is also of great concern in Canada, Germany, Scandinavia[①], and Great Britain. Developing countries that rely heavily on coal, such as China, are facing serious acid rain problems as well.

Causes of Acid Rain

As we have said, sulfur dioxide and nitrogen oxides are the major contributors to acid rain. Amounts of these substances emitted into the environment in the United States are shown in other chapters. As shown earlier in other table, emissions of SO_2 peaked in the 1970s and declined to about 13 million metric tons per year by 2007; and nitrogen oxides leveled off at about 25 million metric tons per year in the mid-1980s and had dropped to 17 million metric tons by 2007.

In the atmosphere, reactions with oxygen and water vapor transform SO_2, NO_x into sulfuric and nitric acids, which may travel long distances with prevailing winds and be deposited as acid precipitation—rainfall, snow, or fog. Sulfate and nitrate particles may also be deposited directly on the surface of land as dry deposition and later be activated by moisture to become sulfuric and nitric acids.

Again, sulfur dioxide is emitted primarily by stationary sources, such as power plants that burn fossil fuels, whereas nitrogen oxides are emitted by both stationary and mobile sources, such as automobiles. Approximately 80% of sulfur dioxide and 65% of nitrogen oxides in the United States come from states east of the Mississippi River.

Acid Rain's Effects on Forest Ecosystems

It has long been suspected that acid precipitation damages trees. Studies in Germany led scientists to cite acid rain and other air pollution as the cause of death for thousands of acres of evergreen trees in Bavaria. Similar studies in the Appalachian Mountains of Vermont[②] (where many soils are naturally acidic) suggest that in some locations half the red spruce trees have died in recent years. Some high-elevation forests of the Appalachian Mountains, including the Great Smoky Mountains and Shenandoah National Park[③], have been impacted by acid rain, acid fog, and dry deposition of acid. Symptoms include slowed tree growth, leaves and needles that turn brown and fall off, and in extreme cases the death of tress. The acid rain does not directly kill trees; rather, it weakens them as essential nutrients are leached from soils or stripped from leaves by acid fog. Acidic rainfall also may release toxic chemicals, such as aluminum, that damage trees.

Acid Rain's Effects on Lake Ecosystems

Records from Scandinavian lakes show an increase in acidity accompanied by a decrease in fish. The increased acidity has been traced to acid rain caused by industrial processes in other countries, particularly Germany and Great Britain. Thousands of lakes, ponds, and streams in the eastern United States are sensitive to acidification, including the Adirondacks and Catskill Mountains of New York

State and others in the Mideast and in the mountains of the Western U.S. Little Echo Pond in Franklin, New York, is one of the most acidic lakes, with a measured pH of 4.2.

Acid rain affects lake ecosystems. It damages aquatic species (fish, amphibians, and crayfish) directly by disrupting their life processes in ways that limit growth or cause death. For example, crayfish produce fewer eggs in acidic water, and the eggs produced often grow into malformed larvae. Acid rain dissolves chemical elements necessary for life in the lake. Once in solution, the necessary elements leave the lake with water outflow. Thus, elements that once cycled in the lake are lost. Without these nutrients, algae do not grow, animals that feed on the algae have little to eat, and animals that feed on these animals also have less food. Acid rain leaches metals, such as aluminum, lead, mercury, and calcium, from the soils and rocks in a drainage basin and discharges them into rivers and lakes. Elevated concentrations of aluminum are particularly damaging to fish because the metal can clog the gills and cause suffocation[④]. The heavy metals may pose health hazards to people, too, because the metals may become concentrated in fish and then be passed on to people, mammals, and birds that eat the fish. Drinking water from acidic lakes may also have high concentrations of toxic metals.

Acid rain's Effects on Human Society

Acid rain damages not only our forests and lakes but also many building materials, including steel, galvanized steel, paint, plastics, cement, masonry, and several types of rock, essentially limestone, sandstone, and marble. Classical buildings on the Acropolis in Athens and in other cites show considerable decay (chemical weathering) that accelerated in the 20th century as a result of air pollution. The problem has grown to such an extent that buildings require restoration, and the protective coatings on statues and other monuments must be replaced quite frequently, at a cost of billions of dollars a year. Particularly important statues in Greece and other areas have been removed and placed in protective glass containers, leaving replicas standing in their former outdoor locations for tourists to view.

Stone decays about twice as rapidly in cites as it does in less urban areas. The damage comes mainly from acid rain and humidity in the atmosphere, as well as from corrosive groundwater. This implies that measuring rates of stone decay will tell us something about changes in the acidity of rain and groundwater in different regions and ages. It is now possible, where the ages of stone buildings and other structures are known, to determine whether the acid rain problem has changed over time.

Selected from "*Daniel B. Botkin, Edward A. Keller. Environmental Science, Eight Edition, John Wiley & Sons, Inc. Printed in Asia, 2012*"

Words and Expressions

buffer　*n.* 缓冲物（液）
bedrock　*n.* 基岩

granite *n.* 花岗岩
amphibian *n.* 两栖动物
crayfish *n.* 龙虾
malformed larvae 畸形幼虫
igneous rock 火山岩，岩浆岩
masonry *n.* 砖石建筑
marble *n.* 大理石
Acropolis in Athens 希腊雅典卫城
replicas *n.* 复制品

Notes

① Scandinavia 可译为：斯堪的纳维亚。包括丹麦、瑞典、芬兰、挪威和冰岛北欧5国。
② the Appalachian Mountains of Vermont 佛蒙特州的阿巴拉契山脉。位于美国东部北美洲巨大的山脉，是北美洲东部众多山脉的统称。
③ the Great Smoky Mountains and Shenandoah National Park 大烟山和仙纳度国家公园。位于美国 Virginia（维珍尼亚州）。
④ Elevated concentrations of aluminum are particularly damaging to fish because the metal can clog the gills and cause suffocation 可译为：高浓度的金属铝尤其危害鱼类，因为金属会阻塞鱼鳃，导致鱼类窒息死亡。

Exercises

1. Put the following into Chinese.
   gill, suffocation, malformed larvae, amphibian, crayfish, replicas, decay, humidity, monument, protective glass container
2. Put the first paragraph into Chinese.
3. What is acid rain according to the text?
4. Give a brief summary of causes of acid rain.

Reading Material: Reactions of Atmospheric Sulfur Compounds

Fig. 1 shows the main aspects of the global sulfur cycle. This cycle involves primarily H_2S, SO_2, SO_3, and sulfates. Of these species, SO_2 is the most important because of its high abundance and facile oxidation to highly acidic H_2SO_4.

Sulfur Dioxide Reactions in the Atmosphere

Many factors, including temperature, humidity, light intensity, atmospheric transport, and surface characteristics of particulate matter, may influence the atmospheric chemical reactions of sulfur dioxide. Like many other gaseous pollutants, sulfur dioxide undergoes chemical reactions resulting in the formation of particulate matter. Whatever the process involved, much of the sulfur dioxide in the atmosphere ultimately is oxidized to sulfuric acid

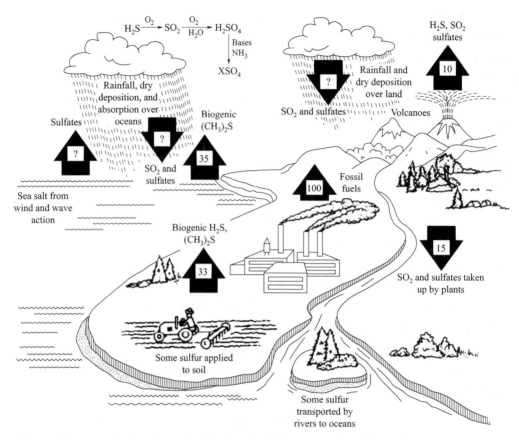

Fig. 1 The atmospheric sulfur cycle. Numbers in the arrows are in millions of metric tons per year.

and sulfate salts, particularly ammonium sulfate and ammonium hydrogen sulfate.

Effect of Atmospheric Sulfur Dioxide

Though not terribly toxic to most people, low levels of sulfur dioxide in air do have some health effects. Sulfur dioxide's primary effect is upon the respiratory tract, producing irritation and increasing airway resistance, especially in people with respiratory weaknesses and in sensitized asthmatics. Therefore, exposure to the gas may increase the effort required to breathe. Mucus secretion is also stimulated by exposure to air contaminated by sulfur dioxide.

Atmospheric sulfur dioxide is harmful to plants. Acute exposure to high levels of the gas kills leaf tissue (leaf necrosis). Chronic exposure of plants to sulfur dioxide causes chlorosis, a bleaching or yellowing of the normally green portions of the leaf. Sulfur dioxide in the atmosphere is converted to sulfuric acid, so that in areas with high levels of sulfur dioxide pollution, plants may be damaged by sulfuric acid aerosols. Such damage appears as small spots where sulfuric acid droplets have impinged upon leaves.

Hydrogen Sulfide, Carbonyl Sulfide, and Carbon Disulfide

Hydrogen sulfide is produced by microbial decay of sulfur compounds and microbial reduction of sulfate, from geothermal steam, as a by-product of wood pulping, and from a number of miscellaneous natural and anthropogenic sources. Most atmospheric hydrogen sulfide is rapidly converted to SO_2 and sulfates. The organic homologisation of hydrogen sulfide, the mercaptans (hydrocarbon groups bonded to the $-SH$ group, such as methyl mercaptan, H_3CSH), enter the atmosphere from decaying organic matter and have particularly objectionable odors.

Hydrogen sulfide pollution from artificial sources is not as much as of an overall air pollution problem as sulfur dioxide pollution. However, there have been several acute incidents of hydrogen sulfide emissions resulting in damage to human health and even fatalities. The most notorious such event occurred in Poza Rica, Mexico, in 1950. Accidental release of hydrogen sulfide from a plant used for the recovery of sulfur from natural gas caused the deaths of at least 22 people and the hospitalization of over 300.

Hydrogen sulfide at levels well above ambient concentrations destroys immature plant tissue. This type of plant injury is readily distinguished from that of other phytoalexins. More sensitive species are killed by continuous exposure to around 3000 ppb H_2S, whereas other species survive with reduction growth, leaf lesions, and defoliation.

Damage to certain kinds of materials is a very expensive effect of hydrogen sulfide pollution. Paints containing basic lead carbonate pigment, $2PbCO_3 \cdot Pb(OH)_2$ (no longer used), were particularly susceptible to darkening by H_2S. A black layer of copper sulfide forms on copper metal exposed to H_2S. Eventually, this layer is replaced by a green coating of basic copper sulfate such as $CuSO_4 \cdot 3Cu(OH)_2$. The green "patina," as it is called, is very resistant to further corrosion. Such layers of corrosion can seriously impair the function of copper contacts on electrical equipment. Hydrogen sulfide also forms a black sulfide coating on silver.

Carbonyl sulfide, COS, is now recognized as a component of the atmosphere at a tropospheric concentration of approximately 500 parts per trillion by volume. It is, therefore, a significant sulfur species in the atmosphere. Both COS and carbon disulfide, CS_2, are oxidized in the atmosphere by reactions initiated by the hydroxyl radical.

Selected from "Stanley E Manahan, Environmental Science and Technology, A Sustainable Approach to Green Science and Technology, Taylor & Franicis Group, USA, 2007"

Unit 7

Text: Introduction to Environmental Chemistry of the Atmosphere

The quality of the air that living organisms breathe, the nature and level of air pollutants, visibility and atmospheric aesthetics, and even climate are dependent upon chemical phenomena that occur in the atmosphere. These, in turn are strongly tied to absorption of solar energy, interactions between the gas phase of the atmosphere and small solid particles suspended in it, and interchange of chemical species with the geosphere[①]. Chemical reactions in the atmosphere and the chemical nature of atmospheric chemical species are the topic of atmospheric chemistry, which is introduced in this text.

Chemistry in the atmosphere involves the unpolluted atmosphere, highly polluted atmospheres, and a wide range of gradations in between. The same general phenomena govern all and produce one huge atmospheric cycle, in which there are numerous subcycles. Gaseous atmospheric chemical species fall into the following somewhat arbitrary and overlapping classifications: inorganic oxides (CO, CO_2, NO_2, SO_2); oxidants (O_2); reductants (CO, SO_2, H_2S); organics (in the unpolluted atmosphere, CH_4 is the predominant organic species, whereas alkanes, alkenes, and aromatic compounds are common around sources of organic pollution); photochemically active species (NO_2, formaldehyde); acids (H_2SO_4); salts (NH_4HSO_4); and unstable reactive species (electronically excited NO_2, $HO\cdot$ radical). In addition, both solid and liquid particles play a strong role in atmospheric chemistry as sources and sinks[②] for gas-phase species, as sites for surface reactions (solid particles), and as bodies for aqueous-phase reactions (liquid droplets). Two constituents of most importance in atmospheric chemistry are radiant energy from the sun, predominantly in the ultraviolet region of the spectrum, and the hydroxyl radical, $HO\cdot$. The former provides a way to pump a high level of energy into individual gas molecules to start a series of atmospheric chemical reactions, and the latter is the most important reactive intermediate and "currency" of daytime atmospheric chemical phenomena.

An important aspect of atmospheric chemistry is that many of the processes occur in the gas phase where molecules are relatively far apart. Therefore, some reactive species can exist for significantly longer times before reacting than they would in water or in solids. This is especially true in the highly rarefied regions of the stratosphere and above.

Photochemical Processes in the Atmosphere

The absorption of electromagnetic solar radiation ("light," usually in the ultraviolet region of the electromagnetic spectrum) by chemical species may cause photochemical reactions to occur. Photochemical reactions give atmospheric chemistry a unique quality and largely determine the nature and ultimate fate of atmospheric chemical species. The ability of electromagnetic radiation to cause photochemical reactions to occur is a function of its energy, E, which increases with increasing frequency (ν) and decreasing wavelength (λ) according to the relationship,

$$E = h\nu \quad (1)$$

Where h is Planck's constant[③], 6.63×10^{-34} joule-seconds (J/sec). In order for a photochemical reaction to occur, a single unit of photochemical energy called a quantum and having an energy of $h\nu$, must be absorbed by the reacting species. If the absorbed light is in the visible region of the sun's spectrum, the absorbing species is colored. Colored NO_2 is a common example of such a species in the atmosphere.

Nitrogen dioxide, NO_2, is one of the most photochemically active species found in a polluted atmosphere. When a molecule such as NO_2 absorbs radiation of energy $h\nu$,

$$NO_2 + h\nu \longrightarrow NO_2^* \qquad (2)$$

an electronically excited molecule designated in the reaction above by an asterisk (*) may be produced. The photochemistry of nitrogen dioxide is discussed in greater detail later.

Electronically excited molecules and atoms are reactive and unstable species in the atmosphere that participate in a wide range of atmospheric chemical processes. Two other generally reactive and unstable species in the atmosphere are free radicals composed of atoms or molecular fragments with unshared electrons, and ions consisting of charged atoms or molecular fragments[④]. The participation of these three kinds of species in atmospheric chemical processes is discussed in later sections.

Electronically excited molecules produced when unexcited ground-state molecules absorb energetic electromagnetic radiation in the ultraviolet or visible regions of the spectrum may possess several possible excited (energized) states. Generally, however, ultraviolet or visible radiation is energetic enough to excite molecules only to several of the lowest energy levels. Because they are energized compared to the ground state, excited chemical species are reactive. Their participation in atmospheric chemical reactions, such as those involved in smog formation, is discussed later.

Electromagnetic radiation absorbed in the infrared region is not sufficiently energetic to break chemical bonds, but does cause the receptor molecules to vibrate and rotate, so it is said that they gain vibrational and rotational energy. The energy absorbed as infrared radiation ultimately is dissipated as heat and raises the temperature of the whole atmosphere. The absorption of infrared radiation is very important in the retention of energy radiated from the Earth's surface.

The reactions that occur following absorption of a photon of light sufficiently energetic to produce an electronically excited species are largely determined by the way in which the excited species loses its excess energy. This may occur by one of several processes divided into two general classes. Of these, photophysical processes are those that do not involve chemical bond breakage or loss of electrons and include loss of energy from the excited molecule by electromagnetic radiation as it returns to the ground state (fluorescence or phosphorescence), transfer of energy to other molecules, or transfer of energy within the absorbing molecule; the last two processes result in dissipation of the excess energy as heat. Photochemical reactions occur as a result of de-excitation processes that involve chemical bond breakage or ion formation, particularly the following:

Photodissociation of the excited molecule (the process responsible for the predominance of atomic oxygen in the upper atmosphere)

$$O_2^* \longrightarrow O + O \qquad (3)$$

Direct reaction with another species

$$O_2^* + O_3 \longrightarrow 2\,O_2 + O \qquad (4)$$

Photoionization through loss of an electron

$$N_2^* \longrightarrow N_2^+ + e^- \tag{5}$$

Insofar as gas-phase reactions in the troposphere are concerned, the most important of the processes listed above is photodissociation. This is because photodissociation converts relatively stable and unreactive molecular species to reactive atoms and free radicals that participate in additional reactions, including chain reactions.

Chain reactions in the Atmosphere: Hydroxyl and Hydroperoxyl Radicals in the Atmosphere

As noted above, energetic electromagnetic radiation in the atmosphere may produce atoms or groups of atoms with unpaired electrons called free radicals.

Free radicals are involved with most significant atmospheric chemical phenomena and are of the most importance in the atmosphere. Because of their unpaired electrons and the strong pairing tendencies of electrons under most circumstances, free radicals are highly reactive; therefore, they generally have short lifetimes. It is important to distinguish between high reactivity and instability. A totally isolated free radical or a single atom, such as an O atom, would be quite stable; it wants to react, but there is nothing around for it to react with. Therefore, free radicals and single atoms from diatomic gases (such as O from O_2) tend to persist under the rarefied conditions of very high altitudes because they can travel long distances before colliding with another reactive species. However, unlike free radicals, electronically excited species have finite, generally very short lifetimes because they can lose energy through emission of photons of electromagnetic radiation without having to react with another species.

A key aspect of chemical processes in the atmosphere is that of chain reactions. These occur when a series of reactions involving particular reactive intermediates, usually free radicals goes through a number of cycles. Most commonly, an atmospheric chain reaction begins with the photochemical dissociation of a molecular species to form free radicals, proceeds through a series of reactions that each generates additional free radicals, and ends with a chain-terminating reaction when free radicals react to form stable species as shown by the following reaction of two methyl radicals:

$$H_3C \cdot + H_3C \cdot \longrightarrow C_2H_6 \tag{6}$$

An example of an important chain reaction sequence occurs with chlorofluorocarbons (Freons) in the stratosphere. Extremely stable chlorofluorocarbons consisting of carbon atoms to which are bonded fluorine and chlorine atoms were once widely used as refrigerant fluids in air conditioners, as aerosol propellants for products such as hair spray, and for foam blowing to make very porous plastic or rubber foams. Dichlorodifluoromethane, CCl_2F_2, was used in automobile air conditioners. Released to the atmosphere, this compound remained as a stable atmospheric gas until it got to very high altitudes in the stratosphere. In this region, ultraviolet radiation of sufficient energy ($h\nu$) is available to break the very strong C-Cl bonds:

$$CCl_2F_2 + h\nu \longrightarrow \cdot CCl_2F_2 + Cl \cdot \tag{7}$$

releasing extremely reactive Cl $\cdot$ atoms in which the dot represents a single unpaired electron remaining with the Cl atom when the bond in the molecule breaks. There are oxygen atoms and molecules of ozone, O_3, also formed by photochemical processes in the stratosphere. A chlorine atom produced by the photochemical dissociation of CCl_2F_2 as shown in reaction (7) can react with a molecule of O_3 to produce O_2 and another reactive free radical species, $ClO \cdot$. This species can react with free O atoms which are present along with the ozone to regenerate Cl atoms, which in turn can react with more O_3 molecules. These reactions are shown below:

$$Cl \cdot + O_3 \longrightarrow O_2 + ClO \cdot \tag{8}$$

$$ClO \cdot + O \longrightarrow O_2 + Cl \cdot \tag{9}$$

These are chain reactions in which $ClO \cdot$ and $Cl \cdot$ are continually reacting and being regenerated, the net result of which is the conversion of O_3 and O in the atmosphere to O_2. One Cl atom can bring about the destruction as many as 10000 ozone molecules! Ozone serves a vital protective function in the atmosphere as a filter for damaging ultraviolet radiation, so its destruction is a very serious problem that has resulted in the banning of chlorofluorocarbon manufacture.

The hydroxyl radical, $HO \cdot$, which is the single most important reactive intermediate species in atmospheric chemical processes, is formed by several mechanisms. At higher altitudes it is produced by photolysis of water:

$$H_2O + h\nu \longrightarrow HO \cdot + H \tag{10}$$

In the presence of organic matter, hydroxyl radical is produced in abundant quantities as an intermediate in the formation of photochemical smog.

Hydroxyl radical is most frequently removed from the troposphere by reaction with methane or carbon monoxide:

$$CH_4 + HO \cdot \longrightarrow H_3C \cdot + H_2O \tag{11}$$

$$CO + HO \cdot \longrightarrow CO_2 + H \tag{12}$$

The reactive methyl radical, $H_3C \cdot$, and the hydrogen atom produced in the preceding reactions undergo additional reactions in the atmosphere. Reaction with hydroxyl radical is the most important means by which greenhouse gas methane and toxic pollutant carbon monoxide are removed from the atmosphere. Other important atmospheric trace species that react with hydroxyl radical and are thus removed from the atmosphere include sulfur dioxide, hydrogen sulfide, nitric oxide, and a variety of hydrocarbons.

Selected from "*Stanley E Manahan, Environmental Science and Technology, A Sustainable Approach to Green Science and Technology, Taylor & Franicis Group, USA, 2007*"

Words and Expressions

aesthetics　　*n.* 美学，美丽，好坏
formaldehyde　　*n.* 甲醛
currency　　*n.* 货币，传媒
rarefied　　*adj.* 稀薄的

stratosphere   *n.* 同温层，平流层
quantum   *n.* 量子
fragment   *n.* 碎片
unshared   *adj.* 独享的
ground-state   基态
fluorescence   *n.* 荧光发射
phosphorescence   *n.* 磷光发射
photodissociation   *n.* 光解
photoionization   *n.* 光电离
insofar as   在……范围，到……程度
hydroperoxyl   *n.* 氢过氧化
methyl radical   甲基自由基
chlorofluorocarbon（Freon）   *n.* 氯氟烃（氟里昂）

Notes

① These, in turn are strongly tied to absorption of solar energy, interactions between the gas phase of the atmosphere and small solid particles suspended in it, and interchange of chemical species with the geosphere. 可译为：这些（化学现象），依此与太阳能吸收，大气的气相和悬浮在大气中固体颗粒以及地圈层化学种类相互作用紧密地联系在一起。

② sources and sinks   源与阱。例如，一个循环过程如卡诺循环必须由热源与热阱构成，热工质从热源中获得热量，向热阱排放热，然后经压缩机做功完成一个循环。

③ Planck's constant   普朗克常数。量子光学中的一个物理常数。

④ Two other generally reactive and unstable species in the atmosphere are free radicals composed of atoms or molecular fragments with unshared electrons, and ions consisting of charged atoms or molecular fragments. 可译为：大气中两个其他常见反应和不稳定的物种是自由基，自由基是由具有不共享电子的原子或分子组成，以及由带电原子或带电分子组成的离子。

Exercises

1. Put the following into Chinese.

   photolysis of water, vital protective function, photochemical smog, hydroxyl radical, electromagnetic radiation, predominant organic species, rarefied, quantum

2. Put the following into English according to the text.

   基态   气溶胶   链反应   质子   光电离   自由基   能见度   氧化剂   还原剂   光分解

3. Give a brief description of Photochemical Processes.
4. Translate "the first paragraph" of the text into Chinese.
5. Give a brief summary of Photochemical Processes in the Atmosphere.

Reading Material: Oxidation Process in the Atmosphere

The 21% (dry basis) by volume content of molecular O_2 makes the atmosphere thermodynamically oxidizing. One prominent manifestation of this condition is the tendency for oxidizable materials to corrode when exposed to the atmosphere. Iron, for example, exposed to moist air tends to rust:

$$4Fe + 3O_2 + xH_2O \longrightarrow 2Fe_2O_3 \cdot xH_2O \tag{1}$$

From the standpoint of atmospheric chemistry, however, the oxidizing tendency of the atmosphere is shown by the conversion of reduced molecular species to oxidized forms. It is this feature of the atmosphere exposed to sunlight that results in the formation of photochemical smog. Among the simple molecular species that enter the atmosphere in relatively reduced forms and that are oxidized are the following:

$$2CO + O_2 \longrightarrow 2CO_2 \tag{2}$$
$$CH_4 + 2O_2 \longrightarrow CO_2 + 2H_2O \tag{3}$$
$$4NO + 3O_2 + 2H_2O \longrightarrow 4HNO_3 \tag{4}$$
$$2SO_2 + O_2 + 2H_2O \longrightarrow 2H_2SO_4 \tag{5}$$
$$H_2S + 2O_2 \longrightarrow H_2SO_4 \tag{6}$$

Although shown here in a very simple form, these reactions actually represent processes that may involve many steps, photochemistry, and reactive intermediates, particularly hydroxyl radical. Oxidation reactions may also occur on particle surfaces and in solution in aqueous aerosol droplets, which are strongly exposed to atmospheric oxygen. Another aspect of these reactions is that the products are often acidic—mildly acidic CO_2 from carbon-containing species, and strongly acidic nitric and sulfuric acid from nitrogen oxides and gaseous sulfur species, respectively.

Reducing agents, such as those shown above, may be quite stable in dry air that is not exposed to sunlight. However, the absorption of photons, from solar radiation starts processes that result in oxidation. As a simple example, the photochemical dissociation of nitrogen dioxide,

$$NO_2 + h\nu \longrightarrow NO + O \tag{7}$$

can produce reactive O atoms that can react with oxidizable molecules,

$$CH_4 + O \longrightarrow H_3C \cdot + HO \cdot \tag{8}$$

to begin the series of reactions that forms the final oxidized products (in this case CO_2 and H_2O). intermediate hydroxyl radical, $HO \cdot$, can abstract H atoms from hydrocarbons,

$$CH_4 + HO \cdot \longrightarrow H_3C \cdot + H_2O \tag{9}$$

or add to molecules such as NO_2,

$$HO \cdot + NO_2 \longrightarrow HNO_3 \tag{10}$$

to bring about oxidations. Chain reactions can be involved, such as the following sequence that regenerates NO_2 from NO:

$$H_3C \cdot (\text{from reaction 8}) + O_2 \longrightarrow H_3COO \cdot \tag{11}$$

$$H_3CO \cdot NO \longrightarrow NO_2 \text{(back to reaction 7)} + H_3CO \cdot \qquad (12)$$

The NO_2 product may undergo photochemical dissociation to produce O atoms and again initiate processes that result in oxidation.

A feature of the photochemical atmosphere, particularly when it is polluted by nitrogen oxides and hydrocarbons, is the generation of strong oxidant molecules. The most common example of a strong organic oxidant species is peroxyacetyl nitrate, PAN, formed from the reaction of $H_3CC(O)OO \cdot$ radical with NO_2:

$$H_3CC(O)OO \cdot + NO_2 + M\text{(energy-absorbing third molecule)} \longrightarrow CH_3C(O)OONO_2 + M \qquad (13)$$

The most prominent inorganic oxidant is ozone, generated by reactions such as,

$$O + O_2 + M\text{(energy-absorbing third molecule)} \longrightarrow O_3 + M \qquad (14)$$

One of the ways in which ozone acts as an oxidant is to add to unsaturated compounds to form reactive ozonide.

Acid-Base Reactions in the Atmosphere

Acid-base reactions occur between acidic and basic species in atmosphere. The atmosphere is normally at least slightly acidic because of the presence of a low level of carbon dioxide, which dissolves in atmospheric water droplets and dissociates slightly:

$$CO_2\text{(g, with water)} \longrightarrow CO_2\text{(aq)} \qquad (15)$$

$$CO_2\text{(aq)} + H_2O \longrightarrow H^+ + HCO_3^- \qquad (16)$$

In terms of pollution, however, strongly acidic HNO_3 and H_2SO_4 formed by the atmospheric oxidation of N oxides, O_2, and H_2S are much more important because they lead to the formation of damaging acid rain.

As reflected by the generally acidic pH of rainwater, basic species are relatively less common in the atmosphere. Particulate calcium oxide, hydroxide, and carbonate can get into the atmosphere from ash and ground rock and can react with acids such as in the following reaction:

$$Ca(OH)_2\text{(s)} + H_2SO_4\text{(aq)} \longrightarrow CaSO_4\text{(s)} + H_2O \qquad (17)$$

The most important basic species in the atmosphere is gas-phase ammonia (NH_3). The greatest source of atmospheric ammonia is from biodegradation of nitrogen-containing biological matter and from bacterial reduction of nitrate:

$$NO_3^-\text{(aq)} + 2\{CH_2O\}\text{(biomass)} + H^+ \longrightarrow NH_3\text{(g)} + 2CO_2 + H_2O \qquad (18)$$

Ammonia is particularly important as a base in the atmosphere because it is the only water-soluble base present at significant levels in the atmosphere. Dissolved in atmospheric water droplets, it plays a strong role in neutralizing atmospheric acids:

$$NH_3^-\text{(aq)} + HNO_3\text{(aq)} \longrightarrow NH_4NO_3\text{(aq)} \qquad (19)$$

$$NH_3^-\text{(aq)} + H_2SO_4\text{(aq)} \longrightarrow NH_4HSO_4\text{(aq)} \qquad (20)$$

These reactions have three effects: (1) They result in the presence of NH_4^+ ion in the atmosphere as dissolved or solid salts, (2) they serve in part to neutralize acidic constituents of the atmosphere, and (3) they produce relatively corrosive ammonium salts such as NH_4NO_3.

Selected from "*Stanley E Manahan, Environmental Science and Technology, A Sustainable Approach to Green Science and Technology, Taylor & Francis Group, USA, 2007*"

Unit 8

Text: Type and Sources of Air Pollutants [I]

What Is Air Pollution? Air pollution is normally defined as air that contains one or more chemicals in high enough concentrations to harm humans, other animals, vegetation or materials. There are two major types of air pollutants. **A primary air pollutant** is a chemical added directly to the air that occurs in a harmful concentration. It can be a natural air component, such as carbon dioxide, that rises above its normal concentration, or something not usually found in the air, such as a lead compound emitted by cars burning leaded gasoline. **A secondary air pollutant** is a harmful chemical formed in the atmosphere through a chemical reaction among air components. Serious air pollution usually results over a city or other area that is emitting high levels of pollutants during a period of air stagnation. The geographic location of some heavily populated cities, such as Los Angeles and Mexico City, makes them particularly susceptible to frequent air stagnation and pollution buildup[①].

We must be careful about depending solely on concentration values in determining the severity air pollutants. By themselves, measured concentrations tell us nothing about the danger caused by pollutants, because threshold levels, synergy, and biological magnification are also determining factors[②]. In addition, we run into the issue of conflicting views of what constitutes harm.

Major air pollutants following are the 11 major types of air pollutants:

1. *Carbon oxides*: carbon monoxide (CO), carbon dioxide (CO_2).

2. *Sulfur oxides*: sulfur dioxide (SO_2), sulfur trioxide (SO_3).

3. *Nitrogen oxides*: nitrous oxide (N_2O), nitric oxide (NO), nitrogen dioxide (NO_2).

4. *Hydrocarbons* (organic compounds containing carbon and hydrogen): methane (CH_4), butane (C_4H_{10}), benzene (C_6H_6).

5. *Photochemical oxidants*: ozone (O_3), PAN (a group of peroxyacylnitrates), and various aldehydes.

6. *Particulates* (solid particles or liquid droplets suspended in air): smoke, dust, soot, asbestos, metallic particles (such as lead, beryllium cadmium), oil, salt spray, sulfate salts.

7. *Other inorganic compounds*: asbestos, hydrogen fluoride (HF), hydrogen sulfide (H_2S), ammonia (NH_3), sulfur acid (H_2SO_4), nitric acid (HNO_3).

8. *Other organic* (carbon-containing) *compounds*: pesticides, herbicides, various alcohols, acids, and other chemicals.

9. *Radioactive substances*: tritium, radon, emissions from fossil fuel and nuclear pow-

er plants.

10. *Heat*.

11. *Noise*.

The following Table summarizes the major sources of these pollutants.

Table major air pollutants and sources

Pollutants	Sources	Pollutants	Sources
Carbon oxides carbon monoxide (CO)	Forest fires and decaying organic matter; incomplete combustion of fossil fuels (about two-thirds of total emissions) and other organic matter in cars and furnaces; cigarette smoke	*Hydrocarbons*	Incomplete combustion of fossil fuels in automobiles and furnaces; evaporation of industrial solvents and oil spills; tobacco smoke; forest fires; plant decay (about 85 percent of emissions)
carbon dioxide (CO_2)	Natural aerobic respiration of living organisms; burning of fossil fuels	Asbestos	Asbestos mining; spraying of fire-proofing insulation in buildings; deterioration of brake linings
Sulfur oxides (SO_2 and SO_3)	Combustion of sulfur-containing coal and oil in homes, industries, and power plants; smelting of sulfur-containing ores; volcanic eruptions	Metals and metal compounds	Mining; industrial processes; coal burning; automobile exhaust
Particulates dust, soot, and oil	Forest fires, wind erosion, and volcanic eruptions; coal burning; farming, mining construction, road building, and other land-clearing activities; chemical reactions in the atmosphere; dust stirred up by automobiles; automobile exhaust; coal-burning electric power and industrial plants	Other inorganic compounds Hydrogen sulfide (H_2S)	Chemical industry; petroleum refining
		Ammonia (NH_3)	Chemical industry; petroleum refining
Pesticides and herbicides	Agriculture; forestry; mosquito control	Sulfuric acid (H_2SO_4)	Reaction of sulfur trioxide and water vapor in atmosphere; chemical industry
Nitrogen oxides (NO and NO_2)	High-temperature fuel combustion in motor vehicles and industrial and fossil fuel power plants; lightning	Nitric acid (HNO_3)	Reaction of sulfur trioxide and water vapor in atmosphere; chemical industry
		hydrogen fluoride (HF)	Petroleum refining; glass etching; aluminum and fertilizer production
Photochemical oxidants	Sunlight acting on hydrocarbons and nitrogen oxides	*Noise*	Automobiles, airplanes, and trains; industry; construction

Selected from "*Tyler G., et al., Living in the environment, Wadsworth publishing company, 1982*"

Words and Expressions

aldehyde *n.* 醛，乙醛

asbestos *n.* 石棉

beryllium *n.* 铍

cadmium *n.* 镉

herbicide *n.* 除草剂，阻碍植物生长的化学剂

peroxyacylnitrate *n.* 过氧酰基硝酸酯

pesticide *n.* 杀虫剂；农药

primary air pollutant *n.* 一次大气污染物

secondary air pollutant *n.* 二次大气污染物

stagnation *n*. 停滞；迟钝；萧条
tritium *n*. 氚，超重氢

Notes

① The geographic location of some heavily populated cities, such as Los Angeles and Mexico City, makes them particularly susceptible to frequent air stagnation and pollution buildup. 可译为：一些人口非常稠密的城市，如洛杉矶和墨西哥城的地理位置，使得它们经常特别易受空气流动停滞和污积物增加的影响。

② By themselves, measured concentrations tell us nothing about the danger caused by pollutants, because threshold levels, synergy, and biological magnification are also determining factors. 可译为：测量的浓度自身并未告诉我们有关污染物造成的危害的信息，因为临界浓度、协同作用和生物放大效应都是决定因素。

Exercises

1. Put the following words or phrase into Chinese.
   primary pollutant, secondary pollutant, air stagnation, nitrous oxide, nitric oxide, nitrogen dioxide, soot, dust, smog, ozone, herbicide, pesticide
2. Put the following into English.
   正常浓度   严重污染的   决定因素   光化学氧化物   液体微滴
   放射性物质 不完全氧化   含硫的   风化   汽车尾气
3. Translate the first paragraph of the text into Chinese.
4. Translate the following passage into Chinese.

    To escape the smog you might go home, close the doors and windows, and breathe in clean air. But a number of scientists have found that the air inside homes and offices is often more polluted and dangerous than outdoor air on a smoggy day. The indoor pollutants include: (1) nitrogen dioxide and carbon monoxide from gas and wood-burning stoves without adequate ventilation; (2) carbon monoxide, soot, and cancer-causing benzopyrene (from cigarette smoke); (3) various organic compounds from aerosol spray cans and cleaning products; (4) formaldehyde (which causes cancer in rats) from urea-formaldehyde foam insulation, plywood, carpet adhesives, and particle board; (5) radioactive radon and some of its decay products from stone, soil, cement, and bricks; and (6) ozone from the use of electrostatic air cleaners.
5. Give a brief summary of sources of air pollution.

Reading Material: Type and Sources of Air Pollutants [Ⅱ]

What are PCBs?

There are 209 possible chlorinated biphenyls, ranging in physical characteristics. The mono- and dichloro biphenyls (27323-18-8), (25512-42-9) are colorless crystalline compounds that when burned in air give rise to soot and hydrogen chloride. The most important products are trichlorobiphenyls, tetrachlorobiphenyls, pentachlorobiphenyls and/or hexachlorobiphenyls.

Chlorinated biphenyls are soluble in many organic solvents and in water only in the ppm range. Although chemically stable (including to oxygen of the air) they can be hydrolyzed to oxybiphenyls under extreme conditions forming toxic polychlorodibenzofurans.

The PCB class of compounds received substantial attention and notoriety when in 1968, in Japan, accidental poisoning occurred by cooking rice in bran oil contaminated by PCBs. Over 1000 patients suffered from various morbid symptoms. A similar poisoning occurred in Taiwan in 1979. Causative agents were considered to be coconaminants of PCBs such as polychlorinated dibenzofurans that are secondarily formed during heating of PCBs congeners in commercial PCB mixtures and require a second look at PCB toxicity.

PCB regulation

Because of concerns regarding PCB's health effects and evidence of presence and persistence in the environment further manufacture of the chemical was banned under The 1976 Toxic Substance Control Act. PCB regulations provide deadlines for removal of most in-use capacitors and transformers containing PCBs and limit time for storage for disposal to one year. EPA has allowed continued use of PCBs in electrical transformers and capacitors when the agency did not pose unreasonable risk. Capacitors, except those in isolated areas should have been removed by October 1988, and transformers of a certain size in or near commercial buildings should be removed by October 1990.

EPA regulations require that PCBs taken out of service be disposed of either by specially designed high temperature incinerators needed to break high concentrations of PCBs to harmless components or by alternate destruction methods approved by the agency. Oils contaminated with low concentrations ($50 \sim 500$ ppm) may be disposed by high efficiency boilers.

The limited number of incinerators approved for PCB incineration and the high cost of building additional incinerators have given incentive for alternate destruction methods. Alternate technologies must be capable of operating as effectively as EPA's incineration efficiency.

PCBs in electrical transformers

There were 304 million lb of PCBs used as electrical fluid in approximately 150000 askarel (non-flammable electrical fluid) transformers in the United States. About 70000 PCB transformers are in or near commercial buildings that are open to the public. About 40000 of these transformers are owned by electrical utilities. Approximately 15000 of these transformers are used in the food and feed industry.

Utilities and other industries must maintain or dispose of approximately 150000 askarel-type transformers that may develop leaks. Each year, an estimated 317 askarel-type transformers can be expected to leak. Each will lose about 5. 3 gal or 66 lb of PCBs.

Transformers classification

The EPA has three classification for transformers: PCB transformers contain 500 or more ppm PCBs. They must be inspected quarterly for leaks, and detailed records must be kept. No maintenance work involving removal of the core is allowed.

At the end of the transformers' useful life, it must be destroyed in an EPA-approved facility, or the transformer liquid must be incinerated and the carcass landfilled. The courts and the EPA have held the original transformer owner liable for leakage that may occur for as long as the carcass remains in the landfill.

PCB-contaminated transformers contain between 50 and 400 ppm PCBs and require annual inspection. The rule concerning disposal, maintenance, and record keeping are less restrictive and less costly than those for PCB transformers.

Non-PCB transformers have less than 50 ppm PCBs and are exempt from the burdensome rules and requirements that apply to PCB and PCBs contaminated transformers. Non-PCB transformers are granted favor under the Toxic Substances Control Act.

Analysis of PCBs

Analytical methods most frequently used for detecting chlorinated biphenyls are capillary column gas chromatography coupled with mass spectrometry in the MID (Multiple Ion Detection) mode and capillary column gas chromatography with ECD (Electron Capture Detector). HPLC (High Pressure Liquid Chromatography) and infrared spectroscopy are applicable to a limited extent. Summary PCB determinations are also possible, though not usual. These require exhaustive chlorination and measurement of the decachlorbipheny content or dechlorination and subsequent biphenyl measurement.

Infrared tested on PCBs

Infrared thermal treatment technology was field tested for effectiveness on PCBs contaminated soil at two Superfund sites during 1986. The transportable pilot system consisted of a primary furnace through which solid/semisolid wastes were conveyed on a wire mesh belt. The heat source was supplied by electric glow bars in lieu of gaseous fuels. Residence times and furnace temperatures could be controlled manually or automatically giving uniform ash. A secondary chamber, heated by electric or gaseous fuels, followed the primary furnace and provided temperatures in excess of $220°F$. The system was designed to comply with all RCRA and TSCA requirements.

Select from "Paul N. C. , *High hazardous pollutants: Asestos, PCBs, dioxins, biomedical wastes, Pollution Engineering, Vol. 21, No. 2, 1989*"

Unit 9

Text: Indoor Air Quality

Fundamentals of Indoor Air Quality

For more than a decade, the U. S. Environmental Protection Agency (U. S. EPA) has ranked indoor air pollution among the top five risks to public health. Potential hazards that may be associated with indoor air include particulates, microbes, and chemicals. Defining the relative impact of specific indoor air pollutants is difficult because of individual genetic susceptibility and exposure to multiple hazards①. Harmful exposure levels are defined for many chemical air pollutants. However, for other indoor pollutants, such as mold, little data is available with respect to acceptable exposure levels. Frequently, the consequences of exposure are evaluated based on retrospective, or even anecdotal, evidence.

In industrialized countries, about 90% of an individual's time is spent indoors, where air quality may be two to five times worse than outdoor air. Approximately one-third of all buildings are expected to have indoor air quality (IAQ) problems at some point during their operational lifespan. Health effects of poor IAQ range from mild and acute (cold and flu-like symptoms, headaches, and nausea), to severe and chronic (allergies, asthma, developmental disorders, cancer, and death)②. While a number of these building-related illnesses have been traced to specific building problems, conditions of complex symptomology related to chemical and/or biological IAQ problems are often vaguely diagnosed as sick building syndrome (SBS) and likely involve multiple pollutants acting collectively or synergistically③. Excessive complaints, related to IAQ, are generated within 30% of new and remodeled buildings, worldwide. In addition, nearly one in four U. S. workers believes there are air quality problems in their work environments, and most thought that these problems affected their work performance.

Sources of Indoor Air Pollutants

1　Volatile Organic Compounds

Volatile organic compounds (VOCs) contain carbon and a variety of other elements, such as hydrogen, oxygen, fluorine, chlorine, bromine, sulfur, or nitrogen. They readily vaporize at room temperature, releasing noxious vapors into the air. Common household products, such as those designed for cleaning, disinfecting, degreasing or waxing, often contain high levels of VOCs. Some of these are potent toxins, capable of persisting in the indoor air space for a long time. Even seemingly innocuous materials, like cosmetics, air fresheners, or dry-cleaned fabrics can emit harmful toxins, such as diethanolamine, paradichlorobenzene, and perchloroethylene, respectively. While many organic chemicals have no known health effects, others range from mild irritants to carcinogens. In addition, research is lacking on the health effects of long-term exposures and individual susceptibilities to different hazards are also unknown.

(1) **Pesticides**

Each year, 75% of all U.S. households report the indoor use of at least one pesticide product. According to the EPA, 80% of a person's exposure to pesticides occurs indoors, and up to 12 different pesticides have been found in single household air samples. Pesticides include insecticides, termiticides, rodenticides, fungicides, and microbial disinfectants, and may be formulated as sprays, foggers, liquids, sticks, or powders. By definition, they aim to kill their designated targets and thus often contain highly toxic compounds. Exposure to pesticides accounts for nearly 80000 poisonings of children each year, primarily due to ingestion, but inhalation and dermal exposures also occur. Health risks include systematic illness, organ damage, respiratory irritation and disease, neurological disorders, and a host of mild symptomology such as headaches, nausea, and dizziness.

(2) **Construction materials and furnishings.**

Materials used in home and building construction and indoor furnishings frequently contain and emit hazardous compounds. Historically, asbestos was used in a variety of construction materials for insulation and as a fire-retardant. Many asbestos products have been removed from buildings and continued use has been banned, but older homes may still contain potentially harmful materials. If in good condition and left undisturbed, asbestos products are generally not a risk; however, aerosolized fibers can be respired, damaging the lungs and abdominal lining and leading to irreversible scarring and cancer.

Carpeting and installation materials, such as adhesives and padding, are known to emit volatile organic compounds. Eye, nose, and throat irritation; headaches; rashes; coughing; fatigue; and shortness of breath have all been reported following new carpet installations. Carpeting may also act as a sink for a multitude of chemical and biological pollutants, including pesticides, dust mites, and molds that may collect in carpet fibers and remain protected from cleaning and vacuuming.

2 Combustion Products

Combustion products such as oil, gas, kerosene, wood, and coal are common to indoor environments due to the use of fuel-burning applications, space heaters, fireplaces, and gas or wood stoves. If not properly vented, harmful pollutants, such as CO, NO_2, and particle or chemical irritants may be released into the air. Improperly installed or maintained chimneys or other ventilation outlets can cause a backdraft of pollutants into the home. Environmental tobacco smoke (ETS) is also considered a combustion product, and exposure via second-hand smoke is a major health concern.

(1) **Carbon monoxide and nitrogen dioxide**

Carbon monoxide (CO) and nitrogen dioxide (NO_2) are both colorless, odorless gases that can cause potent health effects even at low levels. At high concentrations, CO inhalation results in rapid illness and death. About 500 Americans die from unintentional CO poisoning each year. Symptoms of exposure include headache, dizziness, weakness, nausea, vomiting, chest pain, and confusion. Nitrogen dioxide is an irritant of the mucus membranes of the eye, nose, and throat. High-level exposure results in respiratory irritation, shortness of breath, and potentially contributes to respiratory infections and lung diseases like emphysema.

(2) Environmental tobacco smoke (ETS)

According to the U.S. government's Agency for Health Care Policy and Research (AHCPR), 46 million Americans smoke, exposing themselves and others to hazardous products of combustion. About 75% of the nicotine from a cigarette ends up in the atmosphere. In fact, secondhand smoke contains higher concentrations of toxic and cancer-causing chemicals than smoke inhaled directly, with more than 4000 chemicals, including 200 known toxins, and 69 known carcinogens. Each year, an estimated 3000 nonsmoking Americans die of lung cancer, and moer than 35000 die of heart disease from secondhand smoke. Furthermore, an estimated 150000 to 300000 children, below 18 months of age experience respiratory tract infections because of secondhand smoke exposures. Young children exposed to secondhand smoke are also more likely to experience increased incidences of pneumonia, ear infections, bronchitis, coughing, wheezing, and asthma.

3 Lead and Radon

Lead particulates settle on surfaces of indoor environments and are readily redispersed into the air via air currents common to indoor climates. Contaminated lead particles may be inhaled and ingested. Both result in absorption into the blood, where it is then distributed to soft tissues and bone. As lead accumulates over time, it can eventually affect nearly every system in the body.

Radon is a radioactive gas released during the natural decay of uranium, a common component of rocks, soil, and water. Radon enters homes through cracks, drains, and even wells. Once inside, the colorless, odorless gas can become trapped in living spaces, increasing in concentration and remaining undetected. Exposure via breathing the radioactive gas into the lungs results in lung cancer and possible death. Radon gas is listed as second leading cause of lung cancer in the United States. An estimated 21000 deaths per year could be prevented by addressing radon gas exposures, particularly among smokers, who are known to be at increased risk due to synergistic interactions between radon and smoking.

4 Biological Pollutants

The term bioaerosol encompasses any biological agent transmitted by the airborne route, i.e., bacteria, viruses, mold, mites, cockroach particles, pollen, and animal dander and saliva. All of these agents have been associated with adverse health effects, including allergies and asthma, and often coexist in common environments. Allergic diseases have significantly increased worldwide over the last 30 years. More than 50% of all allergic diseases are caused by allergens out of the indoor environment, and nearly one in six persons in the U.S. is effected by hypersensitivity reactions. Indoor molds are a rising concern with ambient air contamination, ranking among the most important allergens of indoor environments. Biological agents can persist in dust particles and animal droppings, or proliferate in humid micro zones until they become aerosolized. Natural breezes, air-conditioning systems, humidifiers, and active movement all create eddies that aid in the aerosolization of spores, microbes, and other toxins. In moist environments, mold and bacterial grow in less than 72 hours, colonizing solid surfaces and subsequently releasing toxins, particulates, and allergens into air spaces.

Select from "*Ian L. Pepper, Charles P. Gerba, Mark L. Brusseau. Environmental Pollution Science, Elserier Inc. USA, 2006*"

Words and Expressions

particulates *n.* 颗粒物
microbes *n.* 微生物，细菌
genetic *n.* 遗传，基因
susceptibility *n.* 敏感性，敏感度，易感受性
retrospective *a.* 追溯的，回顾的
anecdotal *a.* 趣闻的，轶事的
allergy *n.* 过敏症，变态反应
asthma *n.* 哮喘
complex symptomology 综合征
syndrome *n.* 症状
synergistical *a.* 合作的，协同的
termiticides *n.* 杀白蚁剂
rodenticides *n.* 杀鼠剂
fungicides *n.* 杀真菌剂
microbial disinfectants 微生物消毒剂
nausea *n.* 恶心，反感
dizziness *n.* 头晕，混乱
adhesives *n.* 黏合剂
padding *n.* 填充剂
backdraft=backdraught *n.* 倒转，回程
vomiting *n.* 呕吐
pneumonia *n.* 肺炎
radon *n.* 氡
pollen *n.* 花粉
proliferate *v.* 细胞繁殖，激增

Notes

① Defining the relative impact of specific indoor air pollutants is difficult because of individual genetic susceptibility and exposure to multiple hazards. 参考译文：定义特定室内空气污染物对人的相关影响是很难的，由于每个人的基因有其特殊性与敏感性，而人类暴露于多种有害物质中。

② Health effects of poor IAQ range from mild and acute (cold and flu-like symptoms, headaches, and nausea), to severe and chronic (allergies, asthma, developmental disorders, cancer, and death). 可译为：室内空气质量不良对健康的影响从轻微和急性（如感冒症状、头痛和恶心）发展到严重和慢性（过敏、哮喘发展到失调、癌症和死亡）。

③ While a number of these building-related illnesses have been traced to specific building problems, conditions of complex symptomology related to chemical and/or biological IAQ problems are often vaguely diagnosed as sick building syndrome (SBS) and likely involve multiple pollutants acting collectively or synergistically. 可译为：尽管许多与建筑物有关的疾病归结于特定的建筑物问题，但与化学或生物相关的室内空气质量问题综合征通常被误诊为建筑疾病症，这些综合征可能涉及共同或协同作用的多种污染物。

Exercises

1. Put the following words or phrase into Chinese.

   Microbe hypersensitivity reactions allergic disease backdraft of pollutants termiticides multiple hazards asbestos products aerosolized fibers symptoms of exposure rodenticides mucus membranes adverse health effect fungicides microbial disinfectants

2. Put the following into English.

   协同作用 过敏症 敏感性 基因工程 综合征 填充剂（料） 细胞繁殖 肺癌 生物污染物 氡辐射 细菌与病毒 过敏性疾病

3. Translate the first paragraph of the text into Chinese.
4. Give a brief summary of types or sources of biological pollutants.

Reading Material: Indoor Air Pollution

Indoor air pollution from fires for cooking and heating has affected human health for thousands of years. A detailed autopsy of a 4the-century Native American woman, frozen shortly after death, revealed that she suffered from black lung disease from breathing very polluted air over many years. The pollutants included hazardous particles from lamps that burned seal and whale blubber. This same disease has long been recognized as a major health hazard for underground coal miners and has been called "coal miners' disease". As recently as the mid-1970s, black lung disease was estimated to be responsible for about 4000 deaths each year in the United States.

People today spend between 70% and 90% of their time in enclosed places—homes, workplaces, automobiles, restaurants, and so forth—but only recently have we begun to fully study the indoor environment and how pollution of that environment affects our health. The World Health Organization has estimated that as many as one in three people may be working in a building that causes them to become sick, and as many as 20% of public schools in the United States have problems related to indoor air quality. The EPA considers indoor air pollution one of the most significant environmental health hazards people face in the modern workplace.

Hurricane Katrina in 2005 left a great number of people homeless. In response, the Federal Emergency Management Agency (FEMA) provided thousands of trailers for people to live in. That sounded like a great idea until complaints started to come in about health problems of people living in the trailers. A study by the Centers for Disease Control and Prevention (CDC) confirmed that the mobile homes suffered from indoor air pollution by formaldehyde in their construction materials. Formaldehyde is a chemical widely used in the manufacture of building materials, as well as a number of other products. It is considered a probable human carcinogen (a substance that causes or promotes cancer). Common symptoms of exposure to formaldehyde include irritation of the skin, nose, throat, and eyes. People with asthma may be more sensitive to the chemical, and their symptoms may be worse. Since discovery of high levels of formaldehyde in mobile homes in late 2007, plans have gone forward to remove the remaining people, particularly those experiencing symptoms of formaldehyde toxicity.

The history of formaldehyde in the mobile homes provided to Katrina victims is a sad legacy of the entire way our federal government responded to Hurricane Katrina and its aftermath. It is also important because it brings to the public consciousness the potential problems of indoor air pollution, which is often more significant than outdoor air pollution.

The sources of indoor air pollution are incredibly varied and can arise from both human activities and natural processes. Some of common indoor air pollutants, together with guidelines for allowable exposure, are listed as follows:

1. Heating, ventilation, and air-conditioning systems may be sources of indoor air pollutants, including molds and bacteria, if filters and equipment are not maintained properly. Gas and oil furnaces release carbon monoxide, nitrogen dioxide, and particles.

2. Restrooms may have a variety of indoor air pollutants, including secondhand smoke, and also molds and fungi due to humid conditions.

3. Furniture and carpets often contain toxic chemicals (formaldehyde, organic solvents, asbestos) that may be released over time in buildings.

4. Coffee machines, fax machines, computers, and printers can release particles and chemicals, including ozone, which is highly oxidizing.

5. Pesticides can contaminate buildings with cancer-causing chemicals.

6. Fresh-air intake that is poorly located—for example, above a loading dock or first-floor restaurant exhaust fan—can bring in air pollutants.

7. People who smoke indoors, perhaps in restaurants or offices, pollute the indoor envi-

ronment, and even people who smoke outside buildings, particularly near open or revolving doors, may cause pollution as the smoke (secondhand smoke) is drawn into and up through the building by the chimney effect.

8. Remodeling, painting, and other such activities often bring a variety of chemicals and materials into a building. Fumes from such activities may enter the building's heating, ventilation, and air-conditioning system, causing widespread pollution.

9. A variety of cleaning products and solvents used in offices and other parts of buildings contain harmful chemicals whose fumes may circulate throughout a building.

10. People can increase carbon dioxide levels; they can emit bio-effluents and spread bacterial and viral contaminants.

11. Loading docks can be sources of organics from garbage containers, of particulates, and of carbon monoxide from vehicles.

12. Radon gas can seep into a building from soil; rising damp (water), which facilitates the growth of molds, can enter foundations and rise up walls.

13. Dust mites and molds can live in carpets and other indoor places.

14. Pollen can come from inside and outside sources.

Selected from *"Daniel B. Botkin, Edward A. Keller. Environmental Science, Eight Edition, John Wiley & Sons, Inc. Printed in Asia, 2012"*

Unit 10

Text: New Technologies of Air Pollution Control [I]

Biofiltration: An Innovative Air Pollution Control Technology for VOC Emissions

The concept of using microorganisms for the removal of environmentally undesirable compounds by biodegradation has been well established in the area of wastewater treatment for several decades. Not until recently, however, have biological technologies been seriously considered in the United States for the removal of pollutants from other environmental media. Moreover, while bioremediation techniques are now being applied successfully for the treatment of soil and groundwater contaminated by synthetic organics, at present there is very little practical experience with biological systems for the control of air contaminants among environmental professionals in the U. S. In fact, few environmental professionals in this country appear to be aware that "biofiltration," i. e., the biological removal of air contaminants from off-gas streams in a solid phase reactor, is now a well established air pollution control (APC) technology in several European countries, most notably The Netherlands and Germany[①].

In Europe, biofiltration has been used successfully to control odors, and both organic and inorganic air pollutants that are toxic to humans [air toxics, as well as volatile organic compounds (VOC) from a variety of industrial and public sector sources]. The development of biofiltration in West Germany, most of which took place in the late 1970s and the 1980s, was brought about by a combination of increasingly stringent regulatory requirements and financial support from federal and state governments. The experiences in Europe have demonstrated that biofiltration has economic and other advantages over existing APC technologies, particularly if applied to off-gas streams that contain only low concentrations (typically less than 1000 ppm as methane) of air pollutants that are easily biodegraded[2].

The principal reasons why biofiltration is not presently well recognized in the U.S., and has been applied in only a few cases, appear to be a lack of regulatory programs, little governmental support for research and development, and lack of descriptions written in the English language. Specifically, regulatory programs in most U.S. states have not yet addressed, in a comprehensive manner, the control of air toxics, VOC and odors from smaller sources. Moreover, little financial support for investigating the applicability of biofiltration for these sources has been provided by government agencies, Finally, although several important papers on biofiltration have been published in English, most of the technical reports summarizing recent results, were published in German.

Despite these current obstacles, biofiltration is likely to find more widespread application in the U.S. in the near future. In addition to a few existing installations, several full-scale projects are currently in the planning stage or under construction. For example, the first large scale system for VOC control in California, a biofilter to treat ethanol emissions from an investment foundry in Los Angeles area is being planned with co-funding by the South Coast Air Quality Management District (SCAQMD). A detailed description of this system, and an analysis of its performance are provided elsewhere.

The major objective of the present paper is to provide a comprehensive review of important aspects of biofiltration in order to more widely disseminate about this innovative APC technology, and to encourage its implementation where appropriate in the U.S.. Many of the more complex technical and engineering issues related to the development and use of biofiltration cannot be discussed in great depth here. However, we identify and summarize such issues, and refer to more detailed publication. We also note that, in addition to biofitration, other biological APC systems are now in use in Europe for the control of organic off-gas, including "bioscrubbers" and trickling filters. Various articles on these related technologies, which are not discussed here, are available in other literature.

Suggestions to treat odorous off-gases by biological methods can be found in literature as early as 1923 when Bach discussed the basic concept of the control of H_2S emissions from sewage treatment plants. Reports on the application of this concept dating back to the 1950s were published in the U.S. and in West Germany. Pomeroy received U.S. patent No. 2,793,096 in 1957 for a soil bed concept and describes a successful soil bed installation in California. Around 1959 a soil bed was also installed at a municipal sewage treatment plant in Nuremberg, West Germany for the control of odors from an incoming sewer main.

In the U.S., the first systematic research on the biofiltration of H_2S was conducted by Carlson and Leiser in the early 1960s. Their work included the successful installation of several soil filters at a wastewater treatment plant near Seattle and demonstrated that biodegradation rather than sorption accounted for the odor removal.

During the following two decades, several researchers in the U.S. have further studied the soil bed concept and demonstrated its usefulness in several full scale applications. Much of the knowledge about the technology is owed to Hinrich Bohn who has investigated the theory and potential applications of soil beds for more than 15 years. Successful soil bed applications in the U.S. include the control of odors from rendering plants, and the destruction of propane and butane released from an aerosol can filling operation.

While soil beds have been shown to control certain types of odors and VOC efficiently and at fairly low capital and operating cost, their use in the U.S. has been limited by the low biodegradation capacity of soils and the correspondingly large space requirements for the beds[3]. It is estimated that the total number of biofilter and soil bed installations in the U.S. and Canada is currently less than 50 and that they are predominantly used for odor control.

Selected from "*Gero Leson et al., Biofiltration: an innovative air pollution control technology for VOC emission, Air & waste management association, vol. 41, No. 8, 1991*"

Words and Expressions

biodegradation　*n.* 生物降解
bioremediation　*n.* 生物治理
bioscrubber　*n.* 生化洗涤器
butane　*n.* 丁烷
ethanol　*n.* 乙醇
foundry　*n.* 铸工厂；玻璃（制造）厂
full-scale　工业规模
municipal sewage treatment plant　市政污水处理厂
obstacle　*n.* 障碍物
predominantly　*ad.* 主要地
propane　*n.* 丙烷
rendering plants　炼油厂
stringent　*a.* 严格的，严厉的
trickling filter　滴滤池（器）

Notes

① In fact, few environmental professionals in this country appear to be aware that "biofiltration," i.e., the biological removal of air contaminants from off-gas streams in a solid phase reactor, is now a well established air pollution control (APC) technology in several European countries, most notably The Netherlands and Germany. 参考译文：实际上，

这个国家几乎没有环境的专家似乎意识到在好几个欧洲国家（最显著的是荷兰和德国），"生物过滤"，即从固相反应器气流中空气污染物的生物去除法，是一种充分确立的空气污染控制技术。

② The experiences in Europe have demonstrated that biofiltration has economic and other advantages over existing APC technologies, particularly if applied to off-gas streams that contain only low concentrations (typically less than 1000 ppm as methane) of air pollutants that are easily biodegraded. 可译为：欧洲的经验已经证明，生物过滤较已有的空气污染技术有着经济的和其他方面的优越性，特别是用于仅含有低浓度（典型地是甲烷浓度小于 1000ppm）易生物降解的空气污染物的尾气。

③ While soil beds have been shown to control certain types of odors and VOC efficiently and at fairly low capital and operating cost, their use in the U.S. has been limited by the low biodegradation capacity of soils and the correspondingly large space requirements for the beds. 参考译文：虽然已证实土壤床可用相当低的投资和操作费用来控制某些类型的臭气和挥发性有机化合物。但是，土壤的低生物降解能力和相当大的占地要求限制了土壤床在美国的应用。

Exercise

1. Put the following into English.

   尾气　　　可应用性　　　工业规模　　　土壤床　　　生物过滤器
   固定资本　　易生物降解的

2. Put the following into Chinese.

   VOC，APC，regulatory program，financial support，operating cost，biodegradation capacity，environmental media，biological，technologies，inorganic air pollutants

3. Translate the following sentences into Chinese.

   （1）Due to lower operating costs, biofiltration can provide significant economic advantages over other APC technologies if applied to off-gases that contain readily biodegradable pollutants in low concentrations.

   （2）Moreover, while bioremediation techniques are now being applied successfully for the treatment of soil and groundwater contaminated by synthetic organics, at present there is very little practical experience with biological systems for the control of air contaminants among environmental professionals in the U.S. .

   （3）The principal reasons why biofiltration is not presently well recognized in the U.S., and has been applied in only a few cases, appear to be a lack of regulatory programs, little governmental support for research and development, and lack of descriptions written in English language.

4. Fill in each of the following blanks with an appropriate word given.

   Over the last decade, biofiltration has _____ (developed/grew) in Europe into a cost-effective and environmentally benign control technology for gaseous air _____ (contaminants/wastes). If used to treat low concentration _____ (off-gases/sewage) it can provide treatment at significantly lower cost than _____ (competing/concurrent)

APC technologies. Biofiltration is particularly attractive _____ (because of/since more) the savings in operating costs and its low specific energy demand. Current research efforts are primarily targeted at improving the control of essential operating parameters and increasing biodegradation _____ (rates/velocity), in particular _____ (for/of) recalcitrant compounds.

Continued regulatory trends toward more stringent control of VOC, air toxics and odors _____ (similar/contrary) to controls in Germany and several other European countries, will soon generate demand in the U. S. for this technology. Several experienced European biofilter companies have entered the U. S. marketplace, usually _____ (in/for) cooperation with U. S. engineering or equipment firms.

In order to successfully apply biofiltration in the U. S. to appropriate _____ (projects/program), several requirements are necessary. First, All available APC _____ (alternatives/choices) must be thoroughly evaluated in _____ (particular/however), for off-gases with high concentrations of poorly biodegradable organic compounds, other options will usually be more feasible. Second, off-gases _____ (must/can) be accurately characterized and biofilters carefully designed. Pilot testing may be required, as well as the pretreatment of the off-gas by a particulate filters. Finally, a minimum level of attention and maintenance must be _____ (provided/supplied) by the operator of a biofilter.

_____ (However/Moreover), if these requirements can be met, biofiltration is likely to be used in a variety of APC applications over the next few _____ (years/weeks). Agency support of such projects could assist in overcoming the natural reluctance of regulated industries to try _____ (poor/unknown) technologies. It will be the _____ (duty/responsibility) of environmental consulting and engineering firms to identify the APC needs of an industrial operation on a case-by-case basis and, _____ (if/unless) biofiltration is found to be the method of choice, to design and build the appropriate system.

Reading Material: New Technologies of Air Pollution Control [Ⅱ]

Capital and operating costs associated with managing wastes produced by SO_2 emissions control equipment are important factors in evaluating and comparing alternative SO_2 control systems. Over the past eight years, EPRI has conducted a number of studies to provide utilities with cost information on waste management for conventional wet scrubbing. More recently, a comprehensive investigation has been undertaken to assess waste management costs and issues for five alternate sulfur-reduction technologies: spray drying, atmospheric fluidized-bed combustion, limestone furnace injection, dry-sodium injection, and advanced coal cleaning. For each of the five, studies have characterized waste products; developed engineering designs for effective waste handling, disposal, and/or utilization; and estimated waste management costs.

The first study, completed in late 1986, evaluated spray dryer wastes. Results showed that these wastes can be managed without excessive operating and economic problems for utilities or adverse environmental impacts. However, on a dollar-per-ton-disposed basis, spray dryer waste management costs were found to be higher than those for either conventional fly ash or scrubber sludge alone. This finding indicates that cost estimates for new and retrofit spray dryer applications must be revised upward from those produced earlier by EPRI, under which waste management costs from all sulfur-reduction processes were assumed to be equal.

The process of a typical spray dryer waste management system involves five basic activities: (1) transfer of waste material from the spray dryer and particulate control device to a temporary storage facility; (2) storage; (3) conditioning to improve the material's handling characteristics (e.g., adding water to reduce fugitive dust emissions); (4) transportation to a disposal site or to a location where the material is utilized; and (5) for waste material not utilized, placement and containment at a disposal area.

Methodology and Results

The spray dryer waste management study was conducted in four steps: characterizing spray dryer waste, surveying existing and planned spray dryer installations, developing conceptual designs and case studies for new and retrofit spray dryer installations, and evaluating the utilization potential of spray dryer waste.

Waste characterization

Waste material from seven utility spray dryer installations was analyzed to measure physical, chemical, and leachate properties important in the design of a waste management system. Results show that while properties of spray dryer waste are generally similar to those of conventional fly ash, spray dryer waste is finer and more caustic, has a higher heat of hydration, and produces a more alkaline leachate. It can also become tacky when wet, and exhibits self-hardening properties similar to cement. Flowability test results indicate that normal, relatively dry spray dryer waste is generally free to average flowing. With higher moisture contents, however, the material may set up and create serious storage problems. Flowability testing also indicates that spray dryer waste aerates easily and retains air once aerated. Such characteristics are advantageous if controlled fluidized handling is used, but also indicate a potential for flooding, flushing, and flow rate limitation problems. (Flooding and flushing refer to conditions where an aerated bulk solid behaves like a fluid and flows uncontrollably through an outlet or feed mechanism.)

Survey of Waste Management Systems

Data from 18 existing and planned spray dryer waste management systems were collected. Results indicate that waste management methods and equipment for these systems are similar to those used for conventional fly ash. Separate conveyors are typically used to transfer wastes from the spray dryer and particulate control device to a temporary storage silo. To transfer waste from the particulate collector, dilute-phase pressure pneumatic systems are most common, although pneumatic vacuum systems also are being used successfully. To transfer waste from the spray dryer, mechanical conveyers (primarily drag chain, bucket or screw-type) were initially installed in the existing facilities surveyed. However, all but five of these conveyors were converted to pressure or vacuum pneumatic systems due to operational problems caused by excessive wear and abrasion of parts. Most facilities practicing dry disposal methods condition the wastes with a small amount of water prior to transport by truck to an unlined landfill.

Conceptual Designs and Case Studies

Based on results of the waste characterization and survey, and using the EPRI Technical Assessment Guide (TAG), spray dryer waste management conceptual designs were developed, and new and retrofit spray dryer waste management systems were designed for a hypothetical pulverized-coal-fired power plant burning low-ash western subbituminous coal. For both the new and retrofit designs, collected waste was transferred to a common surge silo-spray dryer waste via a pneumatic system and from there to either a slurry preparation area for recycle or to a disposal silo via a dilute-phase pressure pneumatic conveyance system. Waste withdrawn from the disposal silo is assumed to be conditioned with water and transported by truck to landfill.

For both the new and retrofit cases, the power plant was assumed to be a two-unit 1000-MW station. The new facility was assumed to be equipped with a fabric filter, whereas the existing station was assumed to be using an electrostatic precipitator at the time of the retrofit. Estimated total annual levelized costs for the waste management systems up to and including placement in an unlined landfill were $3029000 (0.53 mills/kWh or $14.69/ton) for the new plant, and $2760500 (0.63 mills/kWh or $17.40/ton) for the retrofit installation.

Utilization

Based on spray dryer waste, physical properties, handling characteristics, chemical composition, environmental effects, and processing requirements, current fly ash utilization options were evaluated and ranked in terms of technical feasibility and marketability. In this analysis, the utilization potential for spray dryer wastes was found to be similar to that for Type "C" self-hardening ashes commonly found at plants in the midwestern Unite States. Of the options considered, the following seven applications were determined to be most attractive:

- Structural fill
- Cement replacement
- Stabilized road base
- Synthetic aggregate
- Lightweight aggregate

- Mineral wool, and
- Brick production

Selected from "*Dean Golden, Spray dryer waste management, JAPCA, Vol. 38, No. 3, 1988*"

Unit 11

Text: Effects of Air Pollution

Only a few examples of air pollution effects can be considered in the sections to follow. All of the examples identify local, observable, or measurable impacts, because it is very difficult to develop direct relationships between specific pollutants and effects for exposures over the longer term or at great distances. A more complete description could fill several books. One of the most obvious local effects of particles in the atmosphere is a reduction in visibility. Only a few decades ago, soiling by soot and smoke was clearly apparent in almost every urban center. Reduction in visibility results in a social cost due to slowdown of air traffic and the need for instrument-guided landing systems.

In the air pollution episodes cited, the local presence of gases was objectionable because of odor, taste, or obvious corrosive or chemical effects. Today these gross sensory insults are rarely encountered. However, subtle health effects persist, such as eye or nose irritation or difficult breathing. In the extreme, health effects extend to the brain (CO) and stomach (several pollutants alone or in combination). Damage to vegetation due to chronic exposure to atmospheric pollutants may be one of the more apparent precursor symptoms leading to identification of chronic air pollution[①].

Health effects

Health effects were the dominant considerations in early air pollution episodes for obvious reasons. While the specific pollutant or groups of pollutants generating the observed effects frequently could not be identified, there was sufficient information to implicate certain pollutants as significant contributors. Early research to relate pollutant concentrations and effects was concentrated on these clearly identifiable pollutants.

The human respiratory system is quite efficient in filtering the larger particles out of the air we breathe. Particles smaller than about $5\mu m$, however, can penetrate to the lungs and be deposited in the alveoli. For example, cigarette smoke particles are smaller than $1\mu m$, and they enter and are deposited in the alveoli.

Some particles are particularly damaging because they adsorb gases which cause more intense irritation locally. Gases also penetrate into the deepest lung pockets. Both particles and gases entering the body through the respiratory system can affect the gastrointestinal system, as was reported in the Donora incident. Some chemical, such as lead, can enter the human bloodstream either from the digestive system (ingestion) or by passing through the lung membranes (the respiratory system), and airborne tritium and a few other chemical can enter the bloodstream through the skin.

Each pollutant affects the human body differently, and records of effects have been assembled relating the intensity to the period of exposure for various pollutants[2]. Health effects of carbon monoxide are clearly detectable at 4 percent carboxyhemoglobin (COHb) in the blood, and possibly as low as 2.5 percent COHb.

Effects on plants and animals

The detrimental impacts of air pollution are not limited to those involving human health. Plants and animals are also susceptible. For example, fluorine is emitted from aluminum, glass, phosphate, fertilizer, and some clay-baking operations in significant quantities. Frequently, the plant damage is observed on the fruit or on the flowers, either of which significantly lowers the value of the crop. Fluorine affects plants at concentrations several orders of magnitude below that at which human health is affected[3].

Fluorine also has an effect at even lower concentrations when it is taken up by shrubs, trees, or grass which is subsequently eaten by cattle or other animals. The animals may develop fluorosis, although the plants may not show signs of damage. The animals act as concentrators of the fluorine, resulting in poor animal health and associated lower animal value or survival capability. Some heavy metals, such as mercury and lead, and most radionuclides also become concentrated in plants and animals, frequently in specific organs.

Different plants and animals have different susceptibilities to pollutants. For example, sugar maple can tolerate relatively high concentrations of sulfur dioxide alone, but it is susceptible to damage from exposure to SO_2 and O_3 together. White pine, on the other hand, is very sensitive to damage from either pollutant alone.

Effects on materials

Sulfur and nitrogen oxides react in the atmosphere to form acidic compounds which attack metal surfaces, a problem which has been particularly acute for the communications, switchgear, and computer industries. Fluorine is particularly reactive, and at high atmospheric concentrations etching of glass has been observed. These impacts are taken into consideration when sensitive components are designed, and the required protective measures or design modifications add to the cost of the item being produced. Hydrogen sulfide in the ambient atmosphere reacts with lead oxide in white paint to form lead sulfide, so that white houses have been observed to take on a brownish tint overnight. Accelerated aging of synthetics and rubber due to exposure to atmospheric oxidants has already been noted. (Both the strength and color of the material are affected).

It is extremely difficult to estimate the dollar value of accelerated deterioration of materials and aesthetic items such as buildings, statues or horticultural plantings. It is generally agreed that damage from atmospheric pollution is measured in the billions of dollars.

Ambient air quality standards

Experimental results and research have been prepared for many other pollutants. Using these data, we can establish ambient air quality standards at which pollutant effects will not be detectable. These data have led to the establishment of primary standards to protect health only and secondary standards to protect against effects to other systems. Standards are established for various samples averaging periods, because effects are dependent upon pollutant concentration and exposure time.

Select from "*Mackeziel L. Davis, Davis A Cornwell. Introduction to Environmental Eng., Mcgraw-Hill Inter. Edition, Chem. Series, 2nd Edition. US, 1991*"

Words and Expressions

instrument-guided 导航
episode *n*. 事件，插曲
Insult *n*. 损害
subtle *a*. 敏感的；微妙的；精细的
precursor *n*. 先驱
symptom *n*. 症状；征兆
alveoli *n*. 肺泡
bloodstream *n*. 血流
airborne tritium 空气中的氚
carbon monoxide 一氧化碳
carboxyhemoglobin *n*. 碳氧血红蛋白
fluorine *n*. 氟
phosphate *n*. 磷酸盐
clay-baking 烧瓷，烤瓷
maple *n*. 枫，枫木
sulfur dioxide 二氧化硫
switchgear *n*. 电力设施，接电装置
brownish *a*. 呈褐色的

Notes

① Damage to vegetation due to chronic exposure to atmospheric pollutants may be one of the more apparent precursor symptoms leading to identification of chronic air pollution. 可译为：长期暴露于大气污染物中的植被所遭受的破坏也许是空气长期污染症状的一种较明显的前兆。

② Each pollutant affects the human body differently, and records of effects have been assembled relating the intensity to the period of exposure for various pollutants. 可译为：每种污染物对人体的影响是不同的，而且所收集的各种影响效果的记录都与暴露于各种污

染物期间的强度有关。

③ Fluorine affects plants at concentrations several orders of magnitude below that at which human health is affected. 可译为：氟对人体健康影响的浓度比它对植物影响的浓度低好几个数量级。

Exercises

1. Put the following into Chinese.
   （1） Reduction in visibility results in a social cost due to slowdown of air traffic and the need for instrument-guided landing systems.
   （2） These impacts are taken into consideration when sensitive components are designed, and the required protective measures or design modifications add to the cost of the item being produced.

2. Put the following into English.
   肺囊 氟中毒 煤烟 呼吸系统 过滤 吸附 浓度 硫化氢 硫化铅

Reading material: Control of Air Pollution by Oxidation

The final fate of VOCs is mostly to be oxidized to CO_2 and H_2O, as a fuel either in engines or furnaces in an incinerator in a biological treatment device, or in the atmosphere (forming ozone and fine particles). VOCs-containing gas streams that are too concentrated to be discharged to the atmosphere but not large enough to be concentrated and recovered are oxidized before discharge, either at high temperatures in an incinerator or at low temperatures by biological oxidation.

Combustion

Some air pollutants consist mostly of material that, when burned, produce other materials that are harmless or much less harmless than the original ones. Materials ion in this category are largely compounds of carbon, hydrogen, oxygen, nitrogen, and sulfur. Most VOCs fall into this category, as do some others. Thus the treatment here mainly considers VOCs, but it also applies to other materials. Converting these materials from the harmful or objectionable form to a harmless or less objectionable form by combustion may be the most economical and practical way of solving a VOC emission problem.

Some examples of interest are

$$CO + \frac{1}{2}O_2 \longrightarrow CO_2 \tag{1}$$

$$C_6H_6 + 7\frac{1}{2}O_2 \longrightarrow 6CO_2 + 3H_2O \tag{2}$$

$$H_2S + \frac{3}{2}O_2 \longrightarrow H_2O + SO_2 \tag{3}$$

In Eqs. (1) and (2), carbon monoxide, which has well-known harmful human health effects, and benzene, which is a reactive hydrocarbon, smog precursor, and carcinogen, are converted to harmless materials. Equation (3) may appear not to belong here because SO_2 is a significant air pollutant for which we have a national control program. However, H_2S (hydrogen sulfide) is very toxic at high concentrations and has a strong smell (rotten eggs) that most of us can detect at much lower concentrations than we can detect SO_2 (the smell of burning sulfur or of a wooden match being lighted). Typical estimates of the minimum concentration that average humans can smell are 0.0005 ppm for H_2S and 0.5 ppm for SO_2. Thus, if the problem is caused by the odor of H_2S, it can frequently be alleviated by burning the H_2S to form the less odorous SO_2. This makes sense only for low H_2S concentrations; higher concentrations are treated in an entirely different way.

As an interesting sidelight on the last example, it is common practice to add a strong-smelling sulfur compound-called an odorant-to odorless natural gas or propane to help detect leaks. This is an absolutely necessary safety feature; without the smell for warning, a stove burner accidentally left on but not ignited could lead to a disastrous explosion! Normally this odorant is a mercaptan, which is a near chemical relative of H_2S. In normal combustion this odorant is converted to water, carbon dioxide, and sulfur dioxide. These odorants are effective at such low concentrations that the resulting SO_2 concentration in the combustion gases is below our ability to detect, and we find the combustion products odorless. The concentration is also low enough that the SO_2 produced probably has no health effects.

The nitrogen present in compounds being incinerated normally enters the atmosphere partly as N_2, NO, or NO_2. The latter two are pollutants for which we have control programs. Thus, we would not use incineration to limit the emissions of these materials. However, many organic chemicals contain small amounts of organic nitrogen that will pass through the combustion process and emerge as NO or NO_2, e.g.

$$2(CH_3)_3N + 11\frac{1}{2}O_2 \longrightarrow 6CO_2 + 9H_2O + 2NO \qquad (4)$$

In this case trimethylamine, one of the smelliest compounds known (decaying fish), is oxidized to harmless CO_2 and H_2O plus some NO that will contribute, after further atmospheric oxidation, to any regional NO_2 and ozone problems.

In most incinerators, the chlorine content of the material burned will leave the incinerators as hydrochloric acid, HCl. Municipal waste incinerators generally receive enough polyvinyl chloride plastic that HCl in the exhaust gas can cause corrosion in the incinerator and damage to the neighborhood. Most modern municipal waste incinerators have some kind of acid capture technology to prevent the emission of this HCl.

As discussed previously, at incinerator temperatures some metals become vapors, e.g, mercury, cadmium, zinc. Municipal waste contains some of these materials (mercury in dry cells, cadmium and zinc in metal plating); their emissions can be a problem.

The combustible pollutant can be a gas, a mist droplet, or a solid particulate. The gases most often by burning are CO, hydrocarbons of all kinds, and strong odor producers, which are normally VOCs containing sulfur and/or nitrogen. Almost all VOCs can be destroyed by incineration. The particulates treated by combustion are largely hydrocarbon smokes; examples are the smoke from meat smokehouses, the fumes formed in asphalt processing and paint baking, miscellaneous tars, etc. The principles of dealing with these are the same, although the details may differ.

In all combustion and incineration, incomplete combustion is a permanent problem. Many of the intermediate products produced between the original components and the final carbon dioxide and water are themselves harmful, e.g., aldehydes, dioxins, furans. Incomplete combustion of a waste stream can produce an exhaust gas that is more harmful than the input gas. All air pollution incinerators (and municipal and hazardous waste incinerators) are designed to ensure that combustion is as complete as practical, and that the emissions of products of incomplete combustion are as small as possible.

Biological Oxidation

As discussed above, the ultimate fate of VOCs is to be oxidized to CO_2 and H_2O, either in our engines or furnaces, or incinerators, or in the environment. Many microorganisms will carry out these reactions fairly quickly at room temperature. They form the basis of most sewage treatment plants (oxidizing more complex organic material than the simple VOCs of air pollution interest). Microorganisms can also oxidize the VOCs contained in gas or air streams. The typical biofilter (not truly a filter but commonly called one; better called a highly porous biochemical reactor) consists of the equivalent of a swimming pool, with a set of gas distributor pipes at the bottom, covered with several feet of soil or compost or loam in which the microorganisms live. The contaminated gas enters through the distributor pipes and flows slowly up through soil, allowing time for the VOCs to dissolve in the water contained in the soil, and then to be oxidized by the microorganisms that live there.

Typically these devices have soil depths of 3 to 4 ft, void volumes of 50%, upward gas velocities of 0.005ft/s to 0.5ft/s, and gas residence times of 15 to 60s. They work much better with polar VOCs, which are fairly soluble in water than with HCs whose solubility is much less. The microorganisms must be kept moist, protected from conditions that could injure them, and in some cases given nutrients. Because of the long time the gases must spend in them, these devices are much larger and take up much more ground surface than any of the devices discussed in this chapter. In spite of these drawbacks, there are some applications for which they are economical, and for which they are used industrially.

Select from "*Noel de Never, Air Pollution Control Engineering, 2nd Edition, McGraw-Hill Inc.U.S., 1995*"

PART 3　WATER AND WASTE-WATER TREATMENT

Unit 12

Text: Water Pollution and Pollutants

The relationship between polluted water and disease was firmly established with the cholera epidemic of 1854 in London, England. Protection of public health, the original purpose of pollution control, continues to be the primary objective in many areas. However, preservation of water resources, protection of fishing areas, and maintenance of recreational waters are additional concerns today. Water pollution problems intensified following World War II when dramatic increases in urban density and industrialization occurred. Concern over water pollution reached a peak in the mid-seventies.

Water pollution is an imprecise term that reveals nothing about either the type of polluting material or its source. The way we deal with the waste problem depends upon whether the contaminants are oxygen demanding, algae promoting, infectious, toxic, or simply unsightly. Pollution of our water resources can occur directly from sewer outfalls or industrial discharges (point sources) or indirectly from air pollution or agricultural or urban runoff (nonpoint sources).

Chemically pure water is a collection of H_2O molecules—nothing else. Such a substance is not found in nature—not in wild streams or lakes, not in clouds or rain, not in falling snow, nor in the polar ice caps. Very pure water can be prepared in the laboratory but only with considerable difficulty. Water accepts and holds foreign matter.

Municipal wastewater, also called sewage, is a complex mixture containing water (usually over 99 percent) together with organic and inorganic contaminants, both suspended and dissolved. The concentration of these contaminants is normally very low and is expressed in mg/L, that is, milligrams of contaminant per liter of the mixture. This is a weight-to-volume ratio used to indicate concentrations of constituents in water, wastewater, industrial wastes, and other dilute solutions.

Microorganisms. Wherever there is suitable food, sufficient moisture, and an appropriate temperature, microorganisms will thrive. Sewage provides an ideal environment for a vast array of microbes, primarily bacteria, plus some viruses and protozoa. Most of these microorganisms in wastewater are harmless and can be employed in biological processes to convert organic matter to stable end products. However, sewage may also contain pathogens from the excreta of people with infectious diseases that can be transmitted by contaminated water. Waterborne bacterial diseases such as cholera, typhoid, and tuberculosis, viral diseases such as infectious hepatitis, and the protozoan-caused dysentery, while seldom a problem now in developed countries, are still a threat where properly treated water is not available for public use. Tests for the few pathogens that might be present are difficult and time consuming, and standard practice is to test for other more plentiful organisms that are always present (in the billions) in the intestines of warm-blooded animals, including humans.

Solids. The total solids (organic plus inorganic) in wastewater are, by definition, the residues after the liquid portion has been evaporated and the remainder dried to a constant weight at 103℃. Differentiation between dissolved solids and undissolved, that is, suspended, solids are accomplished by evaporating filtered and unfiltered wastewater samples. The difference in weight between the two dried samples indicates the suspended solids content. To further categorize the residues, they are held at 550℃ for 15 minutes. The ash remaining is considered to represent inorganic solids and the loss of volatile matter to be a measure of the organic content.

Suspended solids (SS) and volatile suspended solids (VSS) are the most useful. SS and BOD (biochemical oxygen demand) are used as measures of wastewater strength and process performance. VSS can be an indicator of the organic content of raw wastes and can also provide a measure of the active microbial population in biological processes.

Inorganic constituents. The common inorganic constituents of wastewater include:

1. *Chlorides and sulphates*. Normally present in water and in wastes from humans.

2. *Nitrogen and phosphorous*[①]. In their various forms (organic and inorganic) in wastes from humans, with additional phosphorous from detergents.

3. *Carbonates and bicarbonates*[②]. Normally present in water and wastes as calcium and magnesium salts.

4. *Toxic substances*. Arsenic, cyanide, and heavy metals such as Cd, Cr, Cu, Hg, Pb, and Zn are toxic inorganics which may be found in industrial wastes.

In addition to these chemical constituents, the concentration of dissolved gases, especially oxygen, and the hydrogen ion concentration expressed as pH are other parameters of interest in wastewater.

Organic matter. Proteins and carbohydrates[③] constitute 90 percent of the organic matter in domestic sewage. The sources of these biodegradable contaminants include excreta and urine from humans; food wastes form sinks; soil and dirt from bathing; washing, and laundering; plus various soaps, detergents, and other cleaning products.

Various parameters are used as a measure of the organic strength of wastewater. One method is based on the amount of organic carbon (total organic carbon, or TOC[④]) present in the waste. TOC is determined by measuring the amount of CO_2 produced when the organic carbon in the sample is oxidized by a strong oxidizer and comparing it with the amount in a standard of known TOC.

Most of the other common methods are based on the amount of oxygen required to convert the oxidizable material to stable end products. Since the oxygen used is proportional to the oxidizable material present, it serves as a relative measure of wastewater strength. The two methods used most frequently to determine the oxygen requirements of wastewater are the COD[5] and BOD[6] tests. The COD, or chemical oxygen demand, of the wastewater is the measured amount of oxygen needed to chemically oxidize the organics present; the BOD, or biochemical oxygen demand, is the measured amount of oxygen required by acclimated microorganisms to biologically degrade the organic matter in the wastewater.

BOD is the most important parameter in water pollution control. It is used as a measure of organic pollution, as a basis for estimating the oxygen needed for biological processes, and as an indicator of process performance.

The amount of organic matter in water or wastewater can be measured directly (as TOC, for example), but this doesn't tell us whether the organics are biodegradable or not. To measure the amount of biodegradable organics[7], we use an indirect method in which we measure the amount of oxygen used by a growing microbial population to convert (oxidize) organic matter to CO_2 and H_2O in a closed system. The oxygen consumed, or biochemical oxygen demand (BOD), is proportional to the organic matter converted, and therefore BOD is a relative measure of the biologically degradable organic matter present in the system. Because biological oxidation continues indefinitely, the test for ultimate BOD has been arbitrarily limited to 20 days, when perhaps 95 percent or more of the oxygen requirement has been met. Even this period, however, is too long to make measurement of BOD useful, so a five-day test, BOD_5[8], carried out at 20℃, has become standard. The rate of the BOD reaction depends on the type of waste present and the temperature and is assumed to vary directly with the amount of organic matter (organic carbon) present.

Selected from "*J. Glynn Henry, Gray W. Heinke. Environmental Science and Engineering, Prentice-Hall International Editions, Pretice. Hall Englewood Cliffs, NJ, USA, 1989*"

Words and Expressions

epidemic　*n*. 流行，传染；*a*. 流行的，传染的
algae　*n*. alga 的复数，海藻，藻类
unsightly　*a*. 难看的，丑的
outfall　*n*. 出口，排出
runoff　*n*. 排水，流放口
sewage　*n*. 污水，废水
organic　*a*. 有机的，机体的
inorganic　*a*. 无机的，人造的
microorganisms　*n*. 微生物
microbe　*n*. 微生物
bacteria　*n*. 细菌（复数）；bacterium（单数）
protozoa　*n*. 原生动物
pathogen　*n*. 病菌，病原体

tuberculosis *n.* 结核病，肺结核
excreta *n.* 排泄物，粪，便
urine *n.* 尿
oxidizer *n.* 氧化剂
acclimate *v.* 服水土，适应环境
intestine *a.* 内部的；*n.* 肠
filter *v.* 过滤
categorize *vt.* 把……分类
chlorides and sulphates 氯化物和硫化物
calcium and magnesium 钙和镁盐
arsenic *n.* 砷
cyanide *n.* 氰
hepatitis *n.* 肝炎
degrade *v.* 降解

Notes

① phosphorous 亚磷的（三价磷）；phosphorous acid 亚磷酸；phosphoric 磷的（五价磷）。
② carbonates，bicarbonates 碳酸盐，碳酸氢盐。
③ carbohydrates 碳水化合物，主要含氢和碳。
④ TOC（total organic carbon） 总有机碳。
⑤ COD（chemical oxygen demand） 化学需氧量。有机物的化学氧化反应所需要的氧量。
⑥ BOD（biological oxygen demand） 生化需氧量。BOD 并不是对某一污染物直接量度，而是测量细菌或微生物分解有机物所需的氧量。
⑦ biodegradable organics 可生物降解的有机物。
⑧ BOD_5 BOD 的测试标准化，即可将水样置于 20℃ 的黑暗环境五天，所得的结果为 BOD_5（前五天的耗氧量）。

Exercises

1. Put the following into Chinese.
   treatment facilities，municipality，population equivalent，basement flooding，per capital per day，runoff，domestic sewage，the type of terrain
2. How many inorganic constituents are there in wastewater?
3. What is the main idea of part of "Stormwater"?
4. What is the definition of "total solids"?
5. Translate part of "Industrial wastewater" into Chinese.

Reading Material：Wastewater

1. Municipal Wastewater

The excreted waste from humans is called **sanitary sewage**, and wastewater from residential areas is referred to as **domestic sewage** and includes kitchen, bath, laundry, and floor drain wastes. These, together with the liquid wastes from commercial and industrial establishments, are termed **municipal wastewater.** This wastewater is normally collected in a public sewerage system (sewers, manholes, pumping stations, etc.) and directed to treatment facilities for safe disposal.

Quantities of municipal wastewater are commonly determined from water use. Because water is consumed by humans, utilized in industrial products, used for cooling, and required for activities such as lawn watering and street washing, only 70 to 90 percent of the water supplied reaches the sewers. However, the assumption is frequently made that the water loss is balanced by infiltration (groundwater leakage into the sewer system through poor joints) or storm water, which enters the sanitary sewer system through illicit connections (roof downspouts and road catchbasins) or through manhole openings.

In North America, municipal water use and the resulting wastewater flow range from about 280 liters (75 gallons) per capita per day for small, mainly residential municipalities to over 900 Lpcd (240 gpcd) for large industrialized cities. These daily averages are based on annual amounts. However, flows vary from day to day. From larger (500000 people) to smaller (10000 people) municipalities, flows as a percentage of the annual daily average might range from maximums of 150 to 200 percent on a daily basis and 200 to 300 percent on an hourly basis and minimums of 70 to 50 percent on a daily basis and 50 to 30 percent on an hourly basis, with the extremes in each case (the second value) applying to the smaller municipality. The quality of municipal wastewater varies with the proportion of residential, commercial, and industrial contributors and the nature of the industrial wastes which the system receives.

The total dry weight of the wastewater constituents from residential areas will be relatively constant, but their concentration will vary with the amount of water used. As the community grows larger and more diversified, the addition of contaminants from commercial and industrial establishments will change the characteristics of the wastewater.

2. Industrial Wastewater

Wastewaters from industries include employees' sanitary wastes, process wastes from manufacturing, wash waters, and relatively uncontaminated water from heating and cooling operations. The wastewaters from processing are the major concern. They vary widely with the type of industry. In some cases, pretreatment to remove certain contaminants or equalization to reduce hydraulic load may be mandatory before the wastewater can be accepted into the municipal system. In contrast with the relatively consistent characteristics of domestic sewage, industrial wastewaters, often have quite different characteristics, even for similar industries. For this reason, extensive studies may be necessary to assess pretreatment requirements and the effect of the wastewater on biological processes.

Wastes are specific for each industry and can range from strong (high BOD_5) biodegradable wastes like those from meat packing, through wastes such as those from plating shops and textile mills, which may be inorganic and toxic and require on-site physical-chemical treatment before discharge to the municipal system. The volume or strength of industrial wastewaters is often compared to that of domestic sewage in terms of a **population equivalent** (PE) based on typical per capital contributions.

3. Stormwater

The runoff from rainfall, snowmelt, and street washing is less contaminated than municipal wastewater. It therefore receives little or no treatment before being discharged into storm sewers (for direct disposal into receiving waters) or before being combined with the municipal wastewater for delivery to the wastewater treatment plant.

The quantity of stormwater which runs off from a municipality varies widely with the time of year, the type of terrain, and the intensity and duration of the storms which occur. North America, excluding desert areas, has a range of 250 to 2000mm of annual precipitation across the continent. In temperate areas with, for example, 750 to 900mm of rainfall per year, stormwater runoff would amount to about 25 percent of the total annual municipal wastewater volume. However, during a storm the rate of stormwater runoff can often be several times, and on occasion can be as high as 100 times, the normal flow to sanitary sewers. This explains why, in order to minimize the possibility of sewer backup and basement flooding, the admission of stormwater into a separate sanitary sewer system should be prohibited.

Of the total global annual rainfall, two-thirds may be lost through evaporation and transpiration, with the balance going to surface water and groundwater storage. Even during storms, not all rainfall becomes runoff. The proportion which does varies from about 20 percent for parks and lawns up to 100 percent for roofs and paved areas. An overall average value for a municipality might range between 30 and 50 percent during fairly intense storms.

Stormwater runoff, particularly in cities, contains dust and other particulates from roads, leaves from trees, grass cuttings from lawns and parks, and fallout from air pollution. The concentration of these contaminants is highest when they are first flushed into the sewer system during the early stages of runoff and then decreases as the rain continues.

Selected from "*J. Glynn Henry, Gray W. Heinke. Environmental Science and Engineering, Prentice-Hall International Editions, Pretice. Hall Englewood Cliffs, NJ, USA, 1989*"

Unit 13

Text: Pollution of Inland Waters and Oceans

1. Pollution of Streams and Rivers

When sewage is discharged into a freshwater stream, the stream becomes polluted. This does not mean that the oxygen content drops instantaneously. But the potential for oxygen depletion exists wherever there is sewage. The measure of this potential is the BOD, which rises as soon as the sewage goes in. Now follow the water downstream from "Polutionville." Three processes are going on, all at the same time.

- Process 1. The bacteria are feasting on the sewage. Because of this action, the amount of sewage in the water is decreasing, so the BOD is going down.
- Process 2. As the bacteria consume the sewage, they also use dissolved oxygen, so that concentration, too, starts to decrease.
- Process 3. Some of the lost oxygen is being replenished from the atmosphere and from photosynthesis by the vegetation in the stream.

For the first 50km or so downstream, the natural ability of the river to recover its oxygen (process 3) simply cannot keep up with the feasting bacteria (process 2), so the dissolved oxygen concentration goes down. The fish begin to die, but it is not the sewage that is killing them. (In fact, the sewage provides food.) Instead, the fish die from lack of oxygen, beginning when the dissolved oxygen concentration falls below about 4mg per liter, depending on the particular species.

The fish kills start about 15km downstream from the introduction of the raw sewage. In time, as the sewage is used up by bacteria, the BOD goes down (process 1), the consumption of oxygen also slows down, and the natural ability of the river to recover (process 3) becomes predominant. The river then begins to repurify itself. About 90km downstream the fish begin to survive again, and at about 140km the oxygen content has increased to its former, unpolluted level.

Of course, if additional sewage is discharged before recovery is complete, as shown in the illustration at 160km, the river becomes polluted again. When sources of pollution are closely spaced, pollution becomes practically continuous. Rivers in such a condition, which unfortunately can be found near densely polluted areas all over the world, support no fish, are high in bacterial content (usually including pathogenic organisms), appear muddily bluegreen from choking algae, and, in extreme cases, stink from putrefaction and fermentation.

2. Pollution of Lakes

Water flows more slowly in lakes than in rivers, and therefore the residence time of water is much longer in a lake. In the summer, the upper waters, called the **epilimnion**[①] (the "surface lake") are warmed by the Sun. These warmer waters, being lighter than the colder ones below, remain on top and maintain their own circulation and oxygen-rich conditions. The lower lake waters (the **hypolimnion**[①]) are cold and relatively airless. Between the two lies a transition layer, the **thermocline**[①], in which both temperature and oxygen content fall off rapidly with depth. As winter comes on, the surface layers cool and become denser. When they become as dense as the lower layers, the entire lake water circulates as a unit and becomes oxygenated. This enrichment is, in fact, enhanced by the greater solubility of oxygen in colder waters. Furthermore, the reduced metabolic rates of all organisms at lower temperatures result in a lesser demand for oxygen. When the lake freezes, then, the waters below support the aquatic life through the winter.

With the spring warmth, the ice melts, the surface water becomes denser, and again the lake "turns over", replenishing its oxygen supply.

Now, what are the effects of oxygen-demanding pollutants on these processes? During the summer, increased supplies of organic matter serve as nutrients in the oxygenated upper waters; the oxygen is replaced as needed by physical contact with the air and from photosynthesis by algae and other water plants. But some organic debris rains down to the lower depths, which are reached neither by air nor by sunlight. Therefore, in an organically rich or eutrophic lake, the bottom suffers first. Fish that live best at low temperatures are therefore the first to disappear from lakes as the cold depths they seek become depleted of oxygen[②] by the increased inflow of nutrients. These fish are frequently the ones most attractive to human diets, such as trout, bass, and sturgeon.

3. Pollution of Groundwater

The accumulation, use, and recharging of groundwater were discussed before. Pollution as well as purification can occur during these transfers. In recent years, however, the pollution of groundwater in some areas has increased to the point where its quality as drinking water is seriously threatened.

If raw sewage is dumped onto the soil, the liquid percolates into the ground. The soil acts as a filter, blocking large solid particles while allowing the liquid to pass through. Smaller particles and even molecules of contaminants, although not physically blocked, adhere to the soil particles, and they, too, are removed. At the same time, however, the percolating water dissolves mineral matter out of the soil and rock. Therefore, all groundwater contains some dissolved inorganic salts. In rare instances, such dissolve matter may be toxic, as for example when the water seeps through areas containing lead or arsenic minerals. Some groundwaters contain natural concentrations of fluorides, which reduce the incidence of dental cavities in children but which in larger concentrations mostle the teeth.

4. Pollution of the Oceans

For many years people were relatively unconcerned about pollution of the oceans because the great mass of the sea can dilute a huge volume of foreign matter to the point where it has little effect. In recent years this attitude has changed for several reasons. For one, pollutants are often added in such relatively high concentrations in local areas that environmental disruption does occur. In addition, the quantity of pollutants dumped into the ocean has grown so large that some scientists fear that global effects may be significant.

Oil. On March 16, 1978, the oil tanker Amoco Cadiz lost her steering off the coast of Brittany in France. High winds blew the ship ashore, and within the next few days she broke apart, spilling approximately 1.6 million barrels (220 thousand tonnes) of crude oil onto the beaches and into the water. It is virtually impossible to estimate the economic damage done by this spill. Over a million sea birds were killed within a few days of disaster. The oil clogged their feathers and respiratory tracts so that they drowned or died from inhaling the oil. A huge oyster fishery was destroyed. Plankton, fish and other sea animals were killed.

Offshore drilling operations also contribute their share of oil to the sea. The largest spill of all started on June 3,1979, at an offshore oil well. Mud, rapidly followed by oil and gas, started to gush through an unsealed drill pipe. The fumes ignited on contact with the pump motors, the drilling tower collapsed, and the spill was out of control. It took nine months to stop it.

Shipwrecks have occurred ever since people first went to sea in ships. In ancient times these accidents were disastrous for the sailors, their families, and ship owners. Today tanker wrecks and discharges represent an even greater hazard because they are threatening the very life of the sea. Can anything be done? The sea has always been an international domain. No nation owns it, no nation can impose laws concerning it.

Crude oil is crude indeed, in the sense that it consists of many thousands of components of widely differing molecular weights. It is usually a dark brown, smelly liquid, about as thick as engine oil. It is largely composed of hydrocarbons, but there is an appreciable proportion of sulfur, and there are trace concentrations of metals such as vanadium and nickel.

Most hydrocarbons are less dense than water, and therefore the major portion of a mixture of hydrocarbons such as crude oil floats. However, some hydrocarbons are dense enough to sink even in seawater, and these materials, together with a portion of the metallic components, may settle to the bottom, where they have the potential to disrupt generations of aquatic organisms. Furthermore, the oxidation of floating oil also yields some products that are denser than seawater. If a typical crude oil is heated to 100℃, some 12 percent of its volume boils off; if it is heated to 200℃, an additional 13 percent boils off. The total (25 percent) may be considered to be the volatile fraction that will evaporate from the floating oil surface within a few days. The remaining oil is slowly metabolized by bacteria, and some of it slowly evaporates. After about three months, practically all the material that can evaporate has evaporated, and all that can be eaten has been eaten. The persistent remainder is an asphaltic residue, representing about 15 percent of the original oil. These leftovers occur as small tarry lumps all over the Earth's seas.

Plastics. Plastics are generally less biodegradable than oil. Sailors in small boats report that the world's oceans are awash with polystyrene cups in a continuous trail across the Atlantic and Pacific Oceans. Polystyrene breaks down into microscopic globules. We have no idea what long-term effects these plastics may have on marine organisms that ingest them. These plastics have mainly been thrown overboard from boats. Most boating associations now urge their members never even to take anything disposable and plastic to sea and encourage ocean liners to follow suit[3]. But much of the damage has, obviously, already been done.

Other Chemical Wastes. There is no known inexpensive and guaranteed safe method of disposing of highly poisonous chemical wastes, such as byproducts from chemical manufacturing, chemical warfare agents, and pesticide residues. It is cheap and therefore tempting to seal such material in a drum and dump it in the sea. But drums rust, and outbound freighters do not always wait to unload until they reach the waters above the sea's depths. As a result, many such drums are found in the fisheries on continental shelves or are even washed ashore. It is estimated that tens of thousands of such drums have been dropped into the sea.

Of course, all the river pollutants enter the same sink—the world ocean. The organic nutrients are recycled in the aqueous food web. But the chemical wastes from factories and the seepages from mines are all carried by the streams and rivers of the world into the sea.

And where do air pollutants go? Airborne lead and other metals from automobile exhaust, mercury vapor, and the fine particles of agricultural sprays ride the winds and fall into the ocean.

Selected from "*Jonathan Turk, Amos Turk. Environmental Science, Third Edition, Saunders College Publishing, The Drylen Press, USA, 1984*"

Words and Expressions

feast *n.* 宴席；*v.* 享受，使……愉快
replenish *vt.* 添满，充足
downstream *ad.* 下游地；*a.* 下游的
putrefaction *n.* 腐烂，腐败
fermentation *n.* 发酵；激动
eutrophic *a.* 发育正常的；营养良好的
metabolic rates 新陈代谢速度
trout *n.* 鳟鱼
sturgeon *n.* 鲟鱼
oyster *n.* 蚝
asphaltic residue 沥青残留物
leftovers 剩余物
tarry *n.* 焦油，煤胶物质；*a.* 柏油状的
polystyrene 聚苯乙烯
globule *n.* 小球，小珠，滴
ocean liner 海轮
outbound *a.* 驶往国外的，处境的
fishery *n.* 渔场，渔业
seepage *v.* 渗透，渗溢

Notes

① epilimnion 表层水；hypolimnion 深水层；thermocline 热变形层。

② become depleted of oxygen　缺氧。

③ Most boating associations now urge their members never even to take anything disposable and plastic to sea and encourage ocean liners to follow suit. 可译为：目前大多数船协会敦促他们的会员再也不要把废弃的物质和塑料扔到海里，并鼓励航海轮船也这样做。

Exercises

1. Put the following into Chinese.

   unpalatable，volatile chemicals，scale in hot pipes，harbor pathogens，precipitate，turbid，intermittent sources，fixtures

2. Put the following into English.

   合成有机化合物　　微生物性能　　普遍接受　　食品加工　　精馏
   蒸发和冷凝　　　　土壤腐蚀　　　过滤作用　　副产品　　　脱盐

3. （1）Translate the part of "Chemical Characteristics" into Chinese.

   （2）Translate the part of "Other Chemical Wastes" into Chinese.

4. Give a brief summary or an abstract of Part "Pollution of Groundwater" in Chinese.

Reading Material：Water Supply

Water Quality Requirements

1. Water Demand

Total water demand on a municipal water supply system is the sum of all the individual demands (from toilet flushing, lawn watering, industrial cooling, street washing, etc.) during a stated period. Demand is not constant, but varies during the day and with the season. Variations decrease as the period over which we measure the demand increases from hourly, to daily, to monthly, to yearly. Consequently, water demand in a particular community is normally specified in terms of **average daily demand**, defined as follows：

Average daily demand＝total water use in one year (volume) /365 days (time)

Units are m^3/d or million m^3/d, or gallons per day (gpd), or million gallons per day (mgd).

It is often convenient to express the rate of demand per person：

Average daily demand ＝ Average daily demand in community/midyear population in community

The units here may be litters per person (capita) per day (Lpcd) or gallons per person (capita) per day (gpcd).

2. Fluctuations in Water Use

The demands on a water system vary not only from year to year and from season to season, but also from day to day and hour to hour. An example of short-term variation in residential

79

water demand during summer and winter is given. Note that during the early hours of summer evenings, a substantial increase in water consumption may result due to lawn watering. It is common practice to express demand fluctuations as a fraction of the average daily demand.

Most of community fire departments obtain water for fire fighting from the nearest fire hydrant connected to the local water distribution system. If there are no fire hydrants, tank trucks or portable pumps and hoses must bring water from the nearest water source. A water distribution system is designed to provide the larger of the maximum hourly demand or the maximum daily demand, plus the fire demand to any group of hydrants in the system. This fire demand is often the governing requirement in establishing pipe sizes, pumping capacity, and reservoir capacity for cities under 200000 people. The flow required to put out or at least contain a fire in an individual group of buildings can be estimated from a following empirical formula,

$$F = 224CA^{1/2}$$

Where　F = the required fire flow in liters per minute;
　　　　C = a coefficient which takes into account the type of construction, existence of automatic sprinklers, and building exposures. Its value is 1.5 for wood frame.

Water Quality Standards

Water contains a variety of chemical, physical, and biological substances which are either dissolved or suspended in it. Form the moment it condenses as rain, water dissolves the chemical compounds of its surroundings as it falls through the atmosphere, runs over ground surfaces, and percolates through the soil. Water also contains living organisms which react with its physical and chemical elements. For these reasons, water must often be treated before it is suitable for use. Water containing certain chemicals or microscopic organisms may be harmful to some industrial processes while being perfectly adequate for others. Disease-causing (pathogenic) microorganisms in water can render it dangerous for human consumption. Groundwater from limestone areas may be very high in calcium carbonates, making it hard and thus requiring softening before use.

Water quality requirements are established in accordance with the intended use of the water. Quality is usually judged as the degree to which water conforms to physical, chemical, and biological standards set by the user. It is not as easy to measure as water quantity because of many tests needed to verify that these standards are being met. Knowing the water quality requirements of each water use is important in order to determine whether treatment of the raw water is required and, if so, what processes are to be used to achieve the desired quality. Water quality standards are also essential in monitoring treatment processes.

Water is evaluated for quality in terms of its physical, chemical, and microbiological properties. It is necessary that the tests, used to analyze the water as regards each of these properties, produce consistent results and have universal acceptance, so that meaningful comparison with water quality standards can be made.

1. Physical Characteristics

Tastes, odors, color, and turbidity are controlled in public water supplies partly because they make drinking water unpalatable, but also because of the use of water in beverages, food processing, and textiles. Tastes and odors are caused by the presence of volatile chemicals and decomposing organic matter. Measurements for these are conducted on the basis of the dilution needed to reduce them to a level barely detectable by human observation. Color in water is caused by minerals such as iron and manganese, organic material, and colored wastes from industries. Color in domestic water may stain fixtures and dull clothes. Testing is done by comparison with a standard set of concentrations of a chemical that produces a color similar to that found in water. Turbidity, as well as being aesthetically objectionable, is a health concern because the particles involved could harbor pathogens. Surface water sources may range in turbidity from 10~1000 units; however, it is possible for very turbid rivers to have 10000 units of turbidity.

2. Chemical Characteristics

The many chemical compounds dissolved in water may be of natural or industrial origin and may be beneficial or harmful depending on their composition and concentration. For example, small amounts of iron and manganese may not just cause color; they can also be oxidized to form deposits of ferric hydroxide and manganese oxide in water mains and industrial equipment. These deposits reduce the capacity of pipes and are expensive to remove.

Hard waters are generally considered to be those waters that require considerable amounts of soap to produce a foam or lather and that also produce scale in hot water pipes, heaters, boilers, and other units in which the temperature of water is increased materially. Water hardness is expressed as equivalent milligrams per liter of calcium carbonate. The bicarbonates of calcium and magnesium precipitate as insoluble carbonates when carbon dioxide is driven off by boiling. This "temporary" hardness, called **carbonate hardness**, should be limited where it causes scale formation in boilers and industrial equipment. Sulfates, chlorides, and nitrates of calcium and magnesium are not removed by boiling. These salts cause **noncarbonate hardness**, sometimes called "permanent" hardness.

Synthetic organic compounds, like DDT, which are products or byproducts of chemicals used in agriculture and industry, can build up to toxic levels in water and living organisms. Measurement techniques have advanced much further than our ability to establish the relationship between synthetic organic compounds now in use and human health. Most governments have set arbitrary limits on the more dangerous of these chemicals until more complete knowledge in this area is available.

Sources of Water

The quality and quantity of water from surface water and groundwater, the two main sources, are influenced by geography, climate, and human activities. Groundwater can normally be used with little or no treatment. Surface water, on the other hand, often needs extensive treatment, particularly if it is polluted.

Groundwater is water that has percolated downward from the ground surface through the soil pores. Formations of soil and rock which have become saturated with water are known as groundwater reservoirs, or **aquifers.** Water is normally withdrawn from these reservoirs by wells. Soil pore size, water viscosity, and other factors combine to limit the speed at which water can move through soil to replenish the well. This flux (velocity) may vary form 1m/day to 1 m/yr. A groundwater reservoir can only support a water withdrawal rate as high as is continually supplied by infiltration. Once this flow is exceeded, the water table will begin to drop, causing existing wells to run dry and requiring expensive deep drilling to locate new wells. There is growing concern that vast stretches of productive farms may lose irrigation water as wells go dry.

Surface waters from rivers and lakes are important sources of public water supplies because of the high withdrawal rates they can normally sustain. One disadvantage of using surface water is that it is open to pollution of all kinds. Contaminants are contributed to lakes and rivers from diverse and intermittent sources, such as industrial and municipal wastes, runoff from urban and agricultural areas, and erosion of soil. Water with variable turbidity, and a variety of substances that contribute to the taste, odor, and color of the water, can necessitate extensive treatment.

Seawater, available in almost unlimited quantities, can be converted into fresh water by a number of processes. However, conversion costs are perhaps two to five times higher than those of treating fresh water.

Desalination is the general term used for the removal of dissolved salts from water. Distillation, the oldest desalination technique, depends upon the evaporation and condensation of water. The process is energy intensive, but using solar energy to evaporate water may make it practical in countries with plentiful sunshine. Another method, freezing, lowers the water temperature until ice crystals free of salt can be separated from the brine. Electrodialysis involves forced migration of charged ions through cation-permeable or anion-permeable membranes by applying an electric potential across a cell containing mineralized water.

Reclaimed Wastewater refers to water that has been treated sufficiently for direct reuse in industry and agriculture, and for limited municipal applications. Such recycling or closed-loop operations may offer the only alternative in areas that cannot obtain enough fresh water. Suspended solids, biodegradable organics, and bacteria can all be removed or degraded by normal wastewater treatment processes, but color, the inorganic salts or magnesium, sodium, and calcium, synthetic organics like pesticides, and other toxic substances must be removed by advanced techniques similar to those used for desalination. Activated carbon is effective in removing many organic pollutants because of its extremely large surface that can trap and adsorb water impurities. Allowing water to cleanse itself by percolating through soil is another technique that removes impurities from water and has wide application in recharging groundwater supplies.

> **Selected from** "*J. Glynn Henry, Gray W. Heinke. Environmental Science and Engineering, Prentice-Hall International Editions, Pretice. Hall Englewood Cliffs, NJ, USA, 1989*"

Unit 14

Text: Water Purification

Water molecules have no memory, and therefore it is silly to talk about the number of times that the water you drink has been polluted and repurified, as if the molecules gradually wore out. All that is important is how pure it is when you drink it.

The purification of water has developed into an elaborate and sophisticated technology. However, the general approaches to purification should be comprehensible, and in some cases even obvious, from a general understanding of the nature of water pollution.

Impurities in water were classified as *suspended*, *colloidal*, or *dissolved*. Suspended particles are large enough to settle out or to be filtered. Colloidal and dissolved impurities are more difficult to remove. One possibility is somehow to make these small particles join together to become larger ones, which can then be treated as suspended matter. Another possibility is to convert them to a gas that escapes from the water into the atmosphere. Whatever the approach, it must be remembered that energy is required to lift water or to pump it through a filter[①].

With these principles in mind, consider the procedures used in purifying municipal waste waters. The first step is the collection system. Waterborne wastes from sources such as homes, hospitals, and schools contain food residues, human excrement, paper, soap, detergents, dirt, cloth, other miscellaneous debris, and, of course, microorganisms. This mixture is called **sanitary** or **domestic sewage**. (The adjective "sanitary" is rather inappropriate because it hardly describes the condition of the sewage; it presumably refers to that of the premises whose wastes have been carried away). These waters, which are sometimes joined by wastes from commercial buildings, by industrial wastes, and by the runoff from rain, flow through a network of sewer pipes. Some systems separate sewage from rainwater, others combine them. The combined piping is cheaper and is adequate in dry weather, but during a storm the total volume is apt to exceed the capacity of the treatment plant, so some is allowed to overflow and pass directly into the receiving stream or river.

Primary Treatment[②]

When the sewage reaches the treatment plant, it first passes through a series of screens that remove large objects such as rats or grapefruits, and then through a grinding mechanism that reduces any remaining objects to a size small enough to be handled effectively during the remaining treatment period. The next stage is a series of settling chambers designed to remove first the heavy grit, such as sand that rainwater brings in from road surfaces, and then, more slowly, any other suspended solids-including organic nutrients that can settle out in an hour or so. Up to this point the entire process, which is called primary treatment, has been relatively inexpensive but has not accomplished much.

Secondary Treatment[3]

The next series of steps is designed to reduce greatly the dissolved or finely suspended organic matter by some form of accelerated biological action. What is needed for such decomposition is oxygen and organisms and an environment in which both have ready access to the nutrients. One device for accomplishing this objective is the **trickling filter**[4]. In this device, long pipes rotate slowly over a bed of stones, distributing the polluted water in continuous sprays. As the water trickles over and around the stones, it offers its nutrients in the presence of air to an abundance of rather unappetizing forms of life.

An alternative technique is the **activated sludge** process. Here the sewage, after primary treatment, is pumped into an aeration tank, where it is mixed for several hours with air and bacteria-laden sludge. The biological action is similar to that which occurs in the trickling filter. The sludge bacteria metabolize the organic nutrients; the protozoa, as secondary consumers, feed on the bacteria. The treated waters then flow to a sedimentation tank, where the bacteria-laden solids settle out and are returned to the aerator. Some of the sludge must be removed to maintain steady-state conditions. The activated sludge process requires less land space than the trickling filters, and, since it exposes less area to the atmosphere, it does not stink so much.

The effluent from the biological action is still laden with bacteria and is not fit for discharge into open waters, let alone for drinking. Since the microorganisms have done their work, they may now be killed. The final step is therefore a disinfection process, usually chlorination. Chlorine gas, injected into the effluent 15 to 30 minutes before its final discharge, can kill more than 99 percent of the harmful bacteria.

Tertiary or "Advanced" Treatment[5]

Although considerable purification is accomplished by the time wastewaters have passed through the primary and secondary stages, these treatments are still inadequate to deal with some complex aspects of water pollution. First, many pollutants in sanitary sewage are not removed. Inorganic ions, such as nitrates and phosphates, remain in the treated waters; these materials, as we have seen, serve as plant nutrients and are therefore agents of eutrophication.

The treatment methods available to cope with these troublesome wastes are necessarily specific to the type of pollutant to be removed, and they are generally expensive. A few of these techniques are described below.

1. Coagulation and Sedimentation

As mentioned earlier in the discussion of biological treatment, it is advantageous to change little particles into big ones that settle faster. So it is also with inorganic pollutants. Various inorganic colloidal particles are waterloving (hydrophilic) and therefore rather adhesive; in their stickiness they sweep together many other colloidal particles that would otherwise fail to settle out in a reasonable time. This process is called flocculation. Lime, alum, and some salts of iron are among these so-called flocculating agents.

2. Adsorption

Adsorption is the process by which molecules of a gas or liquid adhere to the surface of a solid. The process is selective-different kinds of molecules adhere differently to any given solid. To purify water, a solid that has a large surface area and binds preferentially to organic pollutants is needed. The material of choice is activated carbon, which is particularly effective in removing chemicals that produce offensive tastes and odors. These include the biologically resistant chlorinated hydrocarbons.

3. Other Oxidizing Agents

Potassium permanganate ($KMnO_4$) and ozone (O_3) have been used to oxidize waterborne wastes that resist oxidation by air in the presence of microorganisms. Ozone has the important advantage that its only byproduct is oxygen.

$$2O_3 \longrightarrow 3O_2$$

4. Reverse Osmosis

Osmosis is the process by which water passes through a membrane that is impermeable to dissolved ions. In the normal course of osmosis, the system tends toward an equilibrium in which the concentrations on both sides of the membrane are equal. This means that the water flows from the pure side to the concentrated "polluted" side. This is just what we don't want, because it increases the quantity of polluted water. However, if excess pressure is applied on the concentrated side, the process can be reversed, and the pure water is squeezed through the membrane and thus freed of its dissolved ionic or other soluble pollutants.

Selected from *Jonathan Turk, Amos Turk. Environmental Science, Third Edition, Saunders College Publishing, The Drylen Press, USA, 1984*

Words and Expressions

purification　*n.* 净化，纯化
colloidal　*a.* 胶体的，乳化的
debris　*n.* 有机残渣，腐殖质，残骸

sanitary *a.* 有关卫生的，清洁的
waterborne *a.* 水中的
premise *n.* 前提
human excrement 人类排泄物
overflow *v.* 溢出
grapefruit *n.* 朱栾，葡萄柚（一种植物）
grit *n.* 粗砂
activated sludge 活性污泥
let alone 更不用说
coagulation *n.* 凝固，絮凝
sedimentation *n.* 沉淀，沉积
hydrophilic *a.* 亲水的；吸水的
oxidizing agents 氧化剂
ozone *n.* 臭氧
reverse osmosis 反向渗透
membrane *n.* 膜

Notes

① whatever the approach, it must be remembered that energy is required to lift water or to pump it through a filter. 可译为：不论采取哪种方法，要注意提升水或用泵将水输送到过滤床都需要能量。前一句是省略句。

② primary treatment 初级处理。废水处理通常分为三级，初级处理目的是除去悬浮固体物质。

③ secondary treatment 二级处理。二级处理的目的是降低水的 BOD 值。

④ trickling filter 滴滤池。它是二级处理的主要设施之一，是由拳头般大小的岩石填充而成的滤床。运行时水流经滤床，活性大的生物会在岩石上生长，从流经岩石的水获取食物。

⑤ tertiary or advanced treatment 三级或高级处理。

Exercises

1. Put the following into Chinese.

   land disposal, fecal coliform, stringent effluent requirement, assimilation capacity, practical outlets, aquatic life, detrimental to human health, endogenous phase

2. How many ways can be used to remove the suspended, colloidal and dissolved contaminants in wastewater?

3. Give several physical processes to remove suspended solids from wastewater? According to the Text：

4. Translate the part of "Secondary Treatment" into Chinese.

5. Explain the principle of "Reverse Osmosis" in Chinese.

Reading Material: Principles of Wastewater Treatment

Effluent Requirements

The primary objective of wastewater treatment is to remove or modify those contaminants detrimental to human health or the water, land, and air environment. Land disposal, evaporation from ponds, and deep-well injection are occasional options, but usually the only practical outlets for the disposal of treated (or untreated) wastewater are streams, rivers, lakes, and oceans. To protect these water resources, the discharge of pollutants into them must be controlled. This is done in North America by setting effluent requirements for BOD, SS, and fecal coliforms. Secondary treatment is normally necessary to meet these requirements.

In the United States, for example, BOD_5 or SS values in the effluent must not exceed an average of 30 mg/L. The maximum limit for fecal coliforms is 200/100 mL (geometric mean). A further stipulation is that not less than 85 percent removal of BOD and SS must be provided. Canadian requirements are similar, but some provinces use 15 mg/L as the maximum for effluent BOD_5 and SS. This implies a greater than 90 percent removal efficiency in the treatment of "normal" municipal wastewater with a BOD_5 of 200 mg/L.

Where receiving waters have limited assimilation capacity, contain excessive nutrients, provide essential water use, or support valuable aquatic life, then more stringent effluent requirements are warranted. Such situations entail detailed investigations to evaluate the need for additional wastewater treatment.

Treatment Processes

The suspended, colloidal, and dissolved contaminants (both organic and inorganic) in wastewater may be removed physically, converted biologically, or changed chemically.

Physical Processes. Gravity settling is the most common physical process for removing suspended solids from wastewater. It is employed in

- Removing grit (defined as sand particles of 0.2 mm diameter or greater)
- Clarifying raw sewage and concentrating the settled solids (called raw or primary sludge)
- Clarifying biological suspensions and concentrating the settled floc (called biological, activated, or secondary sludge)
- Gravity thickening of primary or secondary sludges.

Biological Processes. Most of the organic constituents in wastewater can serve as food (substrate) to provide energy for microbial growth. This is the principle used in biological waste treatment, where organic substrate is converted by microorganisms, mainly bacteria (with the help of protozoa), to carbon dioxide, water and more new cells. The microorganisms may be aerobic (requiring free oxygen), anaerobic (not requiring free oxygen), or facultative (growing with or without oxygen). Processes in which microorganisms use bound oxygen (from NO_2 for denitrification, for example) are often called anoxic rather than anaerobic.

In a continuous biological process, the system normally operates at some point on the growth curve toward the end of the declining growth phase, or into the endogenous phase where cells utilize their own protoplasm to obtain energy. Utilization of the logarithmic growth phase in wastewater treatment has been impractical, because substrate removal is incomplete and no economical way has been found to separate the microbial population from the liquid.

Aerobic/Anoxic Processes. In aerobic processes (i.e. molecular oxygen is present), heterotrophic bacteria (those obtaining carbon from organic compounds) oxidize about one-third of the colloidal and dissolved organic matter to stable end products ($CO_2 + H_2O$) and convert the remaining two-thirds into new microbial cells that can be removed from the wastewater by settling.

Ever since the importance of wastewater treatment became recognized, municipalities and industries have relied almost exclusively on aerobic rather than anaerobic biological processes for treating their liquid organic wastes. Aerobic treatment has predominated because of its simplicity, stability, efficient and rapid conversion of organic contaminants to microbial cells, and relatively odor-free operation.

Anaerobic Processes. In anaerobic biological processes (i.e. no oxygen is present), two groups of heterotrophic bacteria, in a two-step liquefaction/gasification process, convert over 90 percent of the organic matter present, initially to intermediates (partially stabilized end products including organic acids and alcohols) and then to methane and carbon dioxide gas:

$$\text{Organic matter} \xrightarrow[\text{bacteria}]{\text{acid-forming}} \text{intermediates} + CO_2 + H_2S + H_2O$$

$$\text{Organic acids} \xrightarrow[\text{bacteria}]{\text{methane}} CH_4 + CO_2$$

The process is universally used in heated anaerobic digesters, where primary and biological sludges are retained for approximately 30 days at 35℃ to reduce their volume (by about 30 percent) and their putrescibility, and thus simplify their disposal, usually on agricultural land.

Two major advantages of anaerobic processes over aerobic ones are, that they provide useful energy in the form of methane and that sludge production is only about 10 percent of that from aerobic processes for converting the same amount of organic matter. This is advantageous in the treatment of high-strength wastes, where the handling of large volumes of sludge would be a problem.

Chemical Processes. Many chemical processes, including oxidation, reduction, precipitation, and neutralization, are commonly used for industrial wastewater treatment. For municipal wastewater, precipitation and disinfection are the only processes having wide application.

Chemical treatment alone or with other processes is frequently necessary for industrial wastes that are not amenable to treatment by biological means. The oxidation of toxic cyanide to manageable cyanide (with SO_2) or of hexavalent chromium to the nontoxic trivalent form in the disposal of plating wastes are examples. Chemical processes are also useful in municipal waste treatment, where phosphorus concentrations are reduced and removal of solids increased by precipitation of these contaminants with metallic salts.

Disinfection of the effluent from wastewater treatment plants, generally by chlorination, is desirable where there is a potential health hazard. However, the uncertainty as to when a hazard exists has resulted in wide variation in practices.

Chlorine is the least expensive and most often used chemical for wastewater disinfection, but unfortunately it produces some undesirable side effects. Organic matter present will combine with the chlorine to form chlorinated organics, some of which are known or suspected carcinogens (capable or causing cancer). The fear is that these chlorinated organics are a potential hazard to water supplies. Another concern is the toxicity of chlorine residuals to aquatic life. Residuals as low as 0.05 mg/L are toxic to various species of freshwater fish. Where disinfection is necessary for the protection of public health, but toxicity from chlorine residuals is unacceptable, either the effluent must be dechlorinated or alternatives to chlorination must be considered. Unfortunately, experience with alternatives such as chlorine dioxide, other halogens, ozone, or ultraviolet light is extremely limited. Costs for chlorination and dechlorination or alternative disinfectrion methods are two to three times the cost of chlorination alone. Also, substitutes for chlorine have other disadvantages and unknown long-term effects. There is no ideal disinfectant.

Not all of the available physical, biological, and chemical processes have been described, nor are all of the processes mentioned required in every wastewater facility. The basic units in a typical municipal plant might consist of primary settling tanks, preceded by screening and grit removal; secondary treatment units for oxidation and settling; anaerobic sludge digesters; and chlorination facilities.

Selected from "*J. Glynn Henry, Gray W. Heinke. Environmental Science and Engineering, Prentice-Hall International Editions, Pretice. Hall Englewood Cliffs, NJ, USA, 1989*"

Unit 15

Text: Water Treatment Processes

One of the great achievements of modern technology has been to drastically reduce the incidence of waterborne of diseases such as cholera and typhoid fever. These diseases are no longer the great risks to pubic health that they once were. The key to this advance was the recognition that contamination of pubic water supplies by human wastes was the main source of infection, and that it could be eliminated by more effective water treatment and better waste disposal.

Today's water treatment plants are designed to provide water continuously that meets drinking water standards at the tap. There are four main considerations involved in accomplishing this: source selection, protection of water quality, treatment methods to be used, and prevention of recontamination. Common precautions to prevent groundwater and surface water pollution include prohibiting the discharge of sanitary and storm sewers close to the water reservoir, installing fences to prevent pollution from recreational uses of water, and restrictions on the application of fertilizers and pesticides in areas that drain to the reservoir.

Screening, coagulation/flocculation, sedimentation, filtration, and disinfection are the main unit operations① involved in the treatment of surface water. Water treatment operations fulfill one or more of three key tasks: removal of **particulate** substances such as sand and clay, organic matter, bacteria, and algae; removal of **dissolved** substances such as those causing color and hardness; and removal or destruction of **pathogenic bacteria** and **viruses**. The actual selection of treatment processes depends on the type of water source and the desired water quality.

Occasionally, raw water with low turbidity can be treated by plain sedimentation (no chemicals) to remove larger particles and then filtration to remove the few particles that failed to settle out. Usually, however, particles in the raw water are too small to be removed in a reasonably short time through sedimentation and simple filtration alone. To remedy this, a chemical is added to coagulate/flocculate the small particles, called **colloids**, into large ones, which can then be settled out in sedimentation tanks or removed directly in filters.

Removal of Particulate Matter

The unit operations employed for the removal of particulate matter from water include screening, sedimentation, coagulation/flocculation, and filtration.

Screening to remove large solids such as logs, branches, rags, and small fish is the first stage in the treatment of water. Allowing such debris into the treatment plant could damage pumps and clog pipes and channels. For the same reasons, water intakes are located below the surface of the lake or river in order to exclude floating objects and minimize physical damage from ice.

Sedimentation, the oldest and most widely used form of water and wastewater treatment, uses gravity settling to remove particles from water. It is relatively simple and inexpensive and can be implemented in basins that are round, square, or rectangular. As noted earlier, sedimentation may follow coagulation and flocculation (for highly turbid water) or be omitted entirely (with moderately turbid water). Particulates suspended in surface water can range in size from 10^{-1} to 10^{-7} mm in diameter, the size of fine sand and small clay particles, respectively. Turbidity or cloudiness in water is caused by those particles larger than 10^{-4} mm, while particles smaller than 10^{-4} mm contribute to the water's color and taste.

Coagulation/flocculation is a chemical-physical procedure whereby particles too small for practical removal by plain sedimentation are destabilized and clustered together for faster settling[2]. A significant percentage of particulates suspended in water are so small that settling to the bottom of a tank would take days or weeks. These colloidal particles would never settle by plain sedimentation.

Coagulation is a chemical process used to destabilize colloidal particles. The exact mechanism is not well understood, but the general idea is to add a chemical which has positively charged colloids to water containing negatively charged colloids. This will neutralize the negative change on the colloids and thus reduce the tendency for the colloids to repel each other. Rapid mixing for a few seconds is required to disperse the coagulant. Gentle mixing, called flocculation, of the suspension is then undertaken to promote particle contact. This is achieved by mechanical mixing through the use of slowly rotating paddles inside the coagulation/flocculation tank, or by hydraulic mixing which occurs when flow is directed over and around baffles in the tank. Detention time in the coagulation/flocculation tank is usually between 20~40minutes in tanks 3~4m deep. Through the combined chemical/physical process of coagulation/flocculation, the colloidal particles which would not settle out by plain sedimentation are agglomerated to form larger solids called floc. These appear as fluffy growths of irregular shape that are able to entrap small noncoagulaed particles when settling downward. Aluminum sulfate is the most common coagulant but organic polymers may also be used alone or in combination with alum to improve flocculation. The floc suspension is gently transferred from the coagulation/flocculation tanks to settling tanks, or directly to filters where the flocs are removed.

Disinfection

To ensure that water is free of harmful bacteria it is necessary to **disinfect** it. **Chlorination**[3] is the most common method of disinfecting public water supplies. Sufficient quantities of chlorine from chlorine gas or hypochlorites are added to treated water to kill pathogenic bacteria. Chlorination is a reliable, relatively inexpensive, and easy disinfection method to use. Other disinfectants include chloramines, chlorine dioxide, other halogens, ozone, ultraviolet light, and high temperature. Ozonation, which has been used extensively in France, is now gaining acceptance in North America, especially as an alternative to prechlorination where natural organics are present. Although effective, ozone does not leave a lasting residual for long-term disinfection.

Ozoanation is the disinfection of water by adding ozone, which is a powerful oxidant of inorganic and organic impurities. Its advantages over chlorine are that it leaves no tastes or odors, and unlike chlorine, it apparently does not react with natural organics to form compounds hazardous to humans.

Removal of Dissolved Substances

Aeration is used to remove excessive amounts of iron and manganese from groundwater. These substances cause taste and color problems, interfere with laundering, stain plumbing fixtures, and promote the growth of iron bacteria in water mains. By bubbling air through water, or by creating contact between air and water by spraying, dissolved iron or manganese (Fe^{2+}, Mn^{2+}) is oxidized to a less soluble form (Fe^{3+}, Mn^{4+}), which precipitates out and can be removed in a settling tank or filter. Aeration also removes odors caused by hydrogen sulfide (H_2S) gas.

Softening of water is a process that removes hardness, caused by the presence of divalent metallic ions, principally Ca^{2+} and Mg^{2+}. Hardness in water is the result of contact with soil and rock, particularly limestone, in the presence of CO_2.

Activated Carbon is an extremely adsorbent material used in water treatment to remove organic contaminants. Activated carbon is produced in a two-stage process. First, a suitable base material such as wood, peat, vegetable matter, or bone is carbonized by heating the material in the absence of air. Then the carbonized material is activated by heating it in the presence of air, CO_2, or steam to burn off any tars it has and to increase its pore size. Adsorption of gases, liquids, and solids by activated carbon is influenced by the temperature and pH of the water as well as the complexity of the organics being removed.

In reverse osmosis (RO), fresh water is forced through a semipermeable membrane in the direction opposite to that occurring in natural osmosis. Because the membrane removes dissolved salts, the main application for RO has been in desalination. However, the process also removes organic materials, bacteria, and viruses, and its application in water treatment is growing.

Selected from "*J. Glynn Henry, Gray W. Heinke. Environmental Science and Engineering, Prentice-Hall International Editions, Pretice. Hall Englewood Cliffs, NJ, USA, 1989*"

Words and Expressions

screening *n.*（用拦污栅）隔离
flocculation *n.* 絮凝（沉淀法）
turbidity *n.* 浊度，浑浊性
colloid *n.* 胶体
at the tap 自来水，饮水
plain sedimentation 普通沉淀法
filtration *n.* 过滤
highly turbid water 高度（非常）浑浊的水
moderately turbid water 中等浑浊的水
cluster *v./n.* 集结，成团
agglomerate *v.* 使聚集，成团
entrap *v.* 收集，诱捕
polymer *n.* 聚合物
fluffy *a.* 松散的，易碎的
alum *n.* 铝

aluminum sulfate　硫酸铝

hypochlorite *n.* 次氯酸盐

chloramine *n.* 氯胺

chlorine dioxide　二氧化氯

ultraviolet light　紫外线

halogen *n.* 卤素，如 Br、I、F、Cl

ozonation *n.* 臭氧氧化

chlorine *n.* 氯气

aeration *n.* 曝气

stain *v.* 玷污，弄脏；*n.* 污点

plumbing *n.* 管道

fixture *n.* 装置

limestone *n.* 石灰石

semipermeable *a.* 半渗透的

desalination *n.* （海水）脱盐

divalent *n.* 二价的

Notes

① unit operations　单元操作，如化工原理课程讲的流体输送、传热、蒸发、精馏、吸收等。

② Coagulation/flocculation is a chemical-physical procedure whereby particles too small for practical removal by plain sedimentation are destabilized and clustered together for faster settling. 可译为：混凝/絮凝法是一种化学物理过程，那些太小的用普通沉降法不能去除的颗粒具有在这个过程中失去稳定性能成团的特点，从而能较快地沉淀。

③ chlorination　氯化（消毒）。加氯消毒是目前最常用的自来水和废水消毒法，它主要是消灭残留在水中的微生物，此外，当氯和有机物接触，还可将有机物氧化。但是，该法会和有机物结合，形成卤化物及氯仿等致癌物。目前发达国家已开始使用臭氧氧化法替代氯化法。

Exercises

1. Put the following into Chinese.

   Irrigation, renewable freshwater resources, pathogenic agent, thermocline, pesticides, toxicity, terrestrial, ice caps and glaciers, soil moisture, abyssal

2. Put the following into English.

   农业灌溉　人均年用水量　水力循环　生物圈　水环境
   氧化态　盐化　酸化　放射性　氧化与还原条件　水的硬度

3. Give an abstract of Reading Material in English.

4. What are the advantages of ozonation over chlorine?

5. Describe simply "Coagulation/flocculation process" in Chinese.

Reading Material: Freshwater Systems

"Water, water, everywhere, nor any drop to drink". The well-known line from the poem *The Rime of the Ancient* Mariner describes the situation on our planet well. Water may seem abundant, but water that we can drink is quite rare and limited. About 97.5% of Earth's water resides in the oceans and is too salty to drink or to use to water crops. Only 2.5% is considered **fresh water**, water that is relatively pure, with few dissolved salts. Because most fresh water is tied up in glaciers, icecaps, and underground aquifers, just over 1 part in 10,000 of Earth's water is easily accessible for human use.

Water is renewed and recycled as it moves through the water cycle. Precipitation falling from the sky either sinks into the ground or acts as runoff to form rivers, which carry water to the oceans or large inland lakes. As they flow, rivers can interact with ponds, wetlands, and coastal aquatic ecosystems. Underground aquifers exchange water with rivers, ponds, lakes, and the ocean through the sediments on the bottoms of these water bodies. The movement of water in the water cycle creates a web of interconnected freshwater and marine aquatic systems that exchange water, organisms, sediments, pollutants, and other dissolved substances. What happens in one system therefore affects other system—even those that are far away. Let us examine the freshwater components of the interconnected system, beginning with groundwater. Marine and coastal components of the system will be examined subsequently, but note that all these systems interact extensively.

Groundwater Plays Key Roles in the Hydrologic Cycle

Liquid fresh water occurs either as surface water or groundwater. Surface water is water located atop Earth's surface (such as in a river or lake) and groundwater is water beneath the surface held within pores in soil or rock. Any precipitation reacting Earth's land surface that does not evaporate, flow into waterways, or get taken up by organisms infiltrates the surface. Groundwater makes up one-fifth Earth's fresh water supply and plays a key role in meeting human water needs.

Groundwater flows downward and from areas of high pressure to areas of low pressure. However, a typical rate of flow might be only about 1 m (3 ft) per day, so groundwater can remain underground for a long time. When we pump groundwater through wells, we are drawing up ancient water. The average age of groundwater has been estimated at 1400 years, and some is tens of thousands of years old.

Groundwater is contained with **aquifers**: porous, sponge-like formations of rock, sand, or gravel that hold water. An aquifer's upper layer, or zone of aeration, contains pore spaces partly filled with water. In the lower layer, or zone of saturation, the spaces are completely filled with water. The boundary between these two zones is the **water table.** Picture a sponge resting partly submerged in a tray of water; the lower part of the sponge is saturated, whereas the upper portion contains plenty of air in its pores. Any area where water infiltrates Earth's surface and reaches an aquifer below is known as a recharge zone.

Surface Water Converges in River and Stream Ecosystems

Surface water accounts for just 1% of fresh water, but it is vital for our survival and for the planet's ecological systems. Once water falls from the sky as rain, emerges from springs, or melts from snow or glaciers, it may soak into the ground or may flow downhill over land. Water that flows over land is called runoff.

As it flows downhill, runoff converges where the land dips lowest, forming streams, creeks, or brooks. These small watercourses may merge into rivers, whose water eventually reaches a lake or ocean. A smaller river flowing into a large one is a tributary. The area of land drained by a river system—a river and all its tributaries—is that river's watershed, also called a drainage basin. If you could stand at the mouth of the Mississippi River and trace every drop of water in it back to the spot it first fell as precipitation, then you would have delineated the Mississippi River's watershed, the very area shaded.

Groundwater and surface water interact extensively. Surface water becomes groundwater by infiltration. Groundwater becomes surface water through springs (and from wells drilled by people), often keeping streams flowing or wetlands moist when surface conditions are otherwise dry. Each day in the United States, 1.9 trillion L (492 billion gal) of groundwater are released into bodies of surface water—nearly as much as the daily flow of water in the entire Mississippi River.

Lakes and Ponds are Ecologically Diverse Systems

Lakes and ponds are bodies of standing surface water. Their physical conditions and the types of life within them vary with depth and the distance from shore. As a result, scientists have described several zones typical of lakes and ponds.

Ponds and lakes change over time as streams and runoff bring them sediment and nutrients. Oligotrophic lakes and ponds, which are low in nutrients and high in oxygen, may slowly give way to the high nutrient, low oxygen conditions of eutrophic water bodies. Eventually, water bodies may fill in completely by the process of aquatic succession. As lakes or ponds change over time, species of fish, plants, and invertebrates adapted to oligotrophic conditions may give way to those that thrive in eutrophic conditions. These changes occur naturally, but eutrophication can also result from human-caused nutrient pollution.

The largest lakes are sometimes known as inland seas. North America's Great Lakes are prime examples. Lake Baikal in Asia is the world's deepest lake, at 1637 m (just over 1 mile) deep. The Caspian Sea is the world's largest body of fresh water, covering nearly as much area as Montana or California.

Selected from "*Jay Withgott, Matthew Laposata. , Environment, the Science Behind the Stories, 5th Ed. , Pearson Education Inc. , US. 2014*"

Unit 16

Text: Biological Wastewater Treatment [I]

Application of microbial degradation and removal of undesirable constituents in industrial and municipal wastes is not a new concept. It is commonly used process for general wastewater treatment activities and has been for many years. As the awareness of chemical contamination of the environment, much research on biological degradation of toxic chemicals has occurred. Among the range of treatment technologies, biological degradation ranks among the most effective. Its management application is enhanced by the potential to apply biological treatment in sequence with other chemical and thermal processes.

Activated Biofilter

A biofilter first state is followed by an activated-sludge second state and a settler. Sludge is recycled to the biofilm state and to the activated sludge tank. This variation combines biofilm and suspended-growth characteristics.

Activated sludge

The activated sludge process is a typical type of suspended growth biological treatment system and probably the most widely used biological process for the treatment of organic and industrial waste waters. However, it can only treat aqueous organic wastestreams having less than 1% suspended solid content, and can not tolerate shock loading of concentrated organics. Therefore, the wastestream entering this process will usually have passed through a pretreatment process which includes a clarifier (primary clarifier) and an equalization basin. The primary clarifier is used for removal of grit, oily and fatty material and gross solid material, while the equalization basin is used to dampen wastewater flow variations and to provide more uniform organic loading to the activated sludge system[①].

Activated sludge process are used to treat municipal and industrial wastes since they are versatile, flexible, and can be used to produce an effluent of desired quality by varying process parameters. The process was so-named because it produces an active mass of microorganisms capable of aerobically stabilizing a waste. Many versions of the basic process exist but all are fundamentally similar.

The term activated sludge is applied to both the process and to the biological solids in the treatment unit. The mixed liquor suspended solids or activated sludge contains a variety of heterotrophic microorganisms such as bacteria, protozoa, fungi, and larger microorganisms. The predominance of a particular microbial species depends upon the waste that is treated and the way in which the process is operated.

The activated sludge process is currently the most widely used biological treatment process. This is partly the result of the fact that recirculation of the biomass, which is an integral part of the process, allows microorganisms to adapt to changes in wastewater composition with a relatively short acclimation time and also allows a greater degree of control over the acclimated bacterial population[2].

An activated sludge system consists of an equalization basin, a settling tank, an aeration basin, a clarifier, and a sludge recycle line. Wastewater is homogenized in an equalized basin to reduce variations in the feed, which may cause process upsets of the microorganisms and diminish treatment efficiency. Settleable solids are then removed in a settling tank.

Next, wastewater enters an aeration basin, where an aerobic bacterial population is maintained in suspension and oxygen, as well as nutrients, are provide. The contents of the basin are referred to as the mixed liquor. Oxygen is supplied to the aeration basin by mechanical or diffused aeration, which also aids in keeping the microbial population in suspension. The mixed liquor is continuously discharged from the aeration basin into a clarifier, where the biomass is separated from the treated wastewater. A portion of the biomass is recycled to the aeration basin to maintain an optimum concentration of acclimated microorganisms in the aeration basin. The remainder of the separated biomass is discharged or "wasted." The biomass may be further dewatered on sludge drying beds or by sludge filtration to disposal. The clarified effluent is discharged.

The recycled biomass is referred to as activated sludge. The term "activated" is used because the biomass contains living and acclimated microorganisms that metabolize and assimilate organic material at a higher rate when returned to the aeration basin. This occurs because of the low food-to-microorganism ratio in the sludge from the clarifier.

For the treatment of industrial wastewater, supplemental nutrient sources are often needed to provide sufficient nitrogen and phosphorus. In most cases, nitrogen is added as ammonia and phosphorus as phosphoric acid. A proper pH range (6 to 8) and a sufficient dissolved oxygen concentration (a minimum of 1 to 2mg/L) must also be maintained in the aeration basin to support a healthy and active system.

The aeration basin hydraulic retention time (HRT) and sludge residence time (SRT) are important operational factors. HRT is defined as the ratio of the volume of aeration tank to the influent liquid flow rate, and SRT is the total amount of sludge in the system divided by the rate of sludge leaving the system as waste. Sufficient time must be provided to allow the bacteria to assimilate the organic material in the wastewater. The HRT is usually from 6 to 24 hours and SRT is from 4 to 10 days for the activated sludge process. The optimum operating temperature is in the range of 25℃ to 32℃.

Although organisms present in activated sludge systems range from viruses to multicellular organisms, the predominant and most active are heterotrophic, and to lesser extent, autotrophic bacteria, which are both aggregated in the sludge flocs and dispersed in the liquid. Heterotrophic bacteria utilize organic material as a source of carbon and energy, while autotrophic bacteria generally depend on the oxidation of mineral compounds for energy requirements and utilize carbon dioxide as a carbon source. These bacteria are capable of performing hydrolysis and oxidation reactions.

Complex hydrocarbons are oxidized to lower molecular weights by oxygenase enzymes which incorporate oxygen directed into the long chain or cyclic hydrocarbon molecule. Polysaccharides, fats, and proteins are degraded from their polymeric state to monomeric units via hydrolysis. The end-products, i.e., alcohols and acids, from those reactions will enter the microorganism and be metabolized by oxidation reactions catalyzed by endo-enzymes③. The oxidation follows the chemical sequence of: alcohols oxidized to aldehydes and then to acids. A portion of the acids are oxidized to carbon dioxide and water to obtain the necessary energy to use remaining acids for cell growth.

Generally, the activated sludge process is readily capable of decomposing alcohols, aldehydes, fatty acids, alkanes, alkenes, cycloalkenes and aromatics. Other compounds such as isoalkanes and halogenated hydrocarbons are more resistant to microbial decomposition. Therefore, the degree of treatment and the rate of decomposition are dependent upon the acclimated biomass in the activated sludge system. However, only dilute aqueous wastes can normally be treated, and most hazardous organic wastes are toxic or inhibitory to the process except at very low concentrations. Therefore, treatment of hazardous wastes by this process is often most practical where the aqueous waste can be mixed with a more readily biodegradable wastewater stream.

Dissolved metal ions and fine metal particles produce an adverse effect on microbial metabolism by blinding at the enzyme-active site or causing conformational changes in the enzyme with activated sludge process④. Normally, microorganisms can tolerate only a few milligrams per liter or less of heavy metals. Heavy metals may be kept insoluble by the addition of ferrous sulfate to encourage sulfide precipitation and light metal cations may be detoxified by encouraging formation of carbonates and bicarbonates. In addition to biodegradation, organic materials may be removed by air-stripping, and/or sorption to the sludge.

Selected from "*Robert Noye, Unit Operation in Environmental Engineering, Noye Publication, USA, 1994*"

Words and Expressions

microbial *a.* 微生物的；由细菌引起的
activated sludge *n.* 活性污泥
clarifier *n.* 澄清器，澄清池
equalization basin *n.* 平衡洗涤槽

heterotrophic *a.* 非自养的，异养的
microorganism *n.* 微生物，微小动植物
protozoa *n.* 原生动物，原形动物
fungi *n.* (fungus 的复数形式) 真菌类
acclimation *n.* 服水土，顺应，适应环境
biomass *n.* (单位面积或体积内) 生物质
influent *a.* 流入的；*n.* 支流
autotrophic *a.* 自造营养物质的，自给营养的
oxygenase *n.* (加) 氧酶
enzyme *n.* 酶
polysaccharide *n.* 多糖，聚糖，多聚糖
hydrolysis *n.* 水解
aldehyde *n.* 乙醛
alkane *n.* 链烷，烷烃
alkene *n.* 烯烃，链烯
cycloalkene *n.* 环烯烃
aromatic *n.* 芳香烃
isoalkane *n.* 异烷烃
halogenated *a.* 卤化的，卤代的
carbonate *n.* 碳酸盐
bicarbonate *n.* 重碳酸盐

Notes

① The primary clarifier is used for removal of grit, oily and fatty material and gross solid material, while the equalization basin is used to dampen wastewater flow variations and to provide more uniform organic loading to the activated sludge system. 可译为：预净化池用来处理砂砾、油和脂肪以及总固体物，而均质池则对废水流率的变化进行缓冲同时使活性污泥系统的有机负荷更加均匀。

② This is partly the result of the fact that recirculation of the biomass, which is an integral part of the process, allows microorganisms to adapt to changes in wastewater composition with a relatively short acclimation time and also allows a greater degree of control over the acclimated bacterial population. 可译为：这部分是基于生物质循环这一结果，生物质循环是整个工艺的一部分，它可以使微生物在相对较短的驯化时间内适应废水组分的变化，并能在较大程度上控制驯化菌的数量。

③ The end-products, i.e., alcohols and acids, from those reactions will enter the microorganism and be metabolized by oxidation reactions catalyzed by endo-enzymes. 可译为：通过这些反应的最终产物即乙醇和酸进入微生物体内并经胞内酶催化氧化被代谢。

④ Dissolved metal ions and fine metal particles produce an adverse effect on microbial metabolism by blinding at the enzyme-active site or causing conformational changes in the enzyme with activated sludge process. 可译为：溶解的金属离子和细小的金属微粒通过结合于酶的活性位而对微生物的新陈代谢产生不利的影响，或在活性污泥处理过程中导致酶的构象变化。

Exercises

1. Put the following into Chinese.

    biological degradation, equalization basin, aeration basin, sludge flocs, settling tank, dissolved oxygen, biofilm, suspended-growth

2. Describe the activated sludge process.

3. Translate the following paragraph into Chinese.

    High biomass systems: Many current approaches to high biomass system employ a combination of fixed film suspended biomass in the process. High biomass systems have gained a certain popularity in Europe. During the past few years, a number of investigations undertaken in the Federal Republic of Germany (FRG) have been reported. Among the advantages attributed to such systems have been improvements in nitrification performance, sludge settleability, and effluent quality.

    Reasons for selecting high biomass systems over construction of additional aeration tanks and clarifiers (or other secondary treatment process) include reduced space requirements, increased process stability, and capital/operating cost savings.

    High biomass systems call for installation of supplemental equipment over that contained in a conventional activated sludge plant. More installed equipment generally implies more maintenance, and, to some extent, this is true for some of the systems. In addition, the presence of both suspended and fixed biomass forms and higher biomass concentrations may require a certain level of additional operator time to achieve optimum system performance.

    The presence of inert support media and higher biomass concentrations in these systems can increase overall power consumption. To achieve desired mixing patterns in retrofitted aeration tanks, power input may have to be increased. Also, the presence of additional biomass increases system oxygen requirements which, in turn, requires additional power input. In addition, high biomass systems generally yield higher levels of nitrification, which also can affect overall power consumption. Such factors should be addressed when analyzing operating costs.

Reading Material: Biological Wastewater Treatment [Ⅱ]

An important variation on the activated sludge process is the Powdered Activated Carbon Treatment (PACT) process. This process offers a combined treatment and pretreatment system in which noncompatible and toxic constituents are adsorbed onto activated carbon, while microorganism-compatible waste remains in solution. Powdered activated carbon is added directly to the aeration basin of the activated sludge treatment system. Overall removal efficiency is improved because compounds that are not readily biodegradable or that are toxic to the microorganism are adsorbed onto the surface of the powered activated carbon. The carbon is removed from the wastewater in the clarifier along with the biological sludge. Usually, the activated carbon is recovered, regenerated, and recycled. Limitations of the PACT system include applicability to dilute liquids and residual sludges. The system is susceptible to clogging when there are high solids or high oil content in the wastewater.

A number of advantages have been reported for this combined physical/biological process. These include the removal of non-biodegradable organics, reduced emission of organics to the air, particularly during the period of acclimation, better settling properties of the biomass/powered activated carbon sludge, and protection of the microbial population from toxic shocks. In addition, the powdered activated carbon helps reduce effluent concentrations of organics during an acclimation period. Some evidence also exists for the ability of microorganisms to bioregenerate the powdered activated carbon during periods of low organic loading.

The process has been used successfully with several industrial wastes including those from the manufacture of complex organic chemicals and from oil refining. In one full-scale study, greater than 82% removal of the priority pollutants was achieved with the PACT process. With a carbon dosage of 100mg/L, 99.6% removal of benzene and 84% removal of 2,4-dichlorophenol was achieved.

High biomass systems: Many current approaches to high biomass system employ a combination of fixed film suspended biomass in the process. High biomass systems have gained a certain popularity in Europe. During the past few years, a number of investigations undertaken in the Federal Republic of Germany (FRG) have been reported. Among the advantages attributed to such systems have been improvements in nitrification performance, sludge settleability, and effluent quality.

Reasons for selecting high biomass systems over construction of additional aeration tanks and clarifiers (or other secondary treatment process) include reduced space requirements, increased process stability, and capital/operating cost savings.

High biomass systems call for installation of supplemental equipment over that contained in a conventional activated sludge plant. More installed equipment generally implies more maintenance, and, to some extent, this is true for some of the systems. In addition, the presence of both suspended and fixed biomass forms and higher biomass concentrations may require a certain level of additional operator time to achieve optimum system performance.

The presence of inert support media and higher biomass concentrations in these systems can increase overall power consumption. To achieve desired mixing patterns in retrofitted aeration tanks, power input may have to be increased. Also, the presence of additional biomass increases system oxygen requirements which, in turn, requires additional power input. In addition, high biomass systems generally yield higher levels of nitrification, which also can affect overall power consumption. Such factors should be addressed when analyzing operating costs.

Oxidation ditches: An oxidation ditch is an activated sludge biological treatment process; commonly operated in the extended aeration mode. Typical oxidation ditch treatment systems consist of single channel or concentric, multichannel configurations.

Some forms of preliminary treatment such as bar screens, comminutors, or grit removal normally precede the oxidation ditch. Primary settling prior to an oxidation ditch is sometimes practiced, however, it is not common. Flow to the oxidation ditch is mixed with return sludge from a secondary clarifier and aerated. The aerators may be brush rotors, disc aerators, surface aerators, draft tube aerators, or fine bubble diffusers. The aerators provide mixing and circulation in the ditch, as well as oxygen transfer. A high degree of nitrification occurs in the ditch due to operation in the extended aeration mode. Oxidation ditches are typically designed with a nominal hydraulic detention time at average design flow of greater than 10 hours and a mean cell residence time (sludge age) ranging from 10 to 50 days. Oxidation ditch is usually settled in a separate secondary clarifier, however, intrachannel clarifiers are also used.

Ditches may be constructed of various materials, including concrete, gunite, asphalt, or impervious membranes. Concrete is the most commonly used. The single channel oxidation ditch may be found in a variety of shapes including ovals, horseshoes, or ells, whichever best fits the site. The concentric multichannel ditches may be circular or oval in shape. The addition of an intrachannel clarifier may be incorporated into the ditch design.

An oxidation ditch may be operated with an anoxic zone in the channel to achieve partial denitrification. An anoxic tank upstream of the ditch may be added along with recycle to that tank from the anoxic zone in the channel to achieve higher levels of denitrification. An anaerobic tank may be added prior to the ditch for enhanced biological phosphorus removal.

Oxidation ditches were usually not designed for nitrification or denitrifiction. Design parameters used, however, often ensured that nitrification occurred. Current concern over nutrient discharges to natural water systems has led to interest in upgrading existing oxidation ditches and modifying the oxidation ditch system design to incorporate biological nutrient removal.

Modifications to the basic oxidation ditch design can be made to achieve nitrogen and phosphorus removal. The key to obtaining nitrogen removal is the proper control of dissolved oxygen levels in different sections of the oxidation ditch, and the maintenance of adequate mass of bacteria under aerobic and anoxic conditions. To meet more stringent total nitrogen effluent limits a separate anoxic channel or basin outside the ditch channels may be added.

Holding mixed liquid under anaerobic conditions is required for enhanced biological phosphorus removal. This can be accomplished in either a nonaerated channel or by adding an anaerobic basin before the aerobic oxidation ditch channel.

A vertical loop reactor (VLR) is an aerobic suspended growth activated sludge biological treatment process similar to an oxidation ditch.

Other Variations: A variation to the activated sludge process is the use of high purity oxygen instead of air for aerobic treatment. Oxygen can be supplied from on-site gas generators with liquid oxygen storage as back-up. In addition to oxygen use, the aeration tank is covered which helps to eliminate odors and maintain temperatures in cold-weather periods.

There are many design variations to the conventional activated sludge process besides the use of high purity oxygen. These include: multiple unites with series and/or parallel flow patterns; a tapered distribution of air along the tank length; stepwise addition of raw waste; reaeration of the recycled sludge before mixing with the raw influent; and extended aeration, e.g., 24 hours or longer used for small wastewater flows.

Advantages and Disadvantages: Activated sludge treatment is used extensively in industry. It is probably the most cost-effective manner of destroying organics present in an aqueous waste stream. By using activated sludge modes which ensure complete mixing and high dissolved oxygen levels, high strength organic waste streams can be handled at an industrial waste treatment facility.

Some of the commonly listed disadvantages of the activated sludge process include the high capital investment required, high energy costs, the lengthy start-up time, and sensitivity to toxic and hydraulic shocks. On the other hand, the system can handle high organic loads using relatively short retention times, and can be controlled to achieve various degrees of treatment. Finally, the widespread use of activated-sludge facilities means the process has been well researched and documented.

The treatment requires consistent stable operating conditions. Activated sludge processes are not suitable for removing highly chlorinated organics, aliphatics, amines and aromatic compounds from a waste stream. Some heavy metals and organic chemicals are harmful to the organisms. When utilizing conventional open aeration tanks and clarifiers, this technology can result in the escape of volatile hazardous materials.

The efficiency of this process depends upon the satisfactory functioning of both the biological oxidation and the solids separation processes. Bulking and foaming must be controlled as they inhibit satisfactory separation of sludge solids.

Selected from "*Robert Noye, Unit Operation in Environmental Engineering, Noye Publication, USA, 1994*"

Unit 17

Text: Advanced Oxidation Processes (AOPs)—Wastewater Decontamination by Solar Photo-catalysis

Introduction

Advanced oxidation processes (AOPs)[①] may be used for decontaminating water containing organic pollutants, classified as bio-recalcitrant, and/or for disinfection removing current and emerging pathogens. These methods rely on the formation of highly reactive chemical species that degrade even the most recalcitrant molecules into biodegradable compounds. Although there are different reacting systems, all of them are characterized by the same chemical feature: production of hydroxyl radicals (OH·), which are able to oxidize and mineralize almost any organic molecule, yielding CO_2 and inorganic ions. Reaction rate ($r = -dC/dt = k_{OH}[OH·]C$) for most reactions involving hydroxyl radicals in aqueous solution is usually expressed, and its rate constant is generally on the order of $10^6 \sim 10^9 \text{ M}^{-1} \cdot \text{s}^{-1}$. Rate constants are also characterized by their nonselective attack, which is a useful attribute for wastewater treatment and solution of pollution problems. The versatility of the AOPs is also enhanced by the fact there are different ways of producing hydroxyl radicals, facilitating compliance with specific treatment requirements. Methods based on UV, H_2O_2/UV, O_3/UV, and H_2O_2/O_3/UV combinations use photolysis of H_2O_2 and ozone to produce hydroxyl radicals. Other methods, such as heterogeneous photo-catalysis and homogenous photo-Fenton, are based on the use of a wide band-gap semiconductor and addition of H_2O_2 to dissolved iron salts, respectively, and irradiation with Ultraviolet-Visible light. Both processes of special interest, as sunlight can be used for both.

Publications regarding the photo-catalytic process rose continuously over the last years surpassing a total number of more than 1000 peer-reviewed publications per year, much of it taking into account the possibility of driving the process with solar radiation (source: **www.scopus.com**, 2010. Search terms "Photo-catalysis" and "solar" within these results). Although such a simple search does not necessarily include every single article correctly, it still severs to prove the general trend of an increasing interest of the scientific community. This is due to the fact that, a priori, photo-catalytic process seems to be the most apt of all AOPs to be driven by sunlight. In this chapter, we highlight some of the science and technology being developed to improve the solar photo-catalytic disinfection and decontamination of water.

Solar Heterogeneous Photo-catalysis Main Parameters

The heterogeneous solar photo-catalytic detoxification process consists of use the near-ultraviolet (UV) band of the solar spectrum (wavelengths shorter than 400 nm) to proto-excite a semiconductor catalyst in contact with water and in the presence of oxygen. Oxidizing species (hydroxyl radicals, OH·, produced due to photo-generated holes), which attack oxidizable contaminants, are generated producing a progressive breakup of molecules yielding CO_2, H_2O, and diluted inorganic acids. The process takes place at ambient temperature and without overpressure. The basic principles of this method are well established: (1) semiconductors (e.g., TiO_2, ZnO, Fe_2O_3, CdS, and ZnS) can act as sensitizers for light-induced redox processes due to their electronic structure, which is characterized by a filled valence band and an empty conduction band. (2) Absorption of a proton of energy greater than the band-gap energy leads to the formation of an electron-hole pair. (3) In the absence of suitable scavengers, the stored energy is dissipated within a few nanoseconds by recombination, but if a suitable scavenger or surface defect state is available to trap the electron or hole, recombination is prevented and subsequent redox reactions may occur.② (4) The valence band holes are powerful oxidants (+1.0 to +3.5V vs normal hydrogen electrode, depending on the semiconductor and pH); most organic photo-degradation reactions use the oxidizing power of the holes either directly or indirectly. Absorption of a photon with an energy hv greater or equal to the band-gap energy E_g generally leads to the formation of an electron-hole pair in the semiconductor particle. These charge carries subsequently recombine and dissipate the input energy as heat, get trapped in metastable surface states, or react with electron donors and acceptors adsorbed on the surface or bound within the electrical double layer.

Whenever different semiconductor materials have been tested under comparable conditions for the degradation of the same compounds, TiO_2 has generally been demonstrated to be the most active. The strong resistance of TiO_2 to chemical breakdown and photo-corrosion, its safety, and low cost limit the choice of convenient alternatives. Although there are many different sources of semiconductor particles, TiO_2 has effectively become a standard because it has (1) a reasonably well-defined nature and (2) a substantially higher photocatalytic activity than most other readily available (commercial) semiconductor particles. Other semiconductor particles, for example, CdS or GaP absorb larger fractions of the solar spectrum and can form chemically activated surface-bond intermediates, but, unfortunately, these photo-catalysts are degraded during the repeated catalytic cycles involved in heterogeneous photo-catalysis producing final toxic products. In presence of a fluid (water), a spontaneous adsorption occurs and according to the redox potential of each adsorbate, an electron transfer proceeds toward acceptor molecules, whereas a positive hole is transferred to a donor molecule. Subsequently, activation of process goes through the excitation of the solid, but not through that of the reactants. It is well known that O_2 and water are essential for photo-oxidation.

Selected from "Manij K. Ram, Silvana Andreescu, Hanming Ding, Nanotechnology for Environmental Decontamination, McGraw-Hill Companies, Inc, USA, 2011".

Words and Expressions

biorecalcitrant *a.* 顽强的，反抗的，对抗的
hydroxyl radical 氢氧根自由基
peer-reviewed 同行评议（审）的
priori *n.* 先验，预知
detoxification *n.* 消毒，去毒，戒毒
photoexcite *v.* 光激发
scavenger *n.* 清除剂，清洁工，食腐动物，拾荒者
adsorbate *n.* 吸收质，被吸附物

Notes

① Advanced oxidation processes (AOPs) 高级氧化过程。该过程主要通过不同途径产生一种氧化能力很强的氢氧根自由基，其氧化能力仅次于 F，几乎无选择地将所有有机污染物分子氧化降解或矿化为二氧化碳和水，或无机离子，无需进行后续处理。

② In the absence of suitable scavengers, the stored energy is dissipated within a few nanoseconds by recombination, but if a suitable scavenger or surface defect state is available to trap the electron or hole, recombination is prevented and subsequent redox reactions may occur. 可译为：在缺少合适的清除剂情况下，被储存的能量通过再组合在几纳秒内散失掉，但假如存在合适的清除剂或表面缺陷捕获电子或空穴，可避免再组合，后面的氧化还原反应就可能发生。

Exercises

1. Put the following into English.

清除剂 氢氧根自由基 高级氧化过程 光激发
顽强的 同行评审 消毒 先验

2. Put the following into Chinese.

(1) Although there are different reacting systems, all of them are characterized by the same chemical feature: production of hydroxyl radicals (OH·), which are able to oxidize and mineralize almost any organic molecule, yielding CO_2 and inorganic ions.

(2) Methods based on UV, H_2O_2/UV, O_3/UV, and H_2O_2/O_3/UV combinations use photolysis of H_2O_2 and ozone to produce hydroxyl radicals.

(3) Oxidizing species (hydroxyl radicals, OH·, produced due to photo-generated holes), which attack oxidizable contaminants, are generated producing a progressive

breakup of molecules yielding CO_2, H_2O, and diluted inorganic acids.

(4) The strong resistance of TiO_2 to chemical breakdown and photo-corrosion, its safety, and low cost limit the choice of convenient alternatives.

3. Give a brief description of heterogeneous photocatalysis.

Reading Material: Precipitation

Chemical precipitation is a physicochemical process whereby a substance in solution is transformed into a solid phase and driven out of solution. Precipitation entails altering the chemical equilibrium affecting the solubility of the hazardous component by either adjusting the pH of the solution (often involving a redox reaction and sometimes floc formation), or by adding a substance that will react with the dissolved substance to form a less soluble product.

Precipitation reactions can also follow as a result of a temperature change. Current application of precipitation with respect to hazardous waste treatment include the removal of heavy metals from wastewater by lime treatment, treatment of dye manufacturing wastes, and removal of organic colloids from pulp and paper mill wastewater effluents. Precipitation may be applied to almost any liquid waste stream containing hazardous solids which can be settled out of solution. There are many factors which affect the efficiency of precipitation (pH, nature and concentration of hazardous substances in water, precipitant dosage, temperature, water turbulence, etc.). In practice, the optimum precipitant and dosage for a particular application are determined by a "trial and error" approach using jar tests.

Precipitation can be followed by coagulation and flocculation, in order to enhance sedimentation. Coagulation and flocculation are discussed elsewhere. Xanthate precipitation is discussed under "Ion Exchange."

This technology is used to treat aqueous wastes containing metals. Limitations include the fact that not all metals have a common optimum pH at which they precipitate. Chelating and complexing agents can interfere with the process. Organics are not removed except through adsorptive carryover. The resulting sludge may be hazardous by definition but often may be delisted by specific petition.

The principal precipitation reagents include:

1. Lime—least expensive, generates highest sludge volume.

2. Caustic and carbonates—more expensive than lime, generates smaller amount of sludge, applicable for metals where their minimum solubility within a pH range is not sufficient to meet clean-up criteria.

3. Sulfides—effective treatment for solutions with lower metal concentrations.

4. Sodium borohydride—expensive reagent, produces small sludge volumes which can be reclaimed.

Chemical precipitation normally depends on several variables:

1. Maintenance of an alkaline pH throughout the precipitation reaction and subsequent

settling.

2. Addition of a sufficient excess of treatment ions to drive the precipitation reaction to completion.

3. Addition of an adequate supply of sacrificial ions (such as iron or aluminum) to ensure precipitation and removal of specific target ions.

4. Effective removal of precipitated solids.

Installation of a metals precipitation system inevitably results in the problem of sludge disposal. The cost of hauling the sludge to a licensed hazardous waste landfill will depend on the volume of sludge, the distance hauled, and the sludge composition. Sometimes it is possible to dispose of calcium-based reagent sludges through agricultural or acid pond liming.

Another option is to treat the waste by immobilizing the waste constituents for as long as they remain hazardous. This method of treatment, based on fixation or encapsulation processes, is a possibility for some metals containing wastes. Certain of these residuals could be found hazardous; their heavy metal content may lead to positive tests for EP toxicity. In such cases, encapsulation may be needed to eliminate this characteristic.

Chemical precipitation of metals has long been the primary method of treating metal-laden industrial wastewaters. Due to the success of metals precipitation in such applications, the technology is being considered and selected for use in remediating groundwater containing heavy metals. In groundwater treatment applications, the metal precipitation process is often used as a pretreatment for other treatment technologies (such as chemical oxidation or air stripping) where the presence of metals would interfere with the other treatment processes.

Applicability

Chemical precipitation is a treatment technology applicable to wastewaters containing a wide range of dissolved and other metals, as well as other inorganic substances such as fluorides. This technology removes these metals and inorganics from solution in the form of insoluble solid precipitates. The solids formed are then separated from the wastewater by settling, clarification, and/or polishing filtration.

For some wastewaters, such as chromium plating baths or plating baths containing cyanides, the metals exist in solution in a very soluble form. This solubility can be caused by the metal's oxidation state, for example, high cyanide-containing wastewaters. In both cases, pretreatment, such as hexavalent chromium reduction or oxidation of the metal-cyanide complexes, may be required before the chemical precipitation process can be applied effectively. In the case of arsenic, the arsenic-containing solution is normally treated first with oxidizing agents such as alkali hypochlorite solution to convert the lower-valence arsenic compound to arsenate. The arsenic ion is then typically precipitated out as ferric arsenate. Some compounds must be reduced prior to precipitation. For instance, selinites and selenates are oxidants and are readily reduced to elemental selenium, which is insoluble in aqueous solutions. Sulfur dioxide, sulfides, sulfites, and ferrous ion are all effective for this reduction reaction.

Chemical precipitation may also be applicable to mixed waste for separating radionuclides from other hazardous constituents in wastewaters. Specific conditions of pH, temperature, and precipitating reagent addition are required to selectively remove part or all of the radioactive component as a precipitate.

Principles of Operation

The basic principle of operation of chemical precipitation is that metals and inorganics in wastewater are removed by the addition of a precipitating agent that converts the soluble metals and inorganics to insoluble precipitates. These precipitates are settled, clarified, and/or filtered out of solution, leaving a lower concentration of metals and inorganics in the wastewater. The principal precipitation agents used to convert soluble metal and inorganic compounds to less soluble forms include lime [$Ca(OH)_2$], caustic ($NaOH$), sodium sulfide (Na_2S), and, to a lesser extent, soda ash (Na_2CO_3), phosphate (PO_4^{3-}), and ferrous sulfide (FeS).

The solubility of a particular compound depends on the extent to which the electrostatic forces holding the ions of the compound together can be overcome. The solubility changes significantly with temperature, with most metal compounds becoming more soluble as the temperature increases. Additionally, the solubility is affected by other constituents present in the wastewater, including other ions and complexing agents. Regarding specific ionic forms, nitrates, chlorides, and sulfates are, in general, more soluble than hydroxides, sulfides, carbonates, and phosphates.

Once the soluble metal and inorganic compounds have been converted to precipitates, the effectiveness of chemical precipitation is determined by how successfully the precipitates are physically removed. Removal usually relies on a settling process; that is, a particle of a specific size, shape, and composition will settle at a specific velocity, as described by Stokes' Law. For a batch system, Stokes' Law is a good predictor of settling time because the pertinent particle parameters essentially remain constant. In practice, however, settling time for a batch system is normally determined by empirical testing. For a continuous system, the theory of settling is complicated by such factors as turbulence, short-circuiting of the wastewater, and velocity gradients, thus increasing the importance of empirical tests to accurately determine appropriate settling times.

Chemical Precipitation Process

The equipment and instrumentation required for chemical precipitation vary depending on whether the system is batch or continuous. Both systems are discussed below.

For a batch system, chemical precipitation requires a feed system for the treatment chemicals and a reaction tank where the waste can be treated and allowed to settle. When lime is used, it is usually added to the reaction tank in a slurry form. The supernatant liquid is generally analyzed before discharge to ensure that settling of precipitates is adequate.

For a continuous system, additional tanks are necessary, as well as the instrumentation to ensure that the system is operating properly. In this system, wastewater is fed into an equalization tank, where it is mixed to provide more uniformity, thus minimizing the variability in the type and concentration of constituents sent to the reaction tank.

Following equalization, the wastewater is pumped to a reaction tank where precipitating agents are added. This is done automatically by using instrumentation that senses the pH of the system for hydroxide precipitating agents, or the oxidation-reduction potential (ORP) for nonhydroxide precipitating agents, and then pneumatically adjusts the position of the treatment chemical feed valve until the design pH or ORP value is achieved. (The pH and ORP values are affected by the concentration of hydroxide and nonhydroxide precipitating agents, respectively, and are thus used as indicators of their concentrations in the reaction tank.)

In the reaction tank, the wastewater and precipitating agents are mixed to ensure commingling of the metal and inorganic constituents to be removed and the precipitating agents. In addition, effective dispersion of the precipitating agents throughout the tank is necessary to properly monitor and thereby control the amount added.

Following reaction of the wastewater with the stabilizing agents, coagulating or flocculating compounds are added to chemically assist the settling process. Coagulants and flocculants increase the particle size and density of the precipitated solids, both of which increase the rate of settling. The coagulant or flocculating agent that best improves settling characteristics varies depending on the particular precipitates to be settled.

Settling can be conducted in a large tank by relying solely on gravity or can be mechanically assisted through the use of a circular clarifier or an inclined plate settler. Following the addition of coagulating or flocculating agents, the wastewater is fed to a large settling tank, circular clarifier, or inclined plate settler, where the precipitated solids are removed. These solids are generally further treated in a sludge filtration system to dewater them prior to disposal.

The supernatant liquid effluent can be further treated in a polishing filtration system to remove precipitated residuals both in cases where the settling system is underdesigned and in cases where the particles are difficult to settle.

Selected from "*Robert Noye, Unit Operations in Environmental Engineering, Noye Publication, USA, 1994*"

Unit 18

Text: Oxidation of Wastewater [I]

The process describe in this section are based on chemical oxidation as differentiated from thermal (including wet air and supercritical water oxidation), electrolytic, radiative, and biological oxidation.

Liquids are the primary waste form treatable by chemical oxidation. The most powerful oxidants are relatively non-selective; therefore, any easily oxidizable material in the waste stream will be treated. If for instance an easily oxidation organic solvent were used, little of the chemical effect of the oxidizing agent could be used on the hazardous constituent. This, therefore, essentially limits the use of the most commonly used oxidants to aqueous wastes.

Gases have been treated by scrubbing with oxidizing solutions for the destruction of odorous substances, such as certain amines and sulfur compounds. Potassium permanganate, for instance, has been used in certain chemical processes, in the manufacture of kraft paper and in the rendering industry. Oxidizing solutions are also used for small-scale disposal of certain reactive gases in laboratories.

Oxidation has limited application to slurries, tars, and sludges. Because other components of the sludge, as well as the material to be oxidized, may be attacked indiscriminately by oxidizing agents, careful control of the treatment via multi-staging of reaction, careful control pH, etc. are required.

Chemical oxidation has been found very effective in the treatment of industrial and domestic wastewater. In particular, oxidation offers one of the few methods for removing odor, color, and various potentially toxic organic substances like phenols, pesticides and industrial solvents. It also disinfects drinking water by killing or inactivating pathogenic microorganisms that may be present.

Oxidation reactions can be carried out using simple, readily available equipment; only storage vessels, metering equipment, and contact vessels with agitators are required. However, implementation is complication because every oxidation/reduction reaction system must be designed for the specific application. Laboratory-and/or pilot-scale testing is essential to determine the appropriate chemical feed rates and reactor retention times in accordance with reaction kinetics. Oxidation and reduction has not been widely used in treating hazardous wastestreams.

A major consideration in electing to utilize oxidation technology is that the treatment chemicals are invariably hazardous, and great care must be taken in their handling[①]. In particular, the handling of many oxidizing agents is potentially hazardous and suppliers' instructions should be carefully followed.

In some cases, undesirable by-products may be formed as a result of oxidation. For example, addition of chlorine can result in formation of bio-resistant end products which can be odorous and more toxic than the original compound. The possibility of this undesirable side reaction needs to be considered when using chlorine for oxidation of wastewaters.

Chemical oxidation can be effective way of pretreating wastes prior to biological treatment; compounds which are refractory to biological treatment can be partially oxidized making them more amendable to biological oxidation.

One of the major limitations with chemical oxidation is that oxidation reactions frequently are not complete (reactions do not proceed to CO_2 and H_2O). Incomplete oxidation may be due to oxidant concentration, pH, oxidation potential of the oxidant, or formation of a stable intermediate. The danger of incomplete oxidation is that more toxic oxidation products could be formed. Chemical oxidation is not well suited to high-strength, complex waste streams. The most powerful oxidants are relatively non-selective and any oxidizable organics in the waste stream will be treated. For highly concentrated waste streams this will result in the need to add large concentrations of oxidizing agents in order to treat target compounds. Some oxidant such as potassium permanganate can be decomposed in the presence of high concentrations of alcohol and organic solvents.

Chemical oxidation should be considered as a first treatment step when the waste contains cyanide. Chemical oxidation should be evaluated as a first treatment for wastes that have constituents or concentrations of constituents that are not amendable to other treatment methods. Chemical oxidation should be considered as a final polishing step for residual traces of contaminants remaining after certain other treatments.

Chemical oxidation is a treatment technology used to treat wastes containing organics. It is also used to treat sulfide wastes by converting the sulfide to sulfate. The destruction of cyanides in wastes is usually accomplished by chemical oxidation. Chemical oxidation can also be used to change the oxidation state of metallic compounds to valences that are less soluble, such as converting arsenic in wastes to the relatively insoluble pentavalent state.

Chemical oxidation is applicable to dissolved cyanides in aqueous solutions, such as wastewaters from metal plating and finishing operations, or to inorganic sludges from these operations that contain cyanide compounds. For cyanides, chemical oxidation is most applicable to solutions containing less than 500 mg/L of cyanides when the cyanides are in a form that can be easily disassociated in water to yield free cyanide ions. If cyanides are present in water as a tightly bound complex ion, e.g., ferrocyanide, only limited treatment may occur. If the waste contains greater than 500mg/L of cyanide, but no more than about 100000mg/L, electrolytic oxidation may be more appropriate.

Chemical oxidation may also be used for destruction of the organic component of organometallic compounds in wastes, thus freeing the metal component for treatment by chemical precipitation or stabilization[2]. Organic compounds such as EDTA, NTA, citric acid, glutaric acid, lactic acid, and tartrates are often used as chelating agents to prevent metal ions from precipitating out in electroless plating solutions. When these spent plating solutions require treatment for metals removal by chemical precipitation, the organic chelating agents must first be destroyed. Chemical oxidants, potassium permanganate in particular, are effective in releasing metals from complexes with these organic compounds.

The basic principle of operation for chemical oxidation is that in organic cyanides, some dissolved organic compounds, and sulfides can be chemically oxidized to yield carbon dioxide, water, salts, simple organic acids, and, in the case of sulfides, sulfates. Metallic ions such as arsenites can be oxidized to higher, less soluble valences such as arsenates. The principal chemical oxidants used are hypochlorite, chlorine gas, chlorine dioxide, hydrogen peroxide, ozone, and potassium permanganate.

Selected from "Robert Noye, *Unit Operations in Environmental Engineering*, *Noye Publication*, USA, 1994"

Words and Expressions

scrub *v.* 使（气体）净化
slurry *n.* 泥浆，浆
tar *n.* 焦油，柏油
pathogenic *a.* 致病的，病原的，发病的
oxidation/reduction 氧化/还原
pilot-scale 小规模试验，中试
refractory *a.* 难控制的；难熔的
potassium permanganate 高锰酸钾
amenable *a.* 服从的，顺从的
valences *n.* （化合）价，原子价
EDTA 乙二胺四乙酸
NTA 次氨基三乙酸
citric acid 柠檬酸
glutaric acid 戊二酸
lactic acid 乳酸
tartrate *n.* 酒石酸盐
chelating agent 螯合剂
arsenite *n.* 亚砷酸盐
hypochlorite *n.* 次氯酸盐
chlorine dioxide 二氧化氯（面粉漂白剂，又称 Dyox）
hydrogen peroxide 过氧化氢
ozone *n.* 臭氧

Notes

① A major consideration in electing to utilize oxidation technology is that the treatment chemicals are invariably hazardous, and great care must be taken in their handling. 可译为：在选择使用氧化技术时要考虑的一个重要因素就是这些化学处理品都是有毒的，因此在处理过程中要格外小心。

② Chemical oxidation may also be used for destruction of the organic component of organo-metallic compounds in wastes, thus freeing the metal component for treatment by chemical precipitation or stabilization. 可译为：化学氧化还可用来破坏废水中有机金属化合物的有机组分，从而将其中的金属组分离出来以便于用化学沉淀或稳定剂处理。

Exercise

1. Translate the last two paragraphs into Chinese.
2. What is the main limitation for using chemical oxidation?
3. How many methods do you know to treat cyanide in wastewater?
4. Based on the reading material, what are the advantages or disadvantages of chlorine oxidation, chlorine dioxide oxidation, hydrogen peroxide oxidation and ozonation?

Reading Material: Oxidation of Wastewater [Ⅱ]

Catalytic Oxidation

Adding a catalyst that promotes oxygen transfer and thus enhances oxidation has the effect of lowering the necessary reactor temperature and/or improving the level of destruction of oxidizable compounds. For waste constituents that are more difficult to oxidize, catalyst addition may be necessary to effectively destroy the constituent(s) of concern. Catalysts typically used for this purpose include copper bromide and copper nitrate.

One of the earliest investigations of catalytic oxidation was conducted by Battelle Laboratories in 1971 to study the adsorption of free and complexed cyanide onto activated carbon in the presence of copper. Subsequent efforts were undertaken by the Calgon Corp. to also develop a cyanide detoxification method utilizing catalytic oxidation on granular activated carbon. Cupric ions are added to the wastewater along with oxygen prior to passing the cyanide-bearing waste through a granular activated carbon column. According to Calgon, "cupric ions are added to the water to accelerate and increase the efficiency of the catalytic oxidation of the cyanide by granular activated carbon". In addition to improving the catalytic oxidation of the cyanide, "the presence of cupric ions results in the formation of copper cyanides, which have a greater adsorption capacity than copper or cyanide alone."

Hydrogen bromide wastes from pharmaceutical, agrichemical and other plants can be oxidized into bromine, using a solid catalyst from Catalytica Inc. The process uses oxygen instead of chlorine for the oxidation, thus generating water as the only by-product. With chlorine, bromides and sodium chloride brine waste must be disposed of.

Chemical Waste Management, Inc. has developed a process to treat wastewaters, such as leachates, ground water, and process waters, containing mixtures of salts, metals, and organic compounds. The proprietary technology is a combination of evaporation and catalytic oxidation processes. Wastewater is concentrated in an evaporator by boiling off most of the water and the volatile contaminants, both organic and inorganic. Air or oxygen is added to the vapor, and the mixture is forced through a catalyst bed, where the organic and inorganic compounds are oxidized. This stream, composed of mainly steam, passes through a scrubber, if necessary, to remove any acid gases formed during oxidation. The stream is then condensed or vented to the atmosphere. The resulting brine solution is either disposed of or treated further, depending on the nature of the waste.

Chlorine oxidation

Chlorine has been used in drinking water treatment since the turn of the century, however, the formation of trihalomethanes will probably deter further expansion of chlorine for this use. Oxidized organic compounds are produced by the chlorination process. Chlorine is used in a number of ways for detoxification including alkaline chlorination, chlorine dioxide, chloroiodides, hypochlorites, and chlorinolysis.

Alkaline chlorination: When chlorine is added to wastewaters, under alkaline conditions, reactions occur which lead to oxidation(chlorination) of the contaminant. This oxidation process, which is widely used in the treatment of cyanide wastes, is generally referred to as the "alkaline chlorination" process. Cyanides can be oxidized with chlorine to the less toxic cyanates. Additional chlorine will then oxidize the cyanates to nontoxic nitrogen gas, carbon dioxide, and bicarbonates.

Alkaline chlorination is used to treat free cyanides and complex cyanides although combinations with Fe or Ni will take a longer time. Limitations include the exothermic heat of the reaction, nonselective competitions with other species and additional chlorine demands. Fairly close pH control(7.5 to 9.0) is required to avoid toxic volatiles release.

The oxidation of cyanides in wastewater by chlorine has been the most widely acceptable method of cyanide treatment for the past thirty years. In an alkaline solution of pH 8.5 or higher, chlorine reacts with cyanide to form cyanate.

The oxidation of cyanide wastes by chlorine is a widely used process in plants using cyanide in cleaning and metal processing baths. Alkaline chlorination is also used for cyanide treatment in a number of inorganic chemical facilities producing hydrocyanic acid and various metal cyanides.

Chlorine dioxide oxidation: Oxidation by chlorine dioxide has substantial advantages over oxidation by chlorine in the treatment of drinking water. The oxidation power of chlorine dioxide is not impaired over a wide range of pH, and is therefore not very sensitive to fluctuations in pH. The effectiveness of chlorine oxidation is very much dependent on the pH value, and at typical operating values of pH from 6 to 10, chlorine partially loses its oxidizing power. Chlorine dioxide does not react with ammonia or nitrogenous compounds, whereas these react with chlorine to form chloramines. Finally, chlorine dioxide does not form trihalomethanes during oxidation of organic chemicals. All of these factors make oxidation by chlorine dioxide one of the techniques of choice for removing low levels of organic chemicals from water.

Most of the oxidizing capacity of chlorine dioxide comes from the reduction of ClO_2^- to a chlorite ion. As with chlorine, wastewater treatment by chlorine dioxide is usually done at the site where the waste is produced. The chlorine dioxide is usually produced by reacting sodium chlorite with acids, such as, hydrochloric or sulfuric acids. Extreme caution must be exercised when mixing $NaClO_2$ with acids. The reaction between sulfuric acid and solid

$NaClO_2$ is explosive. Therefore, the reaction must be carried out exclusively using solutions of $NaClO_2$ in water.

Hydrogen Peroxide Oxidation

This process is based on the addition of hydrogen peroxide to oxidize organic compounds. Hydrogen peroxide is not the stable oxide of hydrogen and since it readily gives up its extra oxygen, it is an excellent oxidizing agent. The process is a nonspecific reaction. It may be exothermic/explosive or require addition of heat and/or catalysts. Oxidation by hydrogen peroxide is being developed by Steinfeld and Partner in Germany for in situ treatment. It may also be used for soil surface treatment.

The peroxide oxidation process is run under similar conditions, and with similar equipment, to those used in the alkaline chlorination process. Hydrogen peroxide is added as a liquid solution. Oxidation with H_2O_2 is generally performed in the presence of a metal catalyst. Typical catalysts include iron sulfate, iron wool, nickel salts, and aluminum salts. The waste is heated and then treated with H_2O_2 while being agitated. The H_2O_2 oxidation tends to proceed quickly under basic conditions.

Ozone and hydrogen peroxide, can be used together, and has been found by USAE Waterways Experiment Station to be as effective, and less costly than ultraviolet based systems.

Cyanide Oxidation: Hydrogen peroxide oxidation removes both cyanide and metals in cyanide containing wastewaters. In this process, cyanide bearing waters are heated to 49°C to 54°C (120 °F to 130 °F) and the pH is adjusted to 10.5 to 11.8. Formalin (37% formaldehyde) is added while the tank is vigorously agitated. After 2 to 5 min, a proprietary peroxygen compound (41% hydrogen peroxide with a catalyst and additives) is added. After an hour of mixing, the reaction is completed. The cyanide is converted to cyanate, and the metals are precipitated as oxides or hydroxides. The metals are then removed from solution by either settling or filtration.

The main equipment required for this process is two holding tanks equipped with heaters and air spargers or mechanical stirrers. These tanks may be used in a batch or continuous fashion, with one tank being used for treatment while the other is being filled. A settling tank or a filter is needed to concentrate the precipitate.

The hydrogen peroxide oxidation process is applicable to cyanide-bearing wastewaters, especially those containing metal-cyanide complexes. In terms of waste reduction performance, this process can reduce total cyanide to less than 0.1mg/L and the zinc or cadmium to less than 1.0 mg/L.

Chemical costs are similar to those for alkaline chlorination using chlorine, and lower than those for treatment with hypochlorite. All free cyanide reacts and is completely oxidized to the less toxic cyanate state. In addition, the metals precipitate and settle quickly, and they may be recoverable in many instances. However, the process requires energy expenditures to heat the wastewater prior to treatment.

Ozonation

Ozone is a powerful oxidizing agent, having oxidizing potential greater than either chlorine or chlorine dioxide. Oxidation by ozone instead of chlorine or chlorine dioxide has been found to eliminate entirely the formation of undesirable end-products such as organohalides. It is effective for disinfection, odor and color removal, and destruction of cyanides and toxic organic compounds in water. In addition, ozonation coupled with ultraviolet radiation destroys some trihalomethanes and their precursors. The unconsumed dissolved ozone after treatment quickly decomposes to oxygen, such that no secondary pollutant is formed which might directly or indirectly (through side reactions) cause environmental health problems.

Ozone can be used to pretreat wastes to break down refractory organics or as a polishing step after biological or other treatment processes to oxidize untreated organics. Ozone is usually produced by high-voltage ionization of atmospheric oxygen.

Ozone is currently used for treatment of hazardous wastes to destroy cyanide and phenolic compounds. The rapid oxidation of cyanides with ozone offers advantages over the slower alkaline chlorination method. Limitations include the physical form of the waste (i.e., sludges and solids are not readily treated) and nonselective competition with other species.

Ozonation can be conducted in a batch or continuous process. The ozone for treatment is produced on site because of the hazards of transporting and storing ozone as well as its short shelf life. The ozone gas is supplied to the reaction vessels by injection into the wastewater. The batch process uses a single reaction tank. As with alkaline chlorination, the amount of ozone added and the reaction time used are determined by the type and concentration of the oxidizable contaminants, and vigorous mixing should be provided for complete oxidation.

In continuous operation, two separate tanks may be used for reaction. The first tank receives an excess dosage of ozone. Any excess ozone remaining at the outlet of the second tank is recycled to the first tank, thus ensuring that an excess of ozone is maintained and also that no ozone is released to the atmosphere. As with alkaline chlorination, an ORP control system is usually necessary to ensure that sufficient ozone is being added.

Ozone is usually generated at the point of use in a flowing air or oxygen stream by an electric discharge process. Mixtures of 1% to 3% of ozone/air and 3% to 5% ozone/oxygen can be produced. These are then mixed with water in a contactor to ensure efficient transfer into the liquid. Ozone product is energy intensive, with only 10% of the power supplied to the ozone generator producing ozone. The remainder of the power produces light, sound and heat which are undesirable by-products and, hence, represent process inefficiencies. Therefore, the electricity cost for the ozonation process makes up a considerable percentage of the operating cost.

Selected from "*Robert Noye, Unit Operations in Environmental Engineering, Noye Publication, USA, 1994*"

Unit 19

Text: Unit Operations of Pretreatment

Several devices and structures are placed upstream of the primary treatment operation to provide protection to the wastewater treatment plant (WWTP) equipment. These devices and structures are classified as pretreatment because they have little effect in reducing BOD_5.

Bar Racks

Typically, the first device encountered by the wastewater entering the plant is a bar rack. The primary purpose of the rack is to remove trash and large objects that would damage or foul pumps, valves, and other mechanical equipment. Rags, logs, and other objects that find their way into the sewer are removed from the wastewater on the racks[①]. In modern WWTPs, the racks are cleaned mechanically. The solid material is stored in a hopper and removed to a sanitary landfill at regular intervals.

Bar racks placed upstream of grit chambers and pumps are sized to have an open space of 50 to 150mm between each bar. Racks placed downstream of grit chambers normally have openings of 20 to 50mm, with 25mm being a commonly accepted value. Channel approach velocities are held between 0.4 and 0.9m/s. Two channels are provided to allow one to be taken out of service for cleaning and repair.

Grit Chambers

Inert dense material such as sand, broken glass, silt, and pebbles is called grit. If these materials are not removed from the wastewater, they abrade pumps and other mechanical devices, causing undue wear. In addition, they have a tendency to settle in corners and bends, reducing flow capacity and, ultimately, clogging pipes and channels.

There are three basic types of grit removal devices: velocity controlled, aerated, and constant level short-term sedimentation basins. We will discuss only the first two, since they are the most common.

Velocity controlled. This type of grit chamber, also known as a horizontal flow grit chamber, can be analyzed by means of the classical laws of sedimentation for discrete, non-flocculating particles (Type I sedimentation).

Stokes'law may be used for the analysis and design of horizontal flow grit chambers if the horizontal liquid velocity is maintained at about 0.3m/s. Liquid velocity control is achieved by placing a specially designed weir at the end of the channel[②]. A minimum of two channels must be employed so that one can be out of service without shutting down the treatment plant. Cleaning may be either by mechanical devices or by hand. Mechanical cleaning is favored for plants having average flows over 0.04m^3/s. Theoretical detention times are set at about one minute for average flows. Washing facilities are normally provided to remove organic material from the grit.

Aerated grit chambers. The spiral roll of the aerated grit chamber liquid "drives" the grit into a hopper which is located under the air diffuser assembly. The shearing action of the air bubbles is supposed to strip the inert grit of much of the organic material that adheres to its surface[3].

Aerated grit chamber performance is a function of the roll velocity and detention time. The roll velocity is controlled by adjusting the air feed rate. Nominal air flow values are in the range of 0.15 to 0.45 cubic meters per minute of air per meter of tank length [m^3/(min · m)]. Liquid detention times are usually set to be about three minutes at maximum flow. Length-to-width ratios range from 2 : 5 to 5 : 1 with depths on the order of 2 to 5m.

Grit accumulation in the chamber varies greatly, depending on whether the sewer system is a combined type or a separate type, and on the efficiency of the chamber. For combined systems, $90m^3$ of grit per million cubic meters of sewage ($m^3/10^6 m^3$) is not uncommon. In separate systems you might expect something less than $30m^3/10^6 m^3$. Normally the grit is buried in a sanitary landfill.

Comminutors

Devices that are used to macerate wastewater solid (rags, paper, plastic, and other materials) by revolving cutting bars are called comminutors[4]. These devices are placed downstream of the grit chambers to protect the cutting bars from abrasion. They are used as a replacement for the downstream bar rack but must be installed with a hand-cleaned rack in parallel in case they fail.

Equalization

Flow equalization is not a treatment process, but a technique that can be used to improve the effectiveness of both secondary and advanced wastewater treatment processes. Wastewater does not flow into a municipal wastewater treatment plant at a constant rate; the flow rate varies from hour to hour, reflecting the living habits of the area served. In most towns, the pattern of daily activities sets the pattern of sewage flow and strength. Above-average sewage flows and strength occur in mid-morning. The constantly changing amount and strength of wastewater to be treated makes efficient process operation difficult. Also, many treatment units must be designed for the maximum flow condition encountered, which actually results in their being oversized for average conditions. The purpose of flow equalization is to dampen these variations so that the wastewater can be treated at a nearly constant flow rate. Flow equalization can significantly improve the performance of an existing plant and increase its useful capacity. In new plants, flow equalization can reduce the size and cost of the treatment units.

Flow equalization is usually achieved by constructing large basins that collect and store the wastewater flow and from which the wastewater is pumped to the treatment plant at a constant rate. These basins are normally located near the head end of the treatment works, preferably downstream of pretreatment facilities such as bar screens, comminutors, and grit chambers. Adequate aeration and mixing must be provided to prevent odors and solids deposition. The required volume of an equalization basin is estimated from a mass balance of the flow into the treatment plant with the average flow the plant is designed to treat. The theoretical basis is the same as that used to size reservoirs.

With the screening completed and the grit removed, the wastewater still contains light organic suspended solids, some of which can be removed from the sewage by gravity in a sedimentation tank. These tanks can be round or rectangular, are usually about 3.5m deep, and hold the wastewater for periods of two to three hours. The mass of settled solids is called raw sludge. The sludge is removed from the sedimentation tank by mechanical scrapers and pumps. Floating materials, such as grease and oil, rise to the surface of the sedimentation tank, where they are collected by a surface skimming system and removed from the tank for further processing.

Primary sedimentation basins (primary tanks) are characterized by Type II flocculant settling. The Stokes equation cannot be used because the flocculating particles are continually changing in size, shape and, when water is entrapped in the floc, by specific gravity. There is no adequate mathematical relationship that can be used to describe Type II settling. Laboratory tests with settling columns are used to develop design data.

The behavior of the primary tank is analogous to the behavior of clarifiers used in water treatment, with a few important exceptions. The solids concentration is much lower in a primary tank than in a water treatment plant clarifier. Thus, the overflow rates and weir loading rates differ appreciably. Overflow rates are commonly in the range 25 to 60m/d. The Great Lakes-Upper Mississippi River Board of State Sanitary Engineers (GLUMRB) recommends that weir loading (hydraulic flow over the effluent weir) rates not exceed $120m^3/d$ of flow per m of weir length $[m^3/(d \cdot m)]$ for plants with average flows less than $0.04m^3/s$. For larger flows the recommended rate is $190m^3/(d \cdot m)$.

As mentioned previously, approximately 50 to 60 percent of the raw sewage suspended solids and as much as 30 to 35 percent of the raw sewage BOD_5 may be removed in the primary tank.

Selected from "*Mackezil L. David, David A. Cornwen, Introduction to Environmental Engineering, Mcgraw-Hill, Inter. Edition, Chem. Eng. Series, 2nd Edition, USA, 1991*"

Words and Expressions

bar rack　格栅池
rag *n*. 抹布；碎屑；石板瓦
log *n*. 原木；圆形木材，圆木
sewer *n*. 下水道
inert *a*. 无活动的，惰性的；迟钝的

silt *n.* 淤泥；残渣；煤粉；泥沙
pebble *n.* 小圆石，小鹅卵石
abrade *v.* 磨损，擦伤
sedimentation *n.* 沉淀，沉降
shearing *n.* 切应力
sanitary *a.* 公共卫生的，清洁的，清洁卫生的
landfill *n.* 垃圾掩埋场
comminutor *n.* 粉碎机
equalization *n.* 平衡；平等化，同等化
scraper *n.* 刮刀
skimming *n.* 浮渣
flocculant *n.* 凝聚剂
Stokes equation 斯托克斯方程
entrap *v.* 收集；诱捕
overflow *n.* 溢出；泛滥

Notes

① Rags, logs, and other objects that find their way into the sewer are removed from the wastewater on the racks. 可译为：污水中的碎石、木片及其他物体通过格栅滤池从废水中除去。

② Liquid velocity control is achieved by placing a specially designed weir at the end of the channel. 可译为：在流道的末端放置一个专门设计的溢流堰就可以控制液流的速度。

③ The shearing action of the air bubbles is supposed to strip the inert grit of much of the organic material that adheres to its surface. 可译为：气泡的剪切作用能够脱去黏附在惰性砂砾表面大量的有机物质。

④ Devices that are used to macerate wastewater solid (rags, paper, plastic, and other materials) by revolving cutting bars are called comminutors. 可译为：通过旋转的切割刀具来破坏废水中固体物质（碎石、纸张、塑料及其他材料）的装置可称之为粉碎机。

Exercises

1. Translate the following into Chinese.

   Flow equalization is usually achieved by constructing large basins that collect and store the wastewater flow and from which the wastewater is pumped to the treatment plant at a constant rate. These basins are normally located near the head end of the treatment works, preferably downstream of pretreatment facilities such as bar screens, comminutors, and grit chambers. Adequate aeration and mixing must be provided to prevent odors and solids deposition. The required volume of an equalization basin is estimated from a mass balance of the flow into the treatment plant with the average flow the plant is designed to treat. The theoretical basis is the same as that used to size reservoirs.

2. Why are equalization basins required in wastewater treatment?

3. Put the following into Chinese.

预（初）沉池　　生化需氧量　　缓冲池　　砂滤池　　均质池　　溢流堰
城市污水　　　　平衡　　　　　悬浮固体　负荷　　重力沉降　澄清池

Reading Material: Wastewater Treatment

Water used for industrial and municipal purposes is often degraded during use by the addition of suspended solids, salts, nutrients, bacteria, and oxygen-demanding material. In the United States, by law, these waters must be treated before being released back into the environment.

Wastewater treatment—sewage treatment—costs about ﹩20 billion per year in the United States, and the cost keeps rising, but it will continue to be big business. Conventional wastewater treatment includes septic-tank disposal systems in rural areas and centralized wastewater treatment plants in cities. Recently, innovative approaches include applying wastewater to the land and renovating and reusing wastewater. We discuss the conventional methods in this section and some newer methods in later sections.

Septic-Tank Disposal Systems

In many rural areas, no central sewage systems or wastewater treatment facilities are available. As a result, individual septic-tank disposal systems, not connected to sewer systems, continue to be an important method of sewage disposal in rural areas as well as outlying areas of cities. Because not all land is suitable for a septic-tank disposal system, an evaluation of each site is required by law before a permit can be issued. An alert buyer should make sure that the site is satisfactory for septic-tank disposal before purchasing property in a rural setting or on the fringe of an urban area where such a system is necessary.

Wastewater Treatment Plants

In urban areas, wastewater is treated at specially designed plants that accept municipal sewage from homes, businesses, and industrial sites. The raw sewage is delivered to the plant through a network of sewer pipes. Following treatment, the wastewater is discharged into the surface-water environment (river, lake, or ocean) or, in some limited cases, used for another purpose, such as crop irrigation. The main purpose of standard treatment plants is to break down and reduce the BOD and kill bacteria with chlorine. Wastewater treatment methods are usually divided into three categories: primary treatment, secondary treatment, and advanced wastewater treatment. Primary and secondary treatments are required by federal law for all municipal plants in the United States. However, treatment plants may qualify for a waiver exempting them from secondary treatment if installing secondary treatment facilities poses an excessive financial burden. Where secondary treatment is not sufficient to protect the quality of the surface water into which the treated water is discharged—for example, a river with endangered fish species that must be protected—advanced treatment may be required.

Primary Treatment

Incoming raw sewage enters the plant from the municipal sewer line and first passes through a series of screens to remove large floating organic material. The sewage next enters the "grit chamber", where sand, small stones, and grit are removed and disposed of. From there, it goes to the primary sedimentation tank, where particulate matter settles out to form sludge. Sometimes, chemicals are used to help the settling process. The sludge is removed and transported to the "digester" for further processing. Primary treatment removes approximately 30% to 40% of BOD by volume from the wastewater, mainly in the form of suspended solids and organic matter.

Secondary Treatment

There are several methods of secondary treatment. The most common treatment is known as activated sludge, because it uses living organisms—mostly bacteria. In this procedure, the wastewater from primary sedimentation tank enters the aeration tank, where it is mixed with air (pumped in) and with some of the sludge from the final sedimentation tank. The sludge contains aerobic bacteria that consume organic material (BOD) in the wastewater. The wastewater then enters the final sedimentation tank, where sludge settles out. Some of this "activated sludge", rich in bacteria, is recycled and mixed again in the aeration tank with air and new, incoming wastewater acting as a starter. The bacteria are used again and again. Most of the sludge from the final sedimentation tank, however, is transported to the sludge digester. There, along with sludge from the primary sedimentation tank, it is treated by anaerobic bacteria (bacteria that can live and grow without oxygen), which further degrade the sludge by microbial digestion.

Methane gas (CH_4) is a product of the anaerobic digestion and may be used at the plant as a fuel to run equipment or to heat and cool buildings. In some cases, it is burned off. Wastewater from the final sedimentation tank is next disinfected, usually by chlorination, to eliminate disease-causing organisms. The treated wastewater then is discharged into a river, lake, or ocean, or in some limited cases, used to irrigate farmland. Secondary treatment removes about 90% of BOD that enters the treatment plant in the sewage.

Advanced Wastewater Treatment

As noted above, primary and secondary treatments do not remove all pollutants from incoming sewage. Some additional pollutants, however, can be removed by adding more treatment steps. For example, phosphates and nitrates, organic chemicals, and heavy metals can be removed by specifically designed treatments, such as sand filters, carbon filters, and chemicals applied to assist in the removal process. Treated water is then discharged into surface water or may be used for irrigating agricultural lands or municipal properties, such as golf courses, city parks, and grounds surrounding wastewater treatment plants.

Advanced wastewater treatment is used when it is particularly important to maintain good water quality. For example, if a treatment plant discharges treated wastewater into a river and there is concern that nutrients remaining after secondary treatment may cause damage to the river ecosystem (eutrophication), advanced treatment may be used to reduce the nutrients.

Chlorine Treatment

As mentioned, chlorine is frequently used to disinfect water as part of wastewater treatment. Chlorine is very effective in killing the pathogens responsible for outbreaks of serious waterborne diseases that have killed many thousands of people. However, a recently discovered potential is that chlorine treatment also produces minute quantities of chemical by-products, some of which are potentially hazardous to people and other animals. For example, a recent study in Britain revealed that in some rivers, male fish sampled downstream from wastewater treatment plants had testes containing both eggs and sperm. This is likely related to the concentration of sewage effluent and the treatment method used. Evidence also suggests that these by-products in the water may pose a risk of cancer and other human health effects. The degree of risk is controversial and currently being debated.

Selected from " *Daniel B. Botkin, Edward A. Keller. Environmental Science, Eight Edition, John Wiley & Sons, Inc. Printed in Asia, 2012*"

PART 4 SOLID WASTES AND DISPOSAL

Unit 20

Text: Sources and Types of Solid Wastes

Knowledge of the sources and types of solid wastes, along with data on the composition and rates of generation, is basic to the design and operation of the functional elements associated with the management of solid wastes.

Sources of Solid Wastes

Sources of solid wastes are, in general, related to land use and zoning. Although any number of source classifications can be developed, the following categories have been found useful[①]: (1) residential, (2) commercial, (3) municipal, (4) industrial, (5) open areas, (6) treatment plants, and (7) agricultural. Typical solid waste generation facilities, activities, or locations associated with each of these sources are presented in Table 1. The types of wastes generated, which are discussed next, are also identified.

Table 1 Typical Solid Waste Generating Facilities, Activities, and
Locations Associated with Various Source Classifications

Source	Typical facilities, activities, or locations where wastes are generated	Types of solid wastes
Residential	Single-family and multifamily dwellings, low-, medium-, and high-rise apartments, etc.	Food wastes, rubbish, ashes, special wastes
Commercial occasionally	Stores, restaurants, markets, office buildings, hotels, motels, print shops, auto repair shops, medical facilities and institutions, etc.	Food wastes, rubbish, ashes, demolition and construction wastes, special wastes, hazardous wastes
Municipal[①]	As above[①]	As above[①]
Industrial demolition	Construction, fabrication, light and heavy manufacturing, refineries, chemical plants, lumbering, mining power plants, demolition, etc.	Food wastes, rubbish, ashes, and construction wastes, special wastes, hazardous wastes
Open areas	Streets, alleys, parks, vacant lots, playgrounds, beaches, highways, recreational areas, etc.	Special wastes, rubbish
Treatment plant sites principally Agricultural	Water, waste water, and industrial treatment processes, etc. Field and row crops, orchards, vineyards, darries, foodlots, farms, etc.	Treatment plant wastes, composed of residual sludges Spoiled food wastes, agricultural wastes, rubbish, hazardous wastes

① The term municipal normally is assumed to include both the residential and commercial solid wastes generated in the community.

Types of Solid Wastes

The term solid wastes is all-inclusive and encompasses all sources, types of classifications, composition, and properties. Wastes that are discharged may be of significant value in another setting, but they are of little or no value to the possessor who wants to dispose of them. To avoid confusion, the term refuse, often used interchangeably with the term solid wastes, is not used in this text.

As a basis for subsequent discussions, it will be helpful to define the various types of solid wastes that are generated (see Table 1)[2]. It is important to be aware that the definitions of solid waste terms and the classifications vary greatly in the literature. Consequently, the use of published data requires considerable care, judgment, and common sense. The following definitions are intended to serve as a guide and are not meant to be arbitrary or precise in a scientific sense.

Food Wastes Food wastes are the animal, fruit, or vegetable residues resulting from the handling, preparation, cooking, and eating of foods (also called garbage). The most important characteristic of these wastes is that they are highly putrescible and will decompose rapidly, especially in warm weather. Often, decomposition will lead to the development of offensive odors. In many locations, the putrescible nature of these wastes will significantly influence the design and operation of the solid waste collection system. In addition to the amounts of food wastes generated at residences, considerable amounts are generated at cafeterias and restaurants, large institutional facilities such as hospitals and prisons, and facilities associated with the marketing of foods, including wholesale and retail stores and markets.

Rubbish Rubbish consists of combustible and noncombustible solid wastes of household, institutions, commercial activities, etc., excluding food wastes or other highly putrescible material. Typically, combustible rubbish consists of materials such as paper, cardboard, plastics, textiles, rubber, leather, wood, furniture, and garden trimmings. Noncombustible rubbish consists of items such as glass, crockery, tin cans, aluminum cans, ferrous and other nonferrous metals, and dirt.

Ashes and Residues Materials remaining from the burning of wood, coal, coke, and other combustible wastes in homes, stores, institutions, and industrial and municipal facilities for purposes of heating, cooking, and disposing of combustible wastes are categorized as ashes and residues. Residues from power plants normally are not included in this category. Ashes and residues are normally composed of fine, powdery materials, cinders, clinkers, and small amounts of burned and partially burned materials. Glass, crockery, and various metals are also found in the residues from municipal incinerators.

Demolition and Construction Wastes Wastes from razed buildings and other structures are classified as demolition wastes. Wastes from the construction, remodeling, and repairing of individual residences, commercial buildings, and other structures are classified as construction wastes. These wastes are often classified as rubbish. The quantities produced are difficult to estimate and variable in composition, but may include dirt, stones, concrete, bricks, plaster, lumber, shingles, and plumbing, heating, and electrical parts.

Special Wastes　　Wastes such as street sweepings, roadside litter, litter from municipal litter containers, catch-basin debris, dead animals, and abandoned vehicles are classified as special wastes. Because it is impossible to predict where dead animals and abandoned automobiles will be found, these wastes are often identified as originating from nonspecific diffuse sources. This is in contrast to residential sources, which are also diffuse but specific in that the generation of the wastes is a recurring event.

Treatment Plant Wastes　　The solid and semisolid wastes from water, waste water, and industrial waste treatment facilities are included in this classification. The specific characteristics of these materials vary, depending on the nature of the treatment process. At present, their collection is not the charge of most municipal agencies responsible for solid waste management. In the future, however, it is anticipated that their disposal will become a major factor in any solid waste management plan.

Agricultural Wastes　　Wastes and residues resulting from diverse agricultural activities—such as the planting and harvesting of row, field, and tree and vine crops, the production of milk, the production of animals for slaughter, and the operation of feedlots—are collectively called *agricultural wastes*. At present, the disposal of these wastes is not the responsibility of most municipal and county solid waste management agencies. However, in many areas the disposal of animal manure has become a critical problem, especially from feedlots and dairies.

Hazardous Wastes　　Chemical, biological, flammable, explosive, or radioactive wastes that pose a substantial danger, immediately or over time, to human, plant, or animal life are classified as hazardous. Typically, these wastes occur as liquids, but they are often found in the form of gases, solids, or sludges. In all cases, these wastes must be handled and disposed of with great care and caution.

Selected from *"Ven Te Chow, Rolf Eliassen and Ray K. Linsley, Solid Wastes: Engineering Principles and Management Issues, MCGRAW-HILL BOOK COMPANY, 1977"*

Words and Expressions

zoning *n.* 分区，区域划分（商业区、住宅区、工业区等）
residential *a.* 住宅的，居住的
municipal *a.* 市的，市政的
facility *n.* 环境；设备
motel *n.* 汽车旅馆
fabrication *n.* 生产，加工
demolition *n.* 拆除，推翻
lot *n.* 一块地，一块地皮
orchard *n.* 果园

dispose of 处理，处置
inclusive *a.* 包括的，包含的；包括许多或一切的
encompass *v.* 围绕，包围；包含，包括
refuse *n.* 弃物，垃圾，废物
arbitrary *a.* 独裁的，专制的；专横的，任意的
garbage *n.* （丢弃或喂猪等的）剩饭残羹，垃圾
putrescible *a.* 易腐烂的
offensive *a.* 令人不快的，讨厌的
cafeteria *n.* 自助食堂
institutional *a.* 制度上的，慈善机构的
combustible *a.* 容易着火燃烧的
trimming *n.* 装饰物
crockery *n.* 陶（瓦）器；瓦罐
ferrous *a.* 含有铁的
cinder *n.* 煤渣，焦渣
clinker *n.* （煤在火炉、熔炉等中燃烧后所留下的）渣滓，熔渣，熔块
incinerator *n.* 焚化炉
raze *v.* 铲平，拆毁
shingle *n.* 屋顶（墙面）板
litter *n.* 杂乱的废物
catch-basin *n.* 雨水井，沉泥井
in contrast to 与……相反
anticipate *v.* 预料；期望
It is anticipated that 可以预料
diverse *a.* 种类不同的
slaughter *n.* 屠宰
feedlot *n.* 牧场
manure *n.* 粪肥
flammable *a.* 易燃的
with great care and caution 以极细心和谨慎的态度

Notes

① Although any number of source classifications can be developed, the following categories have been found useful. 可译为：虽然已有许多废物来源的分类法，但以下的分类法是很有价值的。

② As a basis for subsequent discussions, it will be helpful to define the various types of solid wastes that are generated. 可译为：作为以下讨论的基础，定义所产生的各种固体废物是有帮助的。

Exercises

1. Write an abstract of the text.
2. Translate the second and the last paragraphs of the text into Chinese.
3. Put the following sentences into English.
（1）饮食废物的重要特点在于它具有易腐烂性，尤其在温暖天气里，会极快地腐烂。
（2）除了在家庭产生饮食废物外，在自助食堂、饭馆、医院和监狱等机构，以及诸如批发、零售等食品市场，均会产生大量的饮食废物。
4. Answer the following questions.
（1）List the sources and types of solid wastes.
（2）List some typical facilities, activities, or locations where wastes are generated.
（3）What kinds of solid wastes do you often see? Where are they from?

Reading Material: Quantities of Wastes

Everyone is familiar with solid wastes, especially those generated in municipalities, such as food wastes and rubbish, abandoned vehicles, demolition and construction wastes, street sweepings, and garden wastes. Far greater amounts, however, result from agricultural, industrial, and mineral sources.

Estimated Total and Per Capita Quantities

Although the data are varied, recent estimates indicate that an average of 4.4 billion tons of solid wastes is generated each year in the United States alone. Of this total, municipal wastes represent approximately 230 million tons; industrial wastes, 140 million tons; and agricultural wastes, 640 million tons. By far the greatest amount of solid wastes comes from mines and minerals and from animal wastes, each with an average of 1.7 billion tons/yr. The total amount generated from all sources by the year 2000 may approach 12 billion tons/yr.

Looking at just the urban and industrial wastes, the generation rate in the United States is approximately 3,600lb/(capita · yr). In comparison, other industrialized countries have lower rates, but they have similar problems. Based on rough estimates, Japan is the closest to the United States with an average of 800lb. The rate in Netherlands is over 600lb; in west Germany, about 500lb. From these figures it can be concluded that in these countries either the rate of consumption of goods is lower or a more serious effort is made to recover and reuse the wastes.

Data from Recent Surveys

Many surveys of solid waste generation have been made by consulting engineers and planners in their studies for municipalities and regional authorities. State and federal agencies, particularly the U.S. Public Health Service (USPHS) and the U.S. Environmental Protection Agency (EPA), have also been active in this area. There are so many reported values that any one for a particular category may be disputed because of the aforementioned impact of technological developments, the marketing of consumer products and their packaging, and commercial and industrial practices. For instance, institutions having large computers, with their long printouts, have had an appreciable impact on the generation of waste paper in certain communities. For these reasons it is imperative that surveys be made for any specific municipality or region to determine the ranges of values of solid wastes generated from municipal and industrial sources.

U. S. Public Health Service Data

In 1968 the USPHS published data obtained in its National Survey of Community Solid Waste Practices. The average generation rates for urban sources in the United States are shown in Table 2.

It should be emphasized that these are yearly averages and that the actual generation rates for a given city vary with the seasons (garden clippings, leaves, Christmas paper and gift containers, etc.). The exact amounts for a given city may also be far above average, such as the vast amount of paper discarded in Washington, D. C., state capitals, and large commercial and industrial centers. Again, beware of averages in the design of facilities, but be guided by them in analyzing the results for a specific location!

Table 2 Average per Capita Quantities of Solid Wastes Collected from Urban Sources in the United States, 1968

Source	lb/(capita·day)	Source	lb/(capita·day)
Combined residential and commercial	4.29	Tree and landscaping	0.18
Industrial	1.90	Park and beach	0.15
Institutional	0.16	Catch basin	0.04
Demolition and construction	0.72	Sewage treatment plant solids	0.50
Street and alley cleanings	0.25	Total	8.19

Note: lb/(capita·day)×0.4536=kg/(capita·day).

Table 3 Components of Municipal Solid Wastes Generated in the United States, 1971

Component	Total generated		Total disposed	
	Tons. millions	Percent	Tons. millions	Percent
Paper	39.1	31.30	47.3	37.8
Glass	12.1	9.7	12.5	10.0
Metal	11.9	9.5	12.6	10.1
Ferrous	10.6	8.5	—	—
Aluminum	0.8	0.6	—	—
Other nonferrous	0.5	0.4	—	—
Plastic	4.2	3.4	4.7	3.8
Rubber and leather	3.3	2.6	3.4	2.7
Textiles	1.8	1.4	2.0	1.6
Wood	4.6	3.7	4.6	3.7
Food	22.0	17.6	17.7	14.2
Subtotal	99.0	79.2	104.80	83.9
Yard wastes	24.1	19.3	18.2	14.6
Miscellaneous inorganics	1.9	1.5	2.0	1.5
Total	125.00	100.00	125.00	100.00

Note: tons×907.2=kg.

Environmental Protection Agency

The EPA has continued the studies of the USPHS and in 1971 published a report to the Congress containing estimates of present and future solid waste generation. The estimate of waste generation in the United States by component is shown in Table 3. Note that the quantities shown as disposed are greater than those shown as generated. The difference is attributed to the increase of moisture content in the disposed wastes and to the measure of the wastes generated on a dry basis. Excluded from this table are treatment plant sludges, demolition and construction wastes, and such special wastes as abandoned automobiles.

Selected from *"Ven Te Chow, Rolf Eliassen and Ray K. Linsley, Solid Wastes: Engineering Principles and Management Issues, MCGRAW-HILL BOOK COMPANY, 1977"*

Unit 21

Text: Everybody's Problems—Hazardous Waste

Every year, billions of tons of solid wastes are discarded in the United States. These wastes range in nature from common household trash to complex materials in industrial wastes, sewage sludge, agricultural residues, mining refuse and pathological wastes from institutions such as hospitals and laboratories.

The U. S. Environmental Protection Agency (EPA) estimated in 1980 that at least 57 million metric tons of the nation's total waste load can be classified as hazardous. Unfortunately, many dangerous materials that society has "thrown away" over recent decades have endured in the environment-making household words of "Love Canal" and "Valley of the Drums."[①] These two incidents are not unique. The EPA has on file hundreds of documented cases of damage to life and the environment resulting from the indiscriminate or improper management of hazardous wastes[②]. The vast majority of cases involve pollution of ground water—the source of drinking water for about half of the U. S. population—from improperly sited or operated landfills and surface impoundments (pits, ponds and lagoons). In addition to polluting ground water, the improper handling or disposal of hazardous waste can cause several other kinds of environmental damage, as illustrated by these case histories (often involving more than one form of damage) from EPA records.

Hazardous Waste Can Pollute Ground Water

- The water supplies of Toone and Teague, Tennessee, were contaminated in 1978 with organic compounds when water leached from a nearby landfill. When the landfill closed, about six years earlier, the site held some 350000 drums, many of them leaking pesticide wastes. Because the towns no longer have access to uncontami-

nated ground water, they must pump water in from other locations.
- Ground water in a 30-square-mile area near Denver was contaminated from disposal of pesticide waste in unlined disposal ponds. The waste, from manufacturing activities of the U. S. Army and a chemical company, dates back to the 1943~1957 period. Decontamination, if possible, could take several years and cost as much as $80 million.

Hazardous Waste Can Contaminate Rivers, Lakes, and Other Surface Water

- At least 1500 drums containing waste, primarily from metal-finishing operations, were buried near Byron, Illinois, for an unknown number of years until about 1972. Surface waters (and soil and ground water as well) were contaminated with cyanides, heavy metals, phenols and miscellaneous other materials. Wildlife, stream life and local vegetation were destroyed. The disposal site suffered long-range damage from the toxic pollutants that drained into the soil.
- About 17000 drums littered a 7-acre site in Kentucky which became known as 'Valley of the Drums'. Some 6000 drums were full, many of them oozing their toxic contents onto the ground. In addition, an undetermined quantity of hazardous waste was buried in drums and subsurface pits. In 1979, EPA analyses of soil and surface water in the drainage area about 25 miles south of Louisville identified about 200 organic chemicals and 30 metals.

Hazardous Waste Can Pollute the Air

- In 1972, waste containing hexachlorobenzene (HCB), one of the family of toxic organic compounds that contains chlorine, was disposed of in a landfill near Darrow and Geismar, Louisiana. The HCB vaporized and subsequently accumulated in cattle over a 100-square-mile area. Some cattle had to be destroyed. This incident represented direct and indirect economic losses of over $380,000. Elevated, although subtoxic, levels of HCB in blood plasma were found in some area residents.
- A truck driver was killed in 1978 as he discharged waste from his truck into one of four open pits at a disposal site in Iberville Parish, Louisiana. He was asphyxiated by hydrogen sulfide produced when liquid wastes mixed in the open pit. The area was surrounded by water and had a history of flooding.

Hazardous Waste Can Burn or Explode

- A fire broke out in 1978 at a disposal site in Chester, Pennsylvania, where 30000 to 50000 drums of industrial waste had been received over a 3-year period. The smoke forced closing of the Commodore Barry Bridge and 45 firemen required medical treatment, mostly as a result of lung and skin irritation from chemical fumes. A number of homes are located within three blocks of the site; drummed waste was kept only 20 feet from a natural gas storage tank, and liquefied natural gas tanks were about 100 yards away. Waste was emptied directly on the soil of the 3-acre site; some probably

drained to the tidal section of the adjacent Delaware River. Waste may even have been dumped into the river.
- A bulldozer operator was killed in a 1975 explosion at a landfill in Edison Township, New Jersey, as he was burying and compacting drums of unidentified chemical waste. Of the 200 truckloads of waste the landfill received daily, about 50 were industrial waste.

Hazardous Waste Can Poison Via the Food Chain

- In 1970, three children in an Alamogordo, New Mexico, family became seriously ill after eating a home-slaughtered pig that had been fed corn treated with a mercury compound. Local health officials found several bags of similarly treated corn in the community dump.
- Over a 4-month period in 1976, an Indiana family consumed milk contaminated with twice the maximum concentration of polychlorinated biphenyls (PCB) considered safe by the Food and Drug Administration. The milk came from the family's cow, which had been grazing in a pasture fertilized with the city of Bloomington's sewage sludge. The sludge contained high levels of PCB from a local manufacturing plant. A federal law passed in 1976 banned production of PCB after January 1, 1979.

Hazardous Waste Can Poison by Direct Contact

- The health of some residents of Love Canal, near Niagara Falls, New York, was seriously damaged by chemical waste buried a quarter of a century ago. As drums holding the waste corroded, their contents percolated through the soil into yards and basements, forcing evacuation of over 200 families in 1978 and 1979. About 80 chemicals, a number of them suspected carcinogens, were identified.
- In 1979, cattle on a Kansas feedlot were contaminated with PCB after waste oil was used in animal backrubbers. The waste oil (from electrical transformers) had been purchased from a salvage yard in 1972, before the effects of PCB were widely known. Inedible byproducts from 54 head of cattle had been shipped to a number of states and had to be traced and disposed of properly; another 112 head had to be destroyed.

These examples provide dramatic evidence of damage to life and the environment from mismanagement of hazardous waste. It was in large part to prevent such tragedies that, in 1976, Congress enacted Subtitle C of the Resource Conservation and Recovery Act (RCRA), Public Law 94-580[3]. This law imposes strict controls over the management of hazardous waste throughout its entire life cycle. The costs for proper environmental controls will be higher than amounts spent in the past to manage these wastes. But the astronomical costs of cleaning up damage caused by poor disposal practices should be eliminated. An ounce of prevention, in this instance, is a sound investment.

A 1979 EPA study indicated that cleaning up abandoned hazardous waste sites and those operating under environmentally unsound conditions could cost as much as $44 billion, only part of which is likely to be paid for by the owners of the sites[4]. The remainder would have to come from other sources. But in many cases it is impossible to assign dollar values to the long-term harm to health and the environment that has resulted from improper management of hazardous waste.

Selected from "Yen-Hsiung Kiang, Amir A. Metry, Hazardous Wastes Processing Technology, ANN ARBOR SCIENCE Publishers, INC. Michigan U.S.A., 1982"

Words and Expressions

trash n. 垃圾；无价值的东西
pathological a. 病理学的；与疾病有关的；有病的
institution n. 公共机构设施；机关
metric ton 吨
indiscriminate a. 不分皂白的；不加选择的
endure v. 忍受；持久
on file 存卷归档
access to 使用或接近的权力，机会
date back to 追溯到
decontamination n. 净化，消除……的污染
unlined a. 无衬里（炉衬、镶衬、衬砌）的
miscellaneous a. （混）杂的，杂项的；各种各样的；多方面的
ooze v./n. （慢慢）渗出（物），徐徐流出（物）
hexachlorobenzene n. 六氯苯
plasma n. 血浆
discharge A from B 从 B 上卸 A
asphyxiate v. 使（人）窒息；闷死
bulldozer n. 推土机
via prep. 经由，通过
polychlorinated biphenyl 多氯代联苯
biphenyl n. 联（二）苯，联苯基
evacuation n. 消除，除清，撤离
carcinogen n. 致癌物，致癌因素
backrubber 摩擦施药器
salvage v. 抢救；废物处理
inedible a. 不适于食用的，不可食的
enact v. 制定
conservation n. 保护，保存
remainder n. 残余物

Notes

① Unfortunately, many dangerous materials that society has "thrown away" over recent decades have endured in the environment-making household words of "Love Canal" and

"Valley of the Drums."可译为：不幸的是，在最近几十年中社会持续扔掉了许多危险物，从而产生了"爱河"和"桶谷"的家用词。

② The EPA has on file hundreds of documented cases of damage to life and the environment resulting from the indiscriminate or improper management of hazardous wastes. 可译为：EPA 有数百起对生命和环境造成损害的案例。他们都是由于不加区别地或不适当地处理有害废物所造成的。

③ It was in large part to prevent such tragedies that, in 1976, Congress enacted Subtitle C of the Resource Conservation and Recovery Act (RCRA), Public Law 94-580. 可译为：正是为了防止这样的悲剧发生，议会在 1976 年制定了资源保护和回收法律附录 C，公共法 94-580。

④ A 1979 EPA study indicated that cleaning up abandoned hazardous waste sites and those operating under environmentally unsound conditions could cost as much as ＄44 billion, only part of which is likely to be paid for by the owners of the sites. 可译为：1979 年的一份 EPA 研究表明，清除废弃有害物场所以及令人不满意的工作场所要花费 440 亿美元，其中仅有部分可由场所雇主支付。这里"those 指代 sites"，"only part of……the site"是定语从句。英语中定语从句有限制性和非限制性定语从句两种。两者区别在于限制意义是否强，且定语从句绝大部分都是后置的。汉语的所有定语都要前置，不涉及限制意义的强弱，故在英译汉时，限制与非限制不起十分重要的作用。汉语不用长定语，若英语中定语从句太长，则不宜译成汉语定语，应用其他方法来处理，如①、④句的翻译。

Exercises

1. Translate the first and second paragraphs of the text into Chinese.
2. Put the following sentences into English.
① 如果对有害废物的处理不恰当，可能会引发其他种类的环境危害。
② 现在用于处理这些废物的费用要远高于过去的费用。
3. Answer the following questions.
① What kind of environmental damage does the improperly handling of hazardous waste cause?
② Can you list some examples to provide dramatic evidences of damage to life and the environment from mismanagement of hazardous waste?

Reading Material: Municipal Solid—Waste Management

Municipal solid—waste management continues to be a problem in the United States and other parts of the world. In many areas, particularly in developing countries, waste management practices are inadequate. These practices, which include poorly controlled open dumps and illegal roadside dumping, can spoil scenic resources, pollute soil and water, and pose health hazards.

Illegal dumping is a social problem as much as a physical one because many people are simply disposing of waste as inexpensively and as quickly as possible, perhaps not seeing their garbage as an environmental problem. If nothing else, this is a tremendous waste of resources, since much of what is dumped could be recycled or reused. In areas where illegal dumping has been reduced, the keys have been awareness, education, and alternatives. Education programs teach people about the environmental problems of unsafe, unsanitary dumping of waste, and funds are provided for cleanup and for inexpensive collection and recycling of trash at sites of origin.

We look next at the composition of solid waste in the United States and then go on to describe specific disposal methods: onsite disposal, compositing, incineration, open dumps and sanitary landfills.

Composition of Solid Waste

The average content of un-recycled solid waste likely to end up at a disposal site in the United States is shown in other chapters. It is no surprise that paper is by far the most abundant component. However, considerable variation can be expected, based on factors such as land use, economic base, industrial activity, climate and time of year.

People have many misconceptions about our waste stream. With all the negative publicity about fast-food packaging, polystyrene foam, and disposable diapers, many people assume that these make up a large percentage of the waste stream and are responsible for the rapid filling of landfills. However, excavations into modern landfills using archaeological tools have cleared up some misconceptions. We now know that fast-food packaging accounts for only about 0.25% of the average landfill; disposable diapers, approximately 0.8%; and polystyrene products about 0.9%. Paper is a major constituent in landfills, perhaps as much as 50% by volume and 40% by weight. The largest single item is newsprint, which accounts for as much as 18% by volume. Newsprint is one of the major items targeted for recycling because big environmental dividends can be expected. However (and this is a value judgment), the need to deal with the major waste products doesn't mean that we need not cut down on our use of disposable diapers, polystyrene, and other paper products. In addition to creating a need for disposal, these products are made from resources that might be better managed.

Onsite Disposal

A common onsite disposal method in unban areas is the garbage—disposal device installed in the wastewater pipe under the kitchen sink to grind garbage and flush it into the sewer system. This effectively reduces the amount of handling and quickly removes food waste. What's left of it is transferred to sewage—treatment plants, where solids remaining as sewage sludge still must be disposed of.

Composting

Composting is a biochemical process in which organic materials, such as lawn clippings and kitchen scraps, decompose to a rich, soil—like material. The process involves rapid partial decomposition of most solid organic waste by aerobic organisms. Although simple backyard compost piles may come to mind, large-scale composing as a waste—management option is generally carried out in the controlled environment of mechanical digesters. This technique is popular in Europe and Asia, where intense farming creates a demand for compost. However, a major drawback of composting is the necessity of separating organic material from other waste. Therefore, it is probably economically advantageous only where organic material is collected separately from other waste. Another negative is that composting plant debris previously treated with herbicides may produce a compost toxic so some plants. Nevertheless, composting is an important component of waste management, and its contribution continues to grow.

Incineration

Incineration burns combustible waste at temperatures high enough (900 to 1000℃ or 1650 to 1830℉) to consume all combustible material, leaving only ash and noncombustible materials to dispose of in a landfill. Under ideal conditions, incineration may reduce the volume of waste by 75% to 95%. In practice, however, the actual decrease in volume is closer to 50% because of maintenance problems as well as waste-supply problems. Besides reducing a large volume of combustible waste to a much smaller volume of ash, incineration has another advantage: It can be used to supplement other fuels and generate electric power.

Incineration of urban waste is not necessarily a clean process; it may produce air pollution and toxic ash. In the United States, for example, incineration is apparently a significant source of environmental dioxin, a carcinogenic toxin. Smokestacks from incinerators also may emit oxides of nitrogen and sulfur, which lead to acid rain; heavy metals, such as lead, cadmium, and mercury; and carbon dioxide, which is related to global warming.

Open Dumps (Poorly Controlled Landfills)

In the past, solid waste was often disposed of in open dumps (now called landfills), where refuse was piled up and left uncovered. Thousands of open dumps have been closed in recent years, and new open dumps are banned in the United States and many other countries. Nevertheless, many are still being used worldwide.

Sanitary Landfills

A sanitary landfill (also called a municipal solid-waste landfill) is designed to concentrate and contain refuse without creating a nuisance or hazard to public health or safety. The idea is to confine the waste to the smallest practical area, reduce it to the smallest practical volume, and cover it with a layer of compacted soil at the end of each day of operation, or more frequently if necessary. Covering the waste is what makes the landfill sanitary. The compacted layer restricts (but does not eliminate) continued access to the waste by insects, rodents, and other animals, such as seagulls. It also isolates the refuse, minimizing the amount of surface water seeping into it and the amount of gas escaping from it.

Selected from "*Daniel B. Botkin, Edward A. Keller. Environmental Science, Eight Edition, John Wiley & Sons, Inc. Printed in Asia, 2012*"

Unit 22

Text: Methods of Waste Disposal

It is inevitable that as there are different types of waste, there will be varying methods of waste disposal. Briefly most solid wastes are deposited on land as tips or spoil heaps, or as land infill to quarries and mine shafts, or as dumps containing a large range of materials. In addition, small quantities of waste are dumped into the sea. Waste is produced continually so there is often a need for some sort of storage facility. In the case of some mineral extractive industries such as deep mined coal, china clay and ironstone, there is storage on the working site as spoil heaps, but this is waste deposition rather than disposal[1]. In other industries the stored waste often has to be transported to disposal areas and tipped or dumped. Alternatively, the stored waste may be treated in various ways before disposal. The treatment may reduce the bulk, or make disposal easier, or extract materials that can be re-used or recycled back into manufacturing processes. In respect of environmental pollution the quantity, the treatment, and disposal methods of waste are of prime importance.

Methods Used by Local Authorities

The Public Health Act 1936 enables Local Authorities to collect, treat, and dispose of all refuse from the domestic sector, and such industrial and trade waste as requested. In 1973, Local Authorities in England dealt with 19.5 M tonnes of waste by various methods (see Table 1). About 15 M tonnes of the tipped waste is household refuse consisting of cinders, ash, dust, vegetable and waste food matter, paper, board, metal, rags, glass, and plastics, whilst the remaining 4.5 M tonnes is from trade sources. About 86% of this waste is not pretreated and is disposed of by land tipping. One quarter of this waste is just dumped in an uncontrolled or non-systematic manner. The other 75% is dealt with by controlled tipping. This means the waste is deposited, spread, and compacted into shallow layers, and covered with soil to assist decomposition and sealing. This method should ensure that loose litter does not blow about, there is no unpleasant odour, and flies and vermin do not breed to produce a health hazard. An alternative to tipping is the use of waste for land in filling. Disused quarries, or land which is derelict by virtue of being low-lying and badly drained, or derelict as a result of spoil tips, can be reclaimed by refuse in-filling.

Table 1 Disposal of Household and Trade Waste, England and Wales 1973

Method	Quantity in M tonnes	Percentage	Method	Quantity in M tonnes	Percentage
Tipping	16.8	86.3	Pulverisation	0.7	3.7
Incineration	1.85	9.5	Composting	0.095	0.5

The shortage of suitable land for tipping purposes has caused some Local Authorities to consider alternative methods of waste disposal. Controlled tipping is usually cheap in respect of capital and labour costs, but it can become costly if highly priced land has to be purchased for future tipping. A way of avoiding this, and assisting land conservation, is to reduce the bulk quantity of the waste before tipping. This allows existing tips to be used for a longer time, reduces the need for new ones, and reduces labour costs for tipping operations. Waste can be pretreated by pulverizing, or mechanically breaking it down into smaller particle sizes, which can reduce the bulk by up to 33% by volume. Whilst the cost of a pulverizing plant is high, the salvaging of materials for recycling, and less waste to deposit can help to off-set the initial capital cost. Another pretreatment waste technique is incineration, which involves combustion in a furnace at a temperature between 950 and 1100℃ to minimize corrosion and the emission of odours[2]. This reduces the waste bulk considerably, and the process can reduce the volume up to 90% and the weight up to 60%, compared to untreated waste. There are environmental advantages to pretreatment. Much less residue has to be disposed of, and it is free from bacteria and wet organic matter that can cause putrefaction odours and gases. Also it is possible to use the heat energy produced for augmenting electrical generation or district heating, and this is already being carried out in some countries[3]. It has been estimated that if all the household and trade refuse collected in the UK in 1974 had been incinerated and the heat used, this could have saved energy equivalent to 6 M tonnes of coal[4].

Methods Used by Industry

An approximate estimate of the annual amount of industrial waste produced in 1973 was 110 M tonnes or 67% from the mining and quarrying industries, 12 M tonnes or 7.3% from CEGB power stations, and 23 M tonnes or 14% from other industrial sectors. There are no published figures for radioactive wastes, but the amount is relatively small. The quantity and type of waste varies from industry to industry, but the major part consists of solid material, liquid slurries and effluent containing a wide range of suspended and dissolved chemical substances. The large amount of solid waste produced by the mining and extractive industries is disposed of by tipping on land or into the sea. Other wastes such as furnace clinker, blast-furnace slag, and copper, tin and zinc-lead slags have been omitted because they are mainly re-used in subsequent production. About 70% of this waste is not treated in any way and is tipped on land, or is used for land in-full and reclamation. The problems associated with this are not discussed in this paper. It should be noted that about 26% of the above wastes are not tipped, but are used to assist in the production of materials such as bricks, concrete blocks, cement, and road and concrete aggregate.

Manufacturing industries produce wastes which are solid, semi-solid, liquid, or gaseous, and each category may contain toxic or non-toxic, flammable, and no-combustible constituents. There are no overall data available across all industries to show the quantities of waste or the methods of disposal. Some limited surveys have been carried out, and these at least provide some detailed information. The Local Government Operational Research Unit (LGORU) conducted a survey of the industrial wastes from 600 firms in the heavily industrialized area of Manchester and Salford in 1970. It was found that one million tonnes of waste per annum was produced, consisting of non-combustible sludge, dust, ash, brick, slag and excavated materials; and combustible paper, rubber, plastics, timber, sawdust, textiles and chemical materials. The quantities and methods of disposal are shown in reference. In this survey, 72% of waste was tipped on land, 16% of the intractable and dangerous waste was dumped at sea in sealed containers, and 8.6% was buried. Only about 3% of the waste was disposed of by the alternative methods of incineration or discharge into sewers.

Selected from "*H. M. Dizc, Environmental Pollution, John Wiley & Sons, Inc., UAS, 1981*"

Words and Expressions

inevitable *a.* 不可避免的
tip *n.* 垃圾场；*v.*（使）倾卸，倾倒，倾翻
heap *v.* 堆积
quarry *n.* 采石场
shaft *n.*（矿）井；通道
china clay 瓷土
bulk *n.* 大小；尺寸
domestic *a.* 生活的；家庭的
whilst *conj.* ＝while
pulverization *n.* 磨碎，粉化
vermin *n.* 害兽
derelict *a.* 被弃的，被遗弃的
low-lying *a.* 低（洼，标高）的，位置很低的
reclaim *a.* 开拓
purchase *v.* 购买，采购
pulverize *v.* 磨碎，研磨
off-set *v.* 补偿
putrefactive *a.*（容易）腐败（烂、朽）的
augment *v.* 增大，加大
aerobic *a.* 需氧的，有氧的
whereby *a.* 借（由，因）此，利用它；凭（为）什么，靠什么
batch *a.* 一次的分量，一批

Notes

① In the case of some mineral extractive industries such as deep mined coal, china clay and ironstone, there is storage on the working site as spoil heaps, but this is waste deposition rather than disposal. 可译为：对某些矿物冶炼工业部门，如深井采煤、采瓷土与采

铁矿，常常在工地设置堆积存放场，但这只是废物存放而不是处理。

② Another pretreatment waste technique is incineration, which involves combustion in a furnace at a temperature between 950 and 1100℃ to minimize corrosion and the emission of odours. 可译为：另一种预处理废物的方法是焚烧，此方法是在温度处于950～1100℃之间的熔炉中燃烧，以减少腐烂和气味的产生。

③ Also it is possible to use the heat energy produced for augmenting electrical generation or district heating, and this is already being carried out in some countries. 可译为：而且可以用产生的热能来增加发电量或用于地区取暖，并且这已经正在某些国家应用。

④ It has been estimated that if all the household and trade refuse collected in the UK in 1974 had been incinerated and the heat used, this could have saved energy equivalent to 6 M tonnes of coal. 可译为：据估计，如果1974年英国所收集的家庭和交易市场的垃圾全部焚烧并且热能都加以利用，则可节约相当于600万吨煤的能量。

Exercises

1. Write an abstract of the text.
2. Translate the first paragraph of the text into Chinese.
3. Put the following sentences into English.
① 在野餐地、公园及其他公共场地随意丢弃纸张、塑料、食品包装盒、瓶子等是另一种常见的污染行为。
② 把废物粉碎成碎块可以使其体积减小33％左右。
4. Answer the following questions.
① What is your present concept of waste disposal?
② How many methods of waste disposal mentioned in this text?
③ Why are the quantity, the treatment, and disposal methods of waste of prime importance?
5. Write an abstract of reading material.

Reading Material: Incineration of Hazardous Waste in the U.S.A.

Incineration is the process of burning flammable hazardous wastes compounds to render them to harmless such as the harmless products of carbon dioxide and water. Incineration has been accepted by many scientists as the only viable alternative to disposal of flammable hazardous wastes. Many others question the safety of such disposal methods. Incineration is probably a good technology for disposal of some wastes it might be inappropriate for other wastes. In either case, the costs of regulatory compliance have caused incineration to be less of a cost effective alternative to waste disposal.

The application of incinerators to the treatment and disposal of hazardous waste has many advantages. The method offers an ultimate disposal technique for flammable and non-flammable substances. In many cases, the technique is the only acceptable method for disposal of many flammable liquids and some organic wastes.

There are, however, several negative aspects to incineration. Many people have a perception that incinerators are unsafe due to technical operating problems. Still others distrust incinerator operators and the regulating governmental authorities. As a result, the sitting of a new incinerator will, in general, result in intense opposition of the installation by the local public.

Federal Regulation

The United States Environmental Protection Agency (EPA) is the single government agency that has control over the location, construction, and operation of incinerators. Since this agency is sensitive to political influence from Congressmen and others, history has shown that incinerator regulation is based on a mixture of sound technical principals and entirely non-technical political whims. Regardless of the source, the present day regulations have driven the cost of regulatory compliance to a point where construction of new facilities is almost cost prohibitive.

The key contribution of the distrust of the public towards incinerator operation is the lack of continuous pollution monitoring equipment. Because of the low concentration of pollutants exiting the flue gas stream, it is impractical to measure the concentrations of these pollutants with conventional on-line analytical equipment.

Theory of Operation

Hazardous chemical are exposed to high temperatures in a combustion furnace. The materials that are combustible are oxidized. Gaseous, liquid, and solid wastes can be destroyed in this manner. The amount by which the hazardous constituent has been reduced is called the Destruction and Removal Efficiency (DRE).

$$DRE = \frac{\text{Waste In} - \text{Waste Out}}{\text{Waste In}} \times 100\%$$

Here, Waste In is the amount of waste fed to the incinerator and Waste Out is the waste lost to the flue gas.

Since most of the waste streams fed to an incinerator system are very complex mixtures of several chemicals (over 200 different chemicals are considered hazardous), it is necessary to pick out one to three different chemicals as representative compounds to use as a measure of the effectiveness of the system. A chemical that is picked as a representative compound is called a Principal Organic Hazardous Constituent (POHC). The POHC of choice is usually a chemical that is present in a large quality, is a particularly insidious compound, or is very difficult to burn because of low heating value. Once the POHCs are chosen, these are the only chemicals that are used in subsequent material balance calculations in describing the efficiency of the incinerator.

Gases and clean liquids are comparatively simple to incinerate because they are easily pumped to the furnace chamber through nozzles that mix them with air in preparation for combustion. High viscosity liquids, sludges, and solids, on the other hand, are difficult to mix with air for proper combustion, hence they are the most difficult materials to incinerate.

Because of mixing difficulties between the air and waste fuel, partial or incomplete combustion may take place. For those chemicals that are partially oxidized, it is possible to form an entirely different chemical species for discharge through the flue gas stream. These alternate chemical products are called products of incomplete combustion (PIC).

PIC's are of even greater concern when chlorine and chloride compounds are present due to the potential for the formation of dioxins and furans. Although the minute concentrations of these chemicals that are present are of negligible hazard, the presence of these chemicals are perceived to be catastrophic by the public. Hence, any new incinerator proposal must deal with the fear of chemicals based on emotion and not the scientific facts.

Another product of incomplete combustion is carbon monoxide (CO). Since continuous monitors are available for CO, CO concentration is often used as an indicator of proper incinerator performance.

Because of the concern that unburned POHC's and PIC's will escape from the combustion apparatus, two combustion furnaces are often employed. The waste is fed to the first or primary combustion chamber (PCC). Here, combustion takes place at a relatively low temperature. The resulting combustion products then are routed to a secondary combustion chamber (SCC) where the gases are further mixed with air and raised to higher temperatures by the combustion of an auxiliary fuel. The design of these two chambers depends largely on the composition and nature of the waste.

For sludges and solid wastes, the primary combustion chamber must be able to move solids through the combustion zone while mixing the waste materials with air. A rotary kiln furnace is generally employed as the PCC for these sludges and solids. Liquid and gaseous wastes, on the other hand, require a much simpler combustion chamber.

Depending on the nature of the wastes being burned, an air pollution control (APC) device may be required. Flue gases from certain types of incinerators may contain particulates, acid gases, or other pollutants such as vaporized heavy metals.

Operating Parameters

In general, the rate of oxidation is controlled by the operating temperature and oxygen concentration. The extent of the oxidation is controlled by the reaction rate and furnace residence time. For a given waste material, an operational test is performed to determine the sets of operating conditions that will guarantee the desired destruction of the POHC. This test operation is given a special name-the trial burn.

The trial burn is a well planned period of operation during which the final operation conditions are proven. The plant management personnel and the EPA personnel negotiate each detail of the plant's operation during the trial burn test. Details include waste source, POHC concentration, waste feed rate, air flow rate, APC operation, etc. The plant is operated at various conditions in an effort to determine the minimum operating temperatures and maximum feed rates that can be employed to operate the plant and still meet the minimum emission requirements. Very accurate measurements of the input and output flows are made and a material balance calculation is performed during the trial burn to determine DRE. The results of the trial bum are used by the EPA to set permit limitations for the operation of the plant.

Performance Standards

Regulations dictate that a DRE of 99.99% of each POHC will be achieved. In addition, 99% of all HCl (or 1.81 kg/hour maximum emission rate) must be removed. Particulate emissions of less than 0.183 grams/cubic meter of flue gas corrected to 7% oxygen must be achieved. Additional requirements for CO concentrations are specified with time weighted CO excursions allowed.

Because of the flue gas emission rate limitations, it is often necessary to install air pollution control devices to control particulates and acid gases. The most popular devices used for hazardous waste incinerators are wet scrubbing equipment such as Venturi scrubbers and ionizing wet scrubbers. In some cases, absorption columns are used in addition to the scrubbers for acid gas removal.

General Requirements

The operator must give special attention to fugitive emission control during the handling of the wastes, automatic waste feed cutoff in the event of an incinerator malfunction, and a rigorous regime of equipment inspection and personnel training must be performed. Any other requirements that might be deemed appropriate are also listed as part of the permit.

Summary

Combustible hazardous wastes can be successfully and completely destroyed in a properly designed incinerator. Regulatory constraints and other considerations make the process very expensive. Because of potential operating problems, incineration should be employed only after other alternatives to waste management such as recycle and waste minimization have been considered.

Selected from *"Clifford E George, Proceedings of International Conference on Environmental Engineering and Chemical Engineering, 1992"*

Unit 23

Text: Disposal of Solid Wastes

Landfilling

Except for the disposal of municipal solid wastes at sea, which is not permitted by most developed countries, solid wastes, or their residues in some form, must go to the land. Landfilling, the most economical and consequently the most common method of solid waste disposal, is used for 90 percent of the municipal solid wastes in the United Kingdom and North America. Even in European countries like West Germany and Switzerland, with massive investments in incineration and composting plants, over 60 percent of domestic and commercial waste is landfilled. Incineration cannot, of course, eliminate landfilling. In fact, it creates a more concentrated residue that may be more hazardous to water supplies than unburned solid wastes. The area needed for landfilling of solid wastes is about 1 ha per year for every 25,000 people (1 Ac /10000 people). This is illustrated in the following Example.

Example

For a population of 25000, estimate the annual area requirements (excluding the buffer zone) for a normally compacted landfill having a refuse depth of 4m excluding cover material[①].

Solution Assuming that per capita waste generation is 2.0 kg/d and that the density of a well-compacted landfill is 450 kg/m^3, the annual area required is

$$\frac{25000 \times 2.0 \text{kg/d} \times 365 \text{d/yr}}{450 \text{kg/m}^3 \times 4\text{m} \times 10000 \text{m}^2/\text{ha}} = 1.0 \text{ha}$$

Area requirements for landfilling can vary considerably with the type of waste and the degree of compaction. Details of the design and operation of sanitary landfills are discussed in reference.

The balance of this section reviews some of the processes that may be used prior to land disposal to reduce waste volume and/or utilize waste components, thus reducing landfill needs.

Incineration

Volume reduction. Large numbers of batch-fed incinerators built during the 1930 and 1940s to reduce waste volume were major contributors to air pollution, performed poorly, and were costly to maintain. Some of these were upgraded, but most were shut down and replaced by land disposal of refuse whenever possible. However, as landfill capacity decreased, volume reduction became more important. At the same time, the fuel value of refuse had been rising steadily. As a result, incineration for reducing waste volume (by about 90 percent) and weight (by 75 percent) with the possibility of energy recovery, became a very popular processing option during the 1970s.

The newer municipal incinerators are usually the continuously burning type, and many have "waterwall" construction in the combustion chamber in place of the older, more common refractory lining[②]. The waterwall consists of joined vertical boiler tubes containing water. The tubes absorb the heat to provide hot water for steam, and they also control the furnace temperature. With waterwall units, costly refractory maintenance is eliminated, pollution control requirements are reduced (because of the reduction in quench water and gas volumes requiring treatment), and heat recovery is simpler. Unfortunately, judging by European experience, corrosion of waterwall units may be a serious problem.

The combustion temperatures of conventional incinerators fueled only by wastes are about 760°C (1400°F) in the furnace proper (insufficient to burn or even melt glass) and in excess of 870°C (1600°F) in the secondary combustion chamber. These temperatures are needed to avoid odor from incomplete combustion. Temperatures up to 1650°C (3,000°F), which would reduce volume by 97 percent and convert metal and glass to ash, are possible with supplementary fuels. Although the first high-temperature pilot installation was built in 1966, application to full-scale units has not followed, presumably because of the high costs involved.

Energy recovery. Mass-burning of solid wastes to produce steam for heating or for use in power generation has been common in Western Europe and Japan for many years. However, until rising fuel prices through the 1970s and early 1980s made the economics of energy recovery attractive, the practice was rare in North America.

Wastes burned solely for volume reduction do not need any auxiliary fuel except for start-up. On the other hand, when the objective is steam production, supplementary fuel (usually gas) must be used with the pulverized refuse, because of the variable energy content of the waste or in the event the quantity of waste available is insufficient. Ferrous metals are normally recovered from the ash.

Markets for steam must be close to the waste-burning incinerators for these combustion systems to be competitive with other heating sources. Wilson (1977) has suggested a maximum distance of 1 mile (1.6 km), but even this may be too far. The city of Chicago (Northwest Plant) and the city of Hamilton, Ontario (SWARU Plant), are two locations where no market for the steam from incineration was available during the first 10 years of operation (APWA, 1979). The incinerator in the city of Montreal, Canada, had no market for the steam from its mass-burning waterwall incinerators until 1983 (about 15 years after it was built).

Of the problems associated with incineration, air pollution control, especially the removal of the fine particulates and toxic gases (including dioxin), are the most difficult. The emission of combustible, carbon-containing pollutants can be controlled by optimizing the combustion process. Oxides of nitrogen and sulfur and other gaseous pollutants have not been a problem because of their relatively small concentration. Other concerns related to incineration include the disposal of the liquid wastes from floor drainage, quench water, and scrubber effluent, and the problem of ash disposal in landfills because of heavy metal residues. Public opposition to incinerators is another serious obstacle to their use. Capital costs (1987 prices) of about $120 million (U.S.) per 1000 ton of daily capacity, and operating costs of $15 to $30 per ton, apply to cities with over 300000 population. Unit costs for smaller centers are much greater, and this tends to limit the use of incinerators to large cities. Even there, however, because of public concern (and considerable evidence) about the toxic gases generated by the burning of solid wastes, incineration is seldom proposed now unless lack of landfill sites leaves no better alternative. This was the situation in Detroit in 1987, where, despite vigorous protests from environmentalists and the neighboring Canadian city of Windsor, Ontario, a $500 million, 4000-ton-per-day waste-to-energy plant, the largest in the U. S., was approved.

Other Conversion Processes

Chemical processes (such as fluidized bed incineration, pyrolysis, and wet oxidation) and biological processes (composting and anaerobic digestion) are other potential methods for reducing municipal waste volumes and/or converting the waste to useful products. Information on many of these processes has been provided in the Handbook of Solid Waste Management and by other investigators. However, of all the chemical and biological conversion processes, only incineration with heat recovery and composting has become widely accepted.

Composting is the aerobic decomposition of organic matter by microorganisms, primarily bacteria and fungi[3]. The reactions generate heat, raising compost temperatures during the composting period. Waste volume is reduced by about 30 percent for wastes with a high proportion of newsprint to perhaps 60 percent for garden debris.

Composting may take place naturally under controlled conditions or in mechanized composting plants. In natural systems, ground garbage, preferably with glass and metals removed, is mixed with a nutrient source (sewage sludge, animal manure, night soil) and a filler (wood chips, ground corn cobs) which permits air to enter the pile. The mixture, maintained at about 50 percent moisture content, is placed in windrows, 2 to 3 m wide, and turned over once or twice a week. In four to six weeks, when the color darkens, the temperature drops, and a musty odor develops, the process is complete. The filler may then be removed and the remaining "humus" used as soil conditioner. With mechanical plants, continual aeration and mixing enable composting time to be reduced by about 50 percent. A short period usually follows the mechanical process to allow the composting material to "mature."

There is a limited market for soil conditioners in North America. Of the 20 or 30 solid waste composting plants built in the U. S. since the first one in Altoona, Pa., in 1951, all but perhaps two or three are closed. Newer composting plants like the Delaware Reclamation Project and others in the northeastern United States combine composting of sewage sludge with municipal solid waste and may indicate a trend toward high-rate composting as one solution to the problems of sludge disposal and solid waste management. By 1985, about 60 composting facilities, primarily for sewage sludge, were operating in 30 states, with the one in Denver, Colo., which opened in 1987, being the largest aerated windrow system[4] in the U. S. The situation is quite different in Western Europe, Israel, Japan, and other advanced countries committed to land reclamation, where many successful solid waste composting plants have been operating for many years. Rotterdam, in Holland, already the location of one of Europe's largest heat recovery incinerators, has a major composting plant to complement its waste management program. The same interest in composting exists in Third-World countries, but in these areas windrow systems are the preferred method.

Selected from "Glynn Henry, Gary W. Heike, Environ. Sci. and Eng.,
Prentiu-Hall Inter. Edition, USA, 1989"

Words and Expressions

landfilling *n.* 土地掩埋
batch-fed 分批投料
upgrade *n.* 升级，上升
fungi *n.*（fungus 的复数形式）真菌类
debris *n.* 碎片，残骸
cob *n.* 玉米棒子
humus *n.* 腐殖质

Notes

① cover material　包装、封面类材料。
② The newer municipal incinerators are usually the continuously burning type, and many have "waterwall" construction in the combustion chamber in place of the older, more common refractory lining. 可译为：较新的城市焚烧炉通常都是连续燃烧型，在燃烧室内很多都带有"水墙"式结构，取代了陈旧的通常难以处理的衬里。
③ Composting is the aerobic decomposition of organic matter by microorganisms, primarily bacteria and fungi. 可译为：堆肥处理是有机物的微生物厌氧降解过程，微生物主要是细菌和真菌。
④ windrow system　风干系统。

Exercises

1. Put the following into Chinese.
   well-compacted landfill, pulverized refuse, fluidized bed incineration, wet oxidation, anaerobic digestion, nutrient source, oil conditioners, mass-burning, municipal waste
2. Answer the following questions.
(1) What is the process of composting?
(2) Where is the largest aerated windrow system?
(3) What is the highest temperature in the incinerator combustion chamber?
3. Write an abstract of the text.
4. Translate the first and second paragraphs in Reading Material into Chinese.

Reading Material: Emerging Technologies in Hazardous Waste Management: An Overview

The goal of most technologies for hazardous waste management, beyond waste avoidance, is to reduce contaminant concentrations to acceptable levels protective of human health and the environment. To achieve such a goal requires considerable innovation, particularly when the technology is applied to remediate contamination in soil and groundwater horizons of heterogeneous constituency and temporally and spatially uncertain source locations and reactivity. Hence, these complexities pose considerable challenges, both in terms of technology development and innovation, and its demonstration and potential verification as well. Correspondingly, such a process requires corroborating data and demonstration from concept through laboratory, pilot and/or field-scale testing, to eventual commercial application.

The status of technology development for remedial options applicable to soils and ground waters has been previously introduced together with technology needs associated with characterization of contaminant source matrices, transport and fate of contaminants in heterogeneous environments, solid/liquid/gas interactions, linkage between hydrometeriological phenomena and geophydrologic response, analytical technology development, biological mediation, and process modeling and field verification.

Exemplary of current activities directed toward improving understanding and development of innovative remediation technologies, supporting customer groups requiring access to information, and technology transfer between developers and the user community are the recently established Ground-Water Remediation Technologies Analysis Center (GWRTAC), an EPA-supported collaboration between the National Environmental Technology Application Center and the University of Pittsburgh, and the Advanced Applied Technology Demonstration Facility, a DOD-supported multi-university/agency/industry collaborative activity at Rice University. Whereas, the former is engaged in collecting and assembling an information database on the Internet and preparing technology status reports such as on treatment walls and the use of surfactants and co-solvents, the latter includes current demonstration of phytoremediation, passive/semi-passive treatment walls, contaminant monitoring, soil vapor extraction, bioventing, hydraulic fracturing, and micro-emulsions, surfactants and co-solvents.

Cleaning and Recycling Methods

Some researchers, like Williams et al. advocate supercritical carbon dioxide for removing contaminants found in precision cleaning of optical components, computer parts, and electronic assemblies. Replicated studies provided a basis for scale up to commercial applications, and information is provided on solvent removal efficiencies, surface interactions, operational cost and waste reduction. Other researchers reported on the successful recovery and reuse of red phosphorus (RP) grenade mix based upon both laboratory and pilot studies. Use of methylene chloride to reclaim RP-butyl rubber mix indicated dependence on consistency factors such as age, amount of talc, and glue used in assembly.

Biodegradation and Bioremediation

Methods for management of pesticide rinsates are often costly and environmentally unsound, with disposal in waste pits leading to groundwater contamination and release of toxic volatiles to the atmosphere. Some researchers, like Woodrow, et al. evaluated a specially fabricated, self-contained device for pesticide degradation using manure as biological digestion medium. Trifluralin, atrazine, carbofuran and diazinon exhibited rapid degradation rates when added to a moist bed of horse manure at an average temperature of 40℃. The build-up and decline of metabolites followed that of the parent compounds. Other scientists, like Boopathy and Manning used a sulfate-reducing bacterial enrichment culture, supplemented with a variety of carbon sources, to examine the degradation patterns of several munitions compounds. With 5% and 10% munitions compounds-contaminated soil, the culture was capable of metabolizing 2,4,6-trinitrotoluen (TNT) under anaerobic conditions using pyruvate as electron donor.

Surfactant and Physicochemical Treatment of Soils

In situ surfactant flushing of DNAPL-contaminated sites may promote the downward migration of hazardous organics into previously uncontaminated zones. Layered-sandpacks can be used to assess the contribution of five different surfactant formulations to the downward mobility of a waste samples from a Surperfund site as contrasted with a model compound, 1,1,2-tricholooethene (TCA). Due to the consequences of the downward mobilization of the waste, none of the surfactants tested were recommended for use at the site. Some scientists also addressed the mobilization of chlorinated hydrocarbons by surfactants, and the requirements for surfactant design that promote relatively low solubilization capacities and/or relative high chlorocarbon/water interfacial tensions. The viability of a number of commercially available as well as laboratory synthesized species was studied for optimum design.

Selected from "*D. William Tedder, Frederick G. Pohland, Emerging Technologies in Hazardous Waste Management VI, Published by American Academy of Environmental Engineers, 1996*"

PART 5　OTHER POLLUTION AND CONTROL TECHNOLOGIES

Unit 24

Text: Noise Control [Ⅰ]

Noise is transmitted from a source to a receiver. To control noise, therefore, we can reduce the source, interrupt the path of transmission, or protect the receiver.

The most obvious method reducing the noise production is changing the source in some way. It might be possible to modify technological approaches so as to accomplish given objectives more quietly. Thus machinery should be designed so that parts do not needlessly hit or rub against each other. Rotary saws instead of jackhammers could be used to break up street pavement. Ultrasonic pile drivers could replace the noisier steam-powered impact-type pile drivers.

Control of noise is a complex and sophisticated technology, and it is most effective when it is applied to the original design of the potentially offending source. All too often a device or a machine or an entire industrial facility is designed with a view only to maximize its capacity to carry out its assigned function. If it turns out to be excessively noisy, an acoustical engineer may be called in to "sound-proof" it. Under such circumstances, the engineer may be forced to accommodate to features of construction that should never have been accepted in the first place. Therefore, much of his effort may necessarily be applied, not to the source, but to the path between sound and receiver.

Sound travels through air by compressions and expansions. It also travels through other elastic media, including solids such as wood. Such solids vibrate in response to sound and therefore do not effectively interrupt its transmission. However, we could use various materials that vibrate very inefficiently, such as wool or lead, and absorb the sound energy, converting it to heat. (Very little heat is involved; the sound power of a symphony orchestra will warm up a room about as much as a-watt electric heater.) Sound absorbing media have been developed extensively; they are called acoustical materials. We could also build interruption of the sound waves mechanically into more kinds of machinery; devices that function in this way are called mufflers. Finally, we may be able to deflect the sound path away from the receiver, as by mechanically directing jet exhaust noise upward instead of down. Such defection is in effect, an interruption between source and receiver.

The final line of defense is strictly personal. We protect ourselves instinctively when we hold our hands over our ears. Alternatively, we can use ear plugs or muffs. A combination of ear plugs and muff can reduce noise by 40 or 50 decibels, which could make a jet plane sound no louder than a vacuum cleaner. Such protection could prevent the deafness caused by combat training, and should also be worn for recreational shooting.

We can also protect ourselves from a noise source by going away from it. In a factory, such reduction of exposure may take the form of rotating assignments so that different workers take their turns at the noisy jobs.

Until 1960's, noise control, in the United States of America, was chiefly handled by state and local governments. The U.S. Department of Labor in 1969 established limits for occupational noise exposure in certain companies under government contract. In 1971 it set an occupational noise standard, and an Executive Order later required federal occupational safety and health programs to be consistent with this standard. These and other government actions led to enactment of the Noise Control Act of 1972 (Public Law 92—574). It declared that the policy of the United States was "to promote an environment for all Americans free from noise that jeopardizes their health or welfare". Under this law, the U.S. Environmental Protection Agency (EPA) coordinates all federal programs relating to noise research and noise control. EPA has the authority to prescribe and amend standard limiting noise generation for products identified as major sources of noise. These include construction and transportation equipment, motors, engines, electrical and electronic equipment. In 1974, the Agency also set noise limits for new medium and heavy trucks, new diesel and diesel-electric locomotives, and railroad cars.

Selected from "Truk A., Wittes J. T., Turk J., Wittes R. E., Environmental Science, 2nd Ed., W. B. Saunders Company, 1978"

Words and Expressions

rotary saw　转锯，带锯
jackhammers $n.$ 汽锤
pile driver　打桩机
accommodate $v.$ 使适应，调节
symphony orchestra　交响乐团
muffler $n.$ 消声器
jeopardize $v.$ 危及，使受危害

Exercises

1. Based on the text, how many ways of noise controls are there for personal use?
2. Put the following into English.
   吸声材料　弹性介质　蒸汽打桩机　耳塞　劳动部　环保局　环保法

Reading Material: Noise Control [Ⅱ]

Source-Path-Receiver Concept

If you have a noise problem and want to solve it, you have to find out something about what the noise is doing, where it comes from, how it travels, and what can be done about it. A straightforward approach is to examine the problem in terms of its three basic elements: that is, sound arises from a source, travels over a path, and affects a receiver or listener.

The source may be one or any number of mechanical devices that radiate noise or vibratory energy. Such a situation occurs when several appliances or machines are in operation at a given time in a home or office.

The most obvious transmission path by which noise travels is simply a direct line-of-sight air path between the source and the listener. For example, aircraft flyover noise reaches an observer on the ground by the direct line-of-sight air path. Noise also travels along structural paths. Noise can travel from one point to another via any one path or a combination of several paths. Noise from a washing machine operating in one apartment may be transmitted to another apartment along air passages such as open windows, doorways, corridors, or duct work. Direct physical contact of the washing machine with the floor or walls sets these building components into vibration. This vibration is transmitted structurally throughout the building, causing walls in other areas to vibrate and to radiate noise.

The receiver may be, for example, a single person, a classroom of students, or a suburban community.

Solution of a given noise problem might require alteration or modification of any or all of these three basic elements:

1. Modifying the source to reduce its noise output.
2. Altering or controlling the transmission path and the environment to reduce noise level reaching the listener.
3. Providing the receiver with personal protective equipment.

Control of Noise Source by Design

Reduce impact forces. Many machines and items of equipment are designed with parts that strike forcefully against other parts, producing noise. Often, this striking action or impact is essential to the machine's function. A familiar example is the typewriter-its keys must strike the ribbon and paper in order to leave an inked impression. But the force of the key also produces noise as the impact falls on the ribbon, paper, and platen.

Several steps can be taken to reduce noise from impact forces. The particular remedy to be applied will be determined by the nature of the machine in question. Not all of the steps listed below are practical for every machine and for every impact-produced noise. But application of even one suggested measure can often reduce the noise appreciably.

Some of the more obvious design modifications are as follows:

1. Reduce the weight, size, or height of fall of the impacting mass.
2. Cushion the impact by inserting a layer of shock-absorbing material between the impacting surfaces. (For example, insert several sheets of paper in the typewriter behind the top sheet to absorb some of the noise-producing impact of the keys.) In some situations,

you could insert a layer of shock-absorbing material behind each of the impacting heads or objects to reduce the transmission of impact energy to other parts of the machine.

3. Whenever practical, one of the impact heads or surfaces should be made of non-metallic material to reduce resonance (ringing) of the heads.

4. Substitute the application of a small impact force over a long time period for a large force over a short period to achieve the same result.

5. Smooth out acceleration of moving parts by applying accelerating forces gradually. Avoid high, jerky acceleration or jerky motion.

6. Minimize overshoot, backlash, and loose play in cams, followers, gears linkages, and other parts. This can be achieved by reducing the operational speed of the machine, better adjustment, or by using spring-loaded restraints or guides. Machines that are well made, with parts machined to close tolerances, generally produce a minimum of such impact noise.

Reduce speeds and pressures. Reducing the speed of rotating and moving parts in machines and mechanical systems results in smoother operation and lower noise output. Likewise, reducing pressure and flow velocities in air, gas, and liquid circulation systems lessens turbulence, resulting in decreased noise radiation. Some specific suggestions that may be incorporated in design are the following:

1. Fans, impellers, rotors, turbines, and blowers should be operated at the lowest bladetip speeds that will still meet job needs. Use large-diameter, low-speed fans rather than small-diameter, high-speed units for quiet operation. In short, maximize diameter and minimize tip speed.

2. All other factors being equal, centrifugal squirrel-cage type fans are less noisy than vane axial or propeller type fans.

3. In air ventilation systems, a 50 percent reduction in the speed of the air flow may lower the noise output by 10 to 20dB, or roughly one-quarter to one-half of the original loudness. Air speeds less than 3m/s measured at a supply or return grille produce a level of noise that usually is unnoticeable in residential or office areas. In a given system, reduction of air speed can be achieved by operating at lower motor or blower speeds, installing a greater number of ventilating grilles, or increasing the cross-sectional area of the existing grilles.

Reduce frictional resistance. Reducing friction between rotating, sliding, or moving parts in mechanical systems frequently results in smoother operation and lower noise output. Similarly, reducing flow resistance in fluid distribution systems results in less noise radiation.

The key to effective noise control in fluid systems is streamline flow. This holds true regardless of whether one is concerned with air flow in ducts or vacuum cleaners, or with water flow in plumbing systems. Streamline flow is simply smooth, nonturbulent, low-friction flow.

The two most important factors that determine whether flow will be streamline or turbulent are the speed of the fluid and the cross-sectional area of the flow path, that is, the pipe or duct diameter. The rule of thumb for quiet operation is to use a low-speed, large-diameter system to meet a specified flow capacity requirement. However, even such a system can inadvertently generate noise if certain aerodynamic design features are overlooked or ignored.

Reduce radiating area. Generally speaking, the larger the vibrating part or surface, the greater the noise output. The rule of thumb for quiet machine design is to minimize the effective radiating surface areas of the parts without impairing their operation or structural strength. This can be done by making parts smaller removing excess material, or by cutting openings, slots, or perforations in the parts. For example, replacing a large, vibrating sheet-metal safety guard on a machine with a guard made of wire mesh or metal webbing might result in a substantial reduction in noise because of the drastic reduction in surface area of the part.

Reduce noise leakage. In many cases, machine cabinets can be made into rather effective soundproof enclosures through simple design changes and the application of some sound-absorbing treatment. Substantial reductions in noise output may be achieved by adopting some of the following recommendations:

1. All unnecessary holes or cracks, particularly at joints, should be caulked.

2. All electrical or plumbing penetrations of the housing or cabinet should be sealed with rubber gaskets or a suitable non-setting caulk.

3. If practical, all other functional or required openings or ports that radiate noise should be covered with lids or shields edged with soft rubber gaskets to effect an airtight seal.

4. Other openings required for exhaust, cooling, or ventilation purposes should be equipped with mufflers or acoustically lined ducts.

5. Openings should be directed away from the operator and other people.

Isolate and damper vibrating elements. In all but the simplest machines, the vibrational energy from a specific moving part is transmitted through the machine structure, forcing other component parts and surfaces to vibrate and radiate sound-often with greater intensity than that generated by the originating source itself.

Generally, vibration problems can be considered in two parts. Firs, we must prevent energy transmission between the source and surfaces that radiate the energy. Second, we must dissipate or attenuate the energy somewhere in the structure. The first part of the problem is solved by isolation. The second part is solved by damping.

The most effective method of vibration isolation involves the resilient mounting of the vibrating component on the most massive and structurally rigid part of the machine. All attachments or connections to the vibrating part, in the form of pipes, conduits, and shaft couplers, must be made with flexible or resilient connectors or couplers. For example, pipe connections to a pump that is resiliently mounted on the structural frame of a machine should be made of resilient tubing and be mounted as close to the pump as possible. Resilient pipe supports or hangers may also be required to avoid bypassing the isolated system.

Damping materials or structures are those that have some viscous properties. This tends to bend or distort slightly, thus consuming part of the noise energy in molecular motion. The use of spring mounts on motors and laminated galvanized steel and plastic in air-conditioning ducts are two examples.

Provide mufflers/silencers. There is no real distinction between mufflers and silencers. They are often used interchangeably. They are in effect acoustical filters and are used when fluid flow noise is to be reduced. The devices can be classified into two fundamental groups: absorptive mufflers and reactive mufflers. An absorptive muffler is one whose noise reduction is determined mainly by the presence of fibrous or porous materials, which absorb the sound. A reactive muffler is one whose noise reduction is determined mainly by geometry. It is shaped to reflect or expand the sound waves with resultant self-destruction.

Although there are several terms used to describe the performance of mufflers, the most frequently used appears to be insertion loss (IL). Insertion loss is the difference between two sound pressure levels that are measured at the same point in space before and after a muffler has been inserted. Since each muffler IL is highly dependent on the manufacturer's selection of materials and configuration, we will not present general IL prediction equations.

Noise Control in the Transmission Path

After you have tried all possible ways of controlling the noise at the source, your next line of defense is to set up devices in the transmission path to block or reduce the flow of sound energy before it reaches your ears. This can be done in several ways: (a) absorb the sound along the path, (b) deflect the sound in some other direction by placing a reflecting barrier in its path, or (c) contain the sound by placing the source inside a sound-insulating box or enclosure.

Selection of the most effective technique will depend upon various factors, such as the size and type of source, intensity and frequency range of the noise, and the nature and type of environment.

Separation. We can make use of the absorptive capacity of the atmosphere, as well as divergence, as a simple, economical method of reducing the noise level. Air absorbs high-frequency sounds more effectively than it absorbs low-frequency sounds. However, if enough distance is available, even low-frequency sounds will be absorbed appreciably.

If you can double your distance from a point source, you will have succeeded in lowering the sound pressure level by 6 dB. It takes about a 10dB drop to halve the loudness. If you have to contend with a line source such as a railroad train, the noise level drops by only 3dB for each doubling of distance from the source. The main reason for this lower rate of attenuation is that line sources radiate, sound waves that are cylindrical in shape. The surface area of such waves only increases two-fold for each doubling of distance from the source. However, when the distance from the train becomes comparable to its length, the noise level will begin to drop at a rate of 6dB for each subsequent doubling of distance.

Indoors, the noise level generally drops only from 3 to 5dB for each doubling of distance in the near vicinity of the source. However, further from the source, reductions of only 1 or 2dB occur for each doubling of distance due to the reflections of sound off hard walls and ceiling surfaces.

Selected from "*Machezil L David, Introduction to Environmental Engineering, Mcgraw-Hill Inter. Edition, Chem. Eng. Series, 2nd Edition, USA, 1991*"

Unit 25

Text: Global Change

In contrast to those human activities that have strictly local effects, **global change** is the term used to describe the effects of human activities on the global environment. Major global change issues are listed in Table 1. The main change events discussed in the following sections, include atmospheric warming and climate change, caused by emission of greenhouse gases; deforestation, caused by overharvesting of timber and clearing of forests for agriculture; desertification, caused primarily by overgrazing of arid and semi-arid shrublands and grasslands; depletion of pelagic (open ocean) marine species by overfishing; and degradation of coastal ecosystems, caused by discharge of pollutants and overdevelopment of shorelines.

Table 1 Global change issues that represent challenges to human welfare and natural ecosystems

Global Change Issue	Impact
1 Air Traffic	Jetliners produce contrails at 8-13km altitude that introduce aerosol pollutants into the upper atmosphere, affecting ozone and greenhouse gas levels.
2 Coral Reef Destruction	Coral reefs are thought to be home to 24% of all marine life, yet 70% of the earth's could be lost over the next few decades.
3 Deforestation	Tropical rainforests are being cleared for agriculture and settlement around the globe, with negative effects on biodiversity. Forest clearing also increases greenhouse warming and increased erosion.
4 Desertification	Over 70% of the world's semi-arid zone has been moderately to severely damaged by overgrazing and unsustainable agricultural practices. Desertification interacts with natural drought cycles to produce starvation and refugee crises in Africa.
5 The Greenhouse Effect and Climate Change	Heat-trapping gases emitted by fossil fuel burning is increasing the earth's surface temperature, with numerous potentially negative impacts on ecosystems and human welfare.
6 Lower Atmosphere Ozone Enrichment	Ozone produced at ground level from pollutants emitted by cars, power plants, and industrial processes causes lung disease in humans and damages plants.
7 Species Extinctions	Poaching, habitat destruction, spread of exotic species, pollution, and global warming are driving many species to extinction. Over 19000 plant species and 5000 animal species are classified as endangered.
8 Upper Atmosphere Ozone Depletion	Chlorofluorocarbons (CFCs) injected into the atmosphere from aerosol spray cans and air conditioners have lowered ozone levels, allowing damaging ultraviolet light to penetrate to the earth's surface.

The physical, chemical, and biological factors that produce global change are called forcing agents. For example, carbon dioxide is a forcing agent that leads to global warming because it can trap heat in the atmosphere. Global change issues are interrelated in both cause and effect through the global biogeochemical cycle, the process by which matter and energy are transformed across the biosphere (the living organisms within earth), the pedosphere (the soil layer), the atmosphere, and the oceans. For example, the burning of tropical forests to clear the land for agriculture leads to a regional loss of biodiversity, the number of species an ecosystem can support. In addition, it also adds carbon dioxide to the atmosphere, which contributes to global warming. Furthermore, forest clearing along tropical coasts leads to soil erosion, which can result in transport of suspended sediment into the sea ad resultant smothering of coral reefs. As temperatures rise, the already damaged coral reef ecosystem becomes susceptible to heat stress, leading to further loss of corals. As coral reefs and tropical forests decline, their capacity to fix carbon dioxide into biomass diminishes, increasing the rate at which carbon dioxide accumulates in the atmosphere.

It is important to note that not all global change is due to human activities. The earth has a long history of both gradual and abrupt change in response to forcing events. Volcanoes add gases and particulates to the atmosphere that can perturb climate. Tsunamis and earthquakes can rearrange the landscape in sometimes catastrophic events, and it is suspected that at least one large meteor has struck the earth in past millennia, causing mass extinctions (such as the disappearance of the dinosaurs).

Humans are not the first organisms to have caused widespread global change. Earth's original atmosphere probably lacked oxygen and was rich in carbon dioxide. However, evolution of the so-called blue-green algae, technically known as cyanobacteria, had the capability to use the energy of sunlight to fix carbon dioxide into organic molecules, at the same time producing gaseous oxygen from water as a byproduct[①]. Over a 2-billion-year period, the atmosphere came to contain 20% oxygen and only a trace amount of carbon dioxide. The main difference between natural change and human-induced global change is in the time scale of events. Natural changes usually take place over thousands or millions of years, giving the earth's life forms time to evolve and adjust to new conditions. Human activities are changing the earth's climate and ecosystems in time scales of decades to hundreds of years, and many life forms may not be able to adjust to such abrupt changes.

Global Warming and the Greenhouse Effect

(1) The Greenhouse effect

The atmosphere contains trace gases that efficiently absorb outgoing long wavelength, infrared radiation, and that therefore contribute greatly to warming the atmosphere. The most important of these is water vapor, followed by carbon dioxide, methane, and nitrous oxide. These heat-trapping molecules are called **greenhouse gases** and occur naturally. Opposing the action of these gases are trace gases derived from sulfur compounds emitted by soil, plants, volcanoes, and the oceans. These gases form sulfate aerosols in the atmosphere that reflect incoming solar radiation, leading to a potential lowering of atmospheric temperature. Except for water, the concentrations of all these trace gases in the atmosphere have been greatly perturbed by human activities over the past century, potentially affecting the atmospheric energy balance and global temperatures.

Scientists are very certain that the recent rise in carbon dioxide levels is due to human activities. It is relatively easy to know the amount of fossil fuel that is burned each year, because records are kept on the amount of coal and gas that are burned for electricity and on the production of liquid fuels for the transportation sector. It is also easy to measure the increase in carbon dioxide and other greenhouse gases in the atmosphere.

The carbon dioxide emissions entering the atmosphere today will contribute to global warming for the next several centuries. Experts project that at present rates of emissions, atmospheric carbon dioxide levels will double over the next century and will peak at 1700 ppm in the year 2400 (over five times preindustrial levels)[②]. This amount of carbon dioxide is projected to produce a temperature rise of 2 to 4℃ at the earth's surface over the next 100 years, and a rise of 4-8 ℃ at the peak of carbon dioxide levels. These elevated temperatures, and their effects at sea level, are expected to persist for many centuries after carbon dioxide emissions have stabilized. Currently, we are just at the beginning of the rise in greenhouse gas levels in the atmosphere.

(2) Effects of Greenhouse Gas Emissions on the Global Climate

There is great uncertainty about the amount of atmospheric warming, if any, that has already taken place due to greenhouse gas emissions. There is even more uncertainty about the impacts of future warming on global climate systems. The rise in atmospheric carbon dioxide levels has been documented by careful measurements over time, whereas climate projections are based on models and other indirect methods that are subject to error. Nevertheless, there are some logical consequences that follow from a warming of global temperature. These include shifts in regional weather patterns due to unequal heating at the equator and the poles; partial melting of Antarctic ice shield, resulting in a rise in sea level around the world; and shifts in the distribution of vegetation zones, with impacts on agriculture and natural ecosystems. The evidence for an actual increase in global temperature, and possible consequences over the next 100 years, are discussed briefly.

Despite the uncertainties, combined annual land-surface and sea-surface temperature databases show a temperature rise of 0.65℃ plus or minus 0.15℃ over the past 150 years. Many scientists believe these data sets confirm the existence of a global warming trend with a certainty of 95% or greater. Other scientists are trying to confirm global warming by using a wide variety of surrogate measurements (indirect tests of a warming signal).

Projected future effects of greenhouse warming on climate are made using computer models of the atmosphere and oceans, called Atmosphere-Ocean General Circulation Models (AOGCMs). Over a dozen modeling groups around the world are simulating the effects of temperature rise on climate systems, and they periodically compare and check each other's results[③]. Results of AOGCMs are combined with models of the carbon cycle, of atmospheric chemistry and physics, and of ice sheets to attempt to predict what is in store for the earth under different greenhouse gas emission scenarios.

Recognizing the problem of global climate change, the World Meteorological Organization and the United Nations Environment Program established the Intergovernmental Panel on Climate Change in 1988. It collects the most recent scientific evidence on all aspects of climate change, including output from AOGCMs, and predicts the potential impacts and options relevant to the risks of human-induced climate change.

Overall, the range of the mean global temperature increase over the next 100 years is expected to be in the range of 2 to 4℃. However, a greater mean annual warming at higher latitudes than near the equator is expected, called polar amplification of the warming. This is in part because, as snow melts, more radiation will be absorbed rather than reflected back to space in the polar regions, leading to greater heating of the land surface. Whereas a rise of less than 1℃ is expected at the equator, at 80° N or S latitude a rise of 3℃ or more is anticipated. For similar reasons, greater warming in winter than in summer is expected at high latitudes, whereas greater warming in summer is expected in arid and semi-arid regions, where soils become drier in summer.

The changes in global temperature distribution could have profound effects on the global climate cycles, because climate patterns are driven in part by differences in surface temperature at different latitudes. Although the direction of change is difficult to predict, it can be expected that monsoon rains, tropical cyclones and hurricanes, precipitation patterns over the continents, and the frequency of extreme weather anomalies such as droughts and floods will be affected by global warming. Also, it is expected that sea level will rise around the globe due to melting of the polar ice caps. A mean global temperature rise of 2 to 4℃ is expected to raise the sea level by 25 to 75cm by the year 2100. This would impact large areas of coastal land around the world, including cities, agricultural areas, and natural coastal ecosystems such as coral reefs, salt marshes, and coastal forests.

Selected from "*Ian L. Pepper, Charles P. Gerba, Mark L. Brusseau. Environmental Pollution Science, Elserier Inc. USA, 2006*"

Words and Expressions

global change 全球（地球）变化
deforestation *n.* 砍伐树木，除去森林
desertification *n.* 沙漠化
overharvesting *n.* 过度砍伐
arid *a.* 干旱，缺水
semi-arid shrublands 半干旱灌木地
pelagic *a.* 深海的，大洋的
overfishing *n.* 过度捕捞
overdevelopment *n.* 过度开发

human welfare 人类的福利
pedosphere *n.*（地球）表土层，土壤圈，土界
biodiversity *n.* 生物多样性
tsunamis *n.* 津浪，海啸地震
heat-trapping *a.* 捕热的
World Meteorological Organization 世界气象组织
tropical cyclones and hurricanes 热带气旋和台风

Notes

① Evolution of the so-called blue-green algae, technically known as cyanobacteria, had the capability to use the energy of sunlight to fix carbon dioxide into organic molecules, at the same time producing gaseous oxygen from water as a byproduct. 参考译文：所谓的蓝绿海藻，科技术语称为氰基菌，演变（进化）过程具有利用太阳能将二氧化碳固定为有机分子，同时由水产生氧作为副产物的能力。

② Experts project that at present rates of emissions, atmospheric carbon dioxide levels will double over the next century and will peak at 1700ppm in the year 2400 (over five times preindustrial levels). 可译为：专家们推测，按目前的排放速率，大气中的二氧化碳水平在下个世纪将翻倍，在2400年达到最大值为1700ppm（超过工业化以前的五倍）。

③ Projected future effects of greenhouse warming on climate are made using computer models of the atmosphere and oceans, called Atmosphere-Ocean General Circulation Models (AOGCMs). Over a dozen modeling groups around the world are simulating the effects of temperature rise on climate systems, and they periodically compare and check each other's results. 参考译文：人们利用大气和海洋计算机模型，称为大气-海洋通用循环模型（AOGCMs），对温室效应未来对气候的影响进行预测。全世界超过12个模拟小组正在模拟温升对气候系统的影响，定期比较和检验彼此的结果。

Exercises

1. Put the following words or phrase into Chinese.
   United Nations Environment Program, greenhouse warming, surrogate measurements, databases, global warming trend, global warming trend, species extinctions, ozone enrichment, polar amplification, global temperature distribution

2. Put the following into English.
   海岸过度开发　津浪（海啸地震）　沙漠化　碳循环　全球气候循环　过度放牧　过度砍伐干旱地区　有机分子　温室气体排放　生物圈　生态系统　生物多样性

3. Translate the last paragraph of the text into Chinese.

4. Give a brief summary of Global Change Issue.

Reading Material: Global Warming

Earth's atmosphere—the layer of gases surrounding the Earth—is a complex, dynamic system that is changing continuously. During an approximate 300-year period from A. D. 950 to 1250, Earth's surface was considerably warmer than what climatologists today call normal (meaning the average surface temperature during the past century or some shorter interval, such as 1960—1990). This warm time is known as the Medieval Warm Period (MWP). With all the concerns today about climate change, perhaps we can learn some lessons from that time. Since weather records were not kept then, we do not have a global picture of what it was like. What we do know is that parts of the world, in particular Western Europe and the Atlantic, may have been warmer some of the time than they were in the last decade of the 20th century. However, on a global basis the MWP was not as warm as it is today.

In Western Europe, it was a time of flourishing culture and activity, as well as expansion of the population; a time when harvests were plentiful, people generally prospered, and many of Europe's grand cathedrals were constructed. Sea temperatures evidently were warmer, and there was less sea ice. Viking explorers from Scandinavia traveled widely in the Far North and established settlements in Iceland, Greenland, and even briefly in North America. Near the end of the 10th century, Erik and Red, the famous Viking explorer, arrived at Greenland with his ships and set up settlements that flourished for several hundred years. The settlers were able to raise domestic animals and grow a variety of crops that had never before been cultivated in Greenland. During the same warm period, Polynesian people in the Pacific, taking advantage of winds flowing throughout the pacific, were able to sail to and colonize islands over vast areas of the Pacific, including Hawaii.

Where some prospered in Western Europe and the Pacific during the Medieval Warm Period, other cultures appear to have been less fortunate. Associated with the warming period were long, persistent droughts (think human-generational length) that appear to have been partially responsible for the collapse of sophisticated cultures in North and Central America. The collapses were not sudden but occurred over a period of many decades, and in some cases the people just moved away. These included the people living near Mono Lake on the eastern side of the Sierra Nevada in California, the Chacoan people in what is today Chaco Canyon in New Mexico, and the Mayan civilization in the Yucatan of southern Mexico and Central America.

The Medieval Warm Period was followed by the Little Ice Age (LIA), which lasted from approximately mid-1400 to 1700. The cooling made life more difficult for people in Western Europe and North America. Crop failures occurred in Western Europe, and some mountain glaciers in Swiss Alps advanced to the extent that they filled valleys and destroyed villages. Areas to north that had enjoyed abundant crop production were under ice. The population was devastated by the Black Plague, whose effects may have been exacerbated by poor nutrition as a result of crop failures and by the damp and cold that reached out across Europe and even to Iceland by about 1400.

Travel and trade became difficult in the Far North. Eventually, the Viking colonies in North America were abandoned and those in Greenland declined greatly. Part of the reason for the abandonment in North America, and particularly in Newfoundland, was that the Vikings may not have been able to adapt to the changing conditions, as did the Inuit peoples living there. As times became tough, the two cultures collided, and the Vikings, despite their fierce reputation, were less able than the Inuit to adapt to the cooling climate.

We do not know what caused the Medieval Warm Period, and the details about it are obscured by insufficient climate data to help us estimate temperature during that period. We do know that it was relatively warm (in Western Europe). We can't associate the warming 1000 years ago with burning of fossil fuels. This suggests that more than one factor can cause warming. In this chapter we will explore climate dynamics so you can better understand what may be the causes of climate change and what might be the best estimates of how it could affect life on Earth and civilizations.

Selected from " *Daniel B. Botkin, Edward A. Keller. Environmental Science, Eight Edition, John Wiley & Sons, Inc. Printed in Asia, 2012*"

Unit 26

Text: Soil

As the human population has grown, so have the amounts of land and resources we devote to agriculture. We can define **agriculture** as the practice of raising crops and livestock for human use and consumption. We obtain most of our food and fiber from **cropland**, land used to raise plants for human use, and from **rangeland**, or pasture, land used for grazing livestock. Today we commandeer more than one out of every three acres of land on Earth to produce food and fiber for ourselves. Rangeland covers 26% of Earth's land surface, and cropland covers 12%.

Healthy soil is vital for agriculture, as well as for forests and for the functioning of Earth's natural systems. **Soil** is not merely lifeless dirt; it is a complex system consisting of disintegrated rock, organic matter, water, gases, nutrients, and microorganisms. Productive soil is a renewable resource. Once depleted, soil may renew itself over time, but renewal generally occurs very slowly. If we abuse soil through careless or uninformed practices, we can greatly reduce its ability to sustain life.

For these reasons, healthy soil is key component of **sustainable agriculture**, agriculture that we can practice in the same way in the same place far into the future. Sustainable agriculture allows soil renew its nutrient content and retain its character from one crop to the next. Sustainable agriculture also requires reliable supplies of clean water, minimized use of fossil-fuel-based fertilizers and pesticides, healthy populations of pollinating insects, sustenance of genetic diversity and, arguably, genetic modification (each of which we explore below). Because most farming and grazing that people have practiced so far have depleted soils faster than they form, it is imperative for our civilization's future that we develop sustainable methods of working with soil.

Soil supports agriculture

Our agriculture relies on healthy soil in several ways. Crop plants depend on soil that contains organic matter to provide the nutrients they need for growth. Plants also need soil with a structure and texture suitable for roots to penetrate deeply. And plants need soil that retains water and makes water and dissolved nutrients accessible to their roots. Livestock also depend on soil with these characteristics, because livestock eat plants that have grown in the soil. If soil becomes degraded, then agriculture suffers. Because everyone in our society relies directly on agriculture for the meals we eat and the clothing we wear, the quality of our lives is closely tied to the quality of our soil.

Healthy soil has sustained agriculture for thousands of years. When people first began farming, they were able to take advantage of deep, nutrient-rich topsoil that had built up over vast spans of time. Today we face the challenge of producing immense amounts of food from soil that has been farmed many times, while also conserving its fertility for the future. Before we examine soil closely, let's step back and consider how agriculture came about in the first place and how we got to where we are today[①].

Soil as a System

We generally overlook the startling complexity of soil. Although it is derived from rock, soil is molded by living organisms. By volume, soil consists very roughly of 50% mineral matter and up to 5% organic matter. The rest consists of pore space taken up by air or water. The organic matter in soil includes living and dead microorganisms as well as decaying material derived from plants and animals. A single teaspoon of soil can contain millions of bacteria and thousands of fungi, algae, and protists. Soil provides habitat for earthworms, insects, mites, millipedes, centipedes, nematodes, sow bugs, and other invertebrates, as well as for burrowing mammals, amphibians, and retiles. The composition of region's soil strongly influences its ecosystems. In fact, because soil is composed of living and nonliving components that interact in complex ways, soil itself meets the definition of an ecosystem.

Soil Forms Slowly

The formation of soil plays a key role in terrestrial primary succession, which begins when the lithosphere's parent material is exposed to the effects of the atmosphere, hydrosphere, and biosphere[②]. **Parent material** is the base geologic material in a particular location. It may be hardened lava or volcanic ash; rock or sediment deposited by glaciers; wind-blown dunes; sediments deposited by rivers, in lakes, or in the ocean; or bedrock, the mass of solid rock that makes up Earth's crust. Parent material is broken down by weathering, the physical, chemical, and biological processes that convert large rock particles into smaller particles.

Once weathering has produced fine particles, biological activity contributes to soil formation through the deposition, decomposition, and accumulation of organic matter. As plants, animals, and microbes die or deposit waste, this material is incorporated amid the weathered rock particles, mixing with minerals. For example, the deciduous trees of temperate forests drop their leaves each fall, and detritivores and decomposers break down this leaf little and incorporate its nutrients into the soil[③]. In decomposition, complex organic molecules are broken down into simple ones that plants can take up through their roots. Partial decomposition of organic matter creates humus, a dark, spongy, crumbly mass of material made up of complex organic compounds. Soils with high humus content hold moisture well and are productive for plant life.

Selected from "*Jay Withgott, Matthew Laposata, Environment, the Science Behind the Stories, 5th Ed., Pearson Education Inc., US. 2014*"

Words and Expressions

rangland *n.* 牧场
commandeer *v.* 征用，强占
disintegrated rock 崩解性岩石
mite *n.* 螨，蚤
millipede *n.* 千足虫
centipede *n.* 蜈蚣
centipede *n.* 蜈蚣
sow bug 潮虫
invertebrate *n.* 无脊椎动物
burrowing mammal 穴居哺乳动物
reptile *n.* 爬行动物
lithosphere *n.* 岩石圈，陆界
detritivore *n.* 食碎屑者，食腐质者，食腐动物

Notes

① Before we examine soil closely, let's step back and consider how agriculture came about in the first place and how we got to where we are today. 可译为：在考察土壤前，让我们回顾和考虑一下当初农业是如何出现，如何走到今天的。

② The formation of soil plays a key role in terrestrial primary succession, which begins when the lithosphere's parent material is exposed to the effects of the atmosphere, hydrosphere, and biosphere. 可译为：土壤的形成在陆地初始的进化过程起着关键的作用，它是在岩石圈母质材料暴露在大气圈、水圈和生物圈的影响作用下开始的。

③ For example, the deciduous trees of temperate forests drop their leaves each fall, and detritivores and decomposers break down this leaf little and incorporate its nutrients into the soil. 可译为：例如，温带阔叶林的落叶树每年秋季都会掉落所有树叶，食腐动物和分解物会把树叶分解为很小块，把树叶中的营养物质吸入到土壤中。

Exercises

1. Translate the following into English.

   岩石卷　征用　牧场　螨虫　蜈蚣　潮虫　穴居哺乳动物　无脊椎动物

2. Translate the following passage into Chinese.

   Healthy soil is vital for agriculture, as well as for forests and for the functioning of Earth's natural systems. Soil is not merely lifeless dirt; it is a complex system consisting of disintegrated rock, organic matter, water, gases, nutrients, and microorganisms. Productive soil is a renewable resource. Once depleted, soil may renew itself over time, but renewal generally occurs very slowly. If we abuse soil through careless or uninformed practices, we can greatly reduce its ability to sustain life.

Reading Material: Soil Erosion

The problem of soil erosion has been recognized for a long time. The Soil Conservation Service (SCS), now called the Natural Resources Conservation Service (NRCS) was formed in response to the Dust Bowl catastrophe of the mid-1930s. The SCS established a nationwide partnership of federal agencies, local conservation districts, and communities to provide assistance to the rural and urban sectors for the conservation of soil and other natural resources. Through the efforts of these partnerships soil erosion on agricultural land has declined considerably.

Soil erosion due to overgrazing on pasturelands and rangelands and loss of soil from croplands is still a major problem in the United States. In 1997, an estimated 1.7 Pg of soil was lost from cropland and Conservation Reserve Program (CRP)[1] land in the United States. This translates into an on-farm economic loss of more than \$27 billion each year, of which \$20 billion is for replacement of nutrients and \$7 billion for lost water and soil.

In addition to the economic effect on farmers, erosion can have serious impacts on the environment, including.

(1) Pollution of lakes and streams by nutrients, particularly phosphorus, and agricultural chemicals that are washed away with the soil.

(2) Flooding due to silt build-up in drains and waterways.

(3) Build-up of sediments in wetlands, which can lead to serious environmental problems or even loss of these habitats.

(4) Siltation in reservoirs, which reduces their capacity.

(5) Siltation in harbors and waterways. The cost for dredging of harbors and waterways in the United States is estimated to be \$1 billion per year.

Both water and wind can erode cropland soils. Water erosion results from the removal of

---

[1] The CRP is a federal program to set aside cropland with the goals of reducing soil erosion, reducing production of surplus commodities, providing income support for farmers, improving environmental quality, and enhancing wildlife habitat.

soil material by flowing water. When a raindrop strikes the soil surface it can break up soil aggregates. On a slope, water begins to flow downhill carrying the detached soil grains with it. Wind erosion occurs in regions of low rainfall; it can be widespread, especially during periods of drought. Unlike water erosion, wind erosion is inherently not related to slope gradient.

Water erosion occurs mainly in areas east of the Corn Belt and Southern Plains. Wind erosion takes place mostly in the west, Northern Plains, and Southern Plains. Both water and wind erosion result in major losses of soil from agricultural land. In 1997, erosion due to water on crop and CRP lands was nearly 1.0Pg. each year and about 0.75Pg every year of soil was lost due to wind erosion.

The conservation particles adopted over the past two decades have considerably reduced soil erosion on agricultural lands. The widespread adoption of conservation tillage practices is one of the major contributors to this improvement. Conventional tillage breaks up and buries the crop residue from the previous planting, creating a bare soil surface, which is vulnerable to erosion. Conservation tillage is the practice of leaving some or all of the crop residues on the soil surface to protect it like a mulch. In addition to protecting the soil from erosion the mulch creates an environment that conserves water in the soil. Conservation tillage has some drawbacks, one of which is an increased dependence on herbicides for weed control. Other practices such as terracing, contour tillage, and crop rotation have helped to reduce water erosion. Diversion of water away from fields or other erosion-sensitive areas can also reduce soil loss.

Measures used to control wind erosion on croplands include conservation tillage, planting windbreaks, and tilling at right angles to the prevailing winds, so that furrows act as small windbreaks to capture blowing soil. Erosion on pastureland and rangeland can be reduced by a number of measures, the single most important of which is control of animal numbers. Animal numbers must be controlled, so that as forage is consumed regrowth has a chance to replace it. For maximum sustained production, plants must be given a period of rest to rebuild these reserves. During droughts the numbers of grazing animals must be tightly controlled, as plants can be injured by a grazing intensity that would not injure them during a less vulnerable season. Proper distribution of animals is also an important grazing management tool. Fences can be used to divide up pastures. On open range land one of the most effective techniques is to provide adequate water facilities that are properly distributed. Ranchers also improve livestock distribution by placing salt and mineral supplies in widespread locations to draw the animals to those areas.

Effective management of soil erosion requires the ability to quantitatively predict the amount of soil loss that would occur under alternative management strategies and practices. Various models have been devised to predict soil loss. The model with the greatest acceptance and use is the universal soil loss equation (USLE). The USLE was developed by U.S. Department of Agriculture's Research Service by statistical analyses of rainfall, runoff, and sediment loss data from many small plots located around the United States and has since been modified to improve its predictions.

Selected from *"Mackenzie L Davis, Susan J Masten. Principles of Environmental Engineering and Science, The McGraw-Hill Companies, Inc. USA 2004"*

Unit 27

Text: Thermal Pollution

The amount of heat that must be removed from an electrical generating facility is quite large. A one-million-kilowatt plant running at 40 percent efficiency would heat 10 million liters of water by 35℃ every hour. It is not surprising that such large quantities of heat, added to aquatic systems cause ecological disruptions. The term thermal pollution has been used to describe these heat effects.

The processes of life involve chemical reaction, and as a rough approximation, the rate of a chemical reaction doubles for every rise in temperature of 10℃. If our body temperature rises by as much as 5℃, which would make a body temperature of 42℃, the fever may be fatal. What then happens to our system when the outside air temperature rises or falls by about 10℃? We adjust by internal regulatory mechanisms that maintain a constant body temperature. This ability is characteristic of warm-blooded animals, such as mammals and birds. In contrast, non-mammalian aquatic organisms such as fish are unable to regulate their body temperatures as efficiently as warm-blooded animals. How then does a fish respond to increases in temperature? All its body processes speed up, and its need for oxygen and its rate of respiration therefore rise. The increased need for oxygen is especially serious, since hot water has smaller capacity for holding dissolved oxygen than cold water. Above some maximum tolerable temperature, death occurs from failure of the nervous system, the respiration system, or essential cell processes. Almost no species of fish common to the United States can survive in waters warmer than 34℃.

In general, not only the fish but also entire aquatic ecosystems are rather sensitively affected by temperature changes. Any disruption of the food chain, for example may upset the entire system. Higher temperatures often prove to be more hospitable for pathogenic organisms, and thermal pollution may therefore convert a low incidence of fish disease to a massive fish kill as the pathogens become more virulent and the fish less resistant. Such situations have long been known in the confined environments of farm and hatchery ponds, which can warm up easily because the total amount of water involved is small. As thermal pollution in larger bodies of water increases, so will the potential for increased loss of fish by disease.

Aquatic ecosystems near power facilities are subject not only to the effects of an elevated average temperature but also to the thermal shocks of unnaturally rapid temperature changes. Thus, the development of cold-water species is hindered by hot water, and the development of hot-water species is upset by the unpredictable flow of heat. Power plants are usually located near population center, and many cities dump sewage into rivers. Since sewage decomposition is dependent on oxygen, hot rivers are less able to cleanse themselves than cold ones. The combination of thermal pollution with increased nutrients from undecomposed sewage can lead to rapid and excessive algal growth. Therefore, thermal pollution imposes the unhappy choice of dirtier rivers or more expensive sewage treatment plants.

Synthetic poisons, too, become more dangerous to fish as the water temperature rises. First of all toxic effects are accelerated at higher temperatures. Second warm water favors increased growth of plant varieties such as algae. The algae tend to collect in the power-plants condensers and reduce water flow efficiency. The electric company responds by periodically introducing chemical poisons into the cooling system to clean the pipes. These poisons are then mixed with the downstream effluent. Additionally, domestic and industrial water consumers are more apt to discharge treatment chemicals into water with high algae concentrations than into clean water. Thus in warm water, not only are fish less likely to resist poisons but they are also likely to be exposed to them more.

The Second Law of Thermodynamics assures us that we cannot invent a process to avoid the production of excess heat. We can however reduce the amount of heat wasted or we can put it to good use.

Selected from "*Truk A., Wittes J. T., Turk J., Wittes R. E., Environmental Science, 2nd Ed., W. B. Saunders Company, 1978*"

Words and Expressions

mammals *n.* 哺乳动物
respiration *n.* 呼吸
virulent *a.* 致命的；极毒的
hatchery pond 鱼塘

Exercises

1. Put the following into English.

   温度每升高 10℃   内部调节机制   一百万千瓦   热血动物   神经系统
   食物链   非哺乳动物   冷却系统   热力学第二定律

2. According to text, how are aquatic ecosystems disrupted by thermal pollution?
3. Translate the third paragraph of the text into Chinese.

Reading Material: The Major Greenhouse Gases

The major anthropogenic greenhouse gases are listed in Table 1. The table also lists the recent rate of increase for each gas and its relative contribution to the anthropogenic greenhouse effect.

Table 1 Major Greenhouse Gases

Trace gases	relative contribution(%)	growth rate(%/YR)
CFC	15-25	5
CH_4	12-20	0.4
O_3	8	0.5
N_2O	5	0.2
Total	40-50	
Contribution of CO_2	50-60	0.3~0.5

Carbon Dioxide

Current estimates suggest that approximately 200 billion metric tons of carbon in the form of carbon dioxide (CO_2) enter and leave Earth's atmosphere each year as a result of a number of biological and physical processes: 50% to 60% of the anthropogenic greenhouse effect is attributed to this gas. Measurements of carbon dioxide trapped in air bubbles in the Antarctic ice sheet suggest that 160,000 years before Industrial Revolution the atmospheric concentration of carbon dioxide varied from approximately 200 to 300 ppm. The highest level or concentration of carbon dioxide in the atmosphere, other than today's, occurred during the major interglacial period about 125000 years ago.

About 140 years ago, just before the major use of fossil fuels began as part of the Industrial Revolution, the atmospheric concentration of carbon dioxide was approximately 280 ppm. Since then, and especially in the past few decades, the concentration of CO_2 in the atmosphere has grown rapidly. Today, the CO_2 concentration is about 392 ppm, and at its current rate of increase of about 0.5% per year, the level may rise to approximately 450 ppm by the year 2050—more than 1.5 times the preindustrial level.

Methane

The concentration of methane (CH_4) in the atmosphere more than doubled in the past 200 years and is thought to contribute approximately 12% to 20% of the anthropogenic greenhouse effect. Certain bacteria that can live only oxygenless atmospheres produce methane and release it. These bacteria live in the guts of termites and the intestines of ruminant mammals, such as cows, which produce methane as they digest woody plants. These bacteria also live in oxygenless parts of freshwater wetlands, where they decompose vegetation, releasing methane as a decay product. Methane is also released with seepage from oil fields and seepage from methane hydrates.

Our activities also release methane. These activities include landfills (the major methane source in the United States), the burning of biofuels, production of coal and natural gas, and agriculture, such as raising cattle and cultivating rice. (Methane is also released by anaerobic activity in flooded lands where rice is grown.) As with carbon dioxide, there are important uncertainties in our understanding of the sources and sinks of methane in the atmosphere.

Chlorofluorocarbons

Chlorofluorocarbons (CFCs) are inert, stable compounds that have been used in spray cans as aerosol propellants and in refrigerators. The rate of increase of CFCs in the atmosphere in the recent past was about 5% per year, and it has been estimated that approximately 15% to 25% of the anthropogenic greenhouse effect may be related to CFCs. Because they affect the stratospheric ozone layer and also play a role in the greenhouse effect, the US banned their use as propellants in 1978. In 1987, 24 countries signed the Montreal Protocol to reduce and eventually eliminate production of CFCs and accelerate the development of alternative chemicals. As a result of the treaty, production of CFCs was nearly phased out by 2000.

Potential global warming from CFCs is considerable because they absorb in the atmospheric window, as explained earlier, and each CFC molecule may absorb hundreds or even thousands of times more infrared radiation emitted from Earth than is absorbed by a molecule of carbon dioxide. Furthermore, because CFCs are highly stable, their residence time in the atmosphere is long. Even though their production was drastically reduced, their concentrations in the atmosphere will remain significant (although lower than today's) for many years, perhaps for as long as a century.

Nitrous Oxide

Nitrous oxide (N_2O) is increasing in the atmosphere and probably contributes as much as 5% of the anthropogenic greenhouse effect. Anthropogenic sources of nitrous oxide include agricultural application of fertilizers and the burning of fossil fuels. This gas, too, has a long residence time; even if emissions were stabilized or reduced, elevated concentrations of nitrous oxide would persist for at least several decades.

Selected from " *Daniel B. Botkin, Edward A. Keller. Environmental Science, Eight Edition, John Wiley & Sons, Inc. Printed in Asia, 2012*"

PART 6 ENVIRONMENTAL MANAGEMENT AND POLICY

Unit 28

Text: Summary of Environmental Impact Assessment (EIA)

1. Definitions

An action is used in this text in the sense of① any engineering of industrial project, legislative proposal, policy, programme or operational procedure with environmental implications. An environmental impact assessment (EIA)② is an activity designed to identify and predict the impact of an action on the biogeophysical environment and on man's health and well-being, and to interpret and communicate information about the impacts.

2. Operational Procedures

(1) Environmental impact assessments should be an integral part of all planning for major actions, and should be carried out at the same time as engineering, economic, and socio-political assessments.

(2) In order to provide guidelines for environmental impact assessments, national goals and policies should be established which take environmental considerations into account; these goals and policies should be widely promulgated.

(3) The institutional arrangements for the process of environmental impact assessment should be determined and made public. Here it is essential that③ the roles of the various participants (decision-maker, assessor, proponent, reviewer, other expert advisors, the public and inter-national bodies) be designated. It is also important that timetables for the impact assessment process be established, so that proposed actions are not held up④ unduly and the assessor and the reviewer are not so pressed that they undertake only superficial analyses.

(4) An environmental impact assessment should contain the following:

① a description of the proposed action and of alternatives;

② a prediction of the nature and magnitude of environmental effects (both positive and negative);

③ an identification of human concerns;

④ a listing of impact indicators as well as the methods used to determine their scales of magnitude and relative weights;

⑤ a prediction of the magnitudes of the impact indicators and of the total impact, for the project and for alternatives;

⑥ recommendations for acceptance, remedial action, acceptance of one or more of the alternatives, or rejection;

⑦ recommendation for inspection procedures.

(5) Environmental impact assessments should include study of all relevant physical, biological, economic, and social factors.

(6) At a very early stage in the process of⑤ environmental impact assessment, inventories should be prepared of relevant sources of data and of technical expertise.

(7) Environmental impact assessments should include study of alternatives, including that of no action.

(8) Environmental impact assessments should include a spatial frame of reference much larger than the area encompassed by the action, e. g. larger than the 'factory fence' in the case of an engineering project.

(9) Environmental impact assessments should include both mid-term and long-term predictions of impacts. In the case of⑥ engineering projects, for example, the following time-frames should be covered:

① during construction;

② immediately after completion of the development;

③ two to three decades later.

(10) Environmental impacts should be assessed as the difference between the future state of the environment if the action took place and the state if no action occurred.

(11) Estimates of both the magnitude and the importance of environmental impacts should be obtained. (Some large effects may not be very important to society, and vice versa⑦.)

(12) Methodologies for impact assessment should be selected which are appropriate to the nature of the action, the data base, and the geographic setting. Approaches which are too complicated or too simple should both be avoided.

(13) The affected parties should be clearly identified, together with the major impacts for each party.

3. Research

Research should be encouraged in the following areas:

(1) Post-audit reviews environmental impact assessments for accuracy and completeness in order that knowledge of assessment methods may be improved. (No systematic post-audit programme has as yet been initiated in any country with experience in impact assessment.)

(2) Study of methods suitable for assessing the environmental effects of social and insti-

tutional programmes, and of other activities of the non-construction type.

(3) Study of criteria for environmental quality.

(4) Study of quantifying value judgments on the relative worth of various components of environmental quality.

(5) Development of modeling techniques for impact assessments, with special emphasis on combined physical, biological, socio-economic systems.

(6) Study of sociological effects and impacts.

(7) Study of methods for communicating the results of highly technical assessments to the non-specialist.

Selected from "*Environmental Impact Assessment: Principles and Procedures*, edited for SCOPE by R. E. Munn in the university of Toronto, Canada. Copyrightc 1975, 1979, by the Scientific Committee on Problem of the Environment (SCOPE). Toronto"

Words and Expressions

well-being 幸福，福利
legislative *a.* 立法的
integral *a.* 完整的；组成的
socio-political 社会政治的
biogeophysical 生物地球物理的
guideline *n.* 准则，指导路线
promulgate *v.* 颁布，公布
institutional *a.* 惯例的，制度的
inventory *n.* 目录，报表
participant *n.* 参加者，参与者
proponent *n.* 建议者，提议者
unduly *ad.* 过度地，过分地
post-audit *n.* 后检查，后审查
superficial *a.* 肤浅的，表面的
criterion (pl. criteria) *n.* 标准，规范
indicator *n.* 指示物

Notes

① in the sense of 有……的意义
② environmental impact assessment 环境影响评价
③ Here it is essential that …… it 为形式主语，代表 that 从句，意为"……是基本的"
④ hold up 阻挡，使停
⑤ in the process of 在……的过程中
⑥ in the case of 就……来说，至于
⑦ vice versa 反之亦然，反过来也一样

Exercises

1. Put the following Chinese.

   action, biogeophysical environment, surperficial analyses, an identification of human concerns, remedial action, relevant physical and biological factors, a spatial frame of reference

2. Put the following into English.

   地理环境　　影响评价方法　　替代方案的研究　　有关环境质量的各组分相对值
   环境质量标准　　　　模型技术　　　　基础数据

3. Translate the following paragraphs into Chinese.

(1) An environmental impact assessment is an activity designed to identify and predict the impact of an action on the biogeophysical environment and on man's health and well-being, and to interpret and communicate information about the impacts.

(2) The institutional arrangements for the process of environmental impact assessment should be determined and made public. Here it is essential that the roles of the various participants (decision-maker, assessor, proponent, reviewer, other expert advisors, the public, and international bodies) be designated. It is also important that timetables for the impact assessment process be established, so that proposed actions are not held up unduly and the assessor and the reviewer are not so pressed that they undertake only superficial analyses.

Reading Material: Introduction to Methods for Environmental Impact Assessment

Many methods for environmental impact assessment (EIA) have been devised to aid identification, prediction and assessment of impacts and preparation of environmental impact statements (EISs). All these methods focus on impacts. Impact is a word much used in EIA, but little attention has been paid, in the academic and professional literature, to the nature of impacts. What is an impact? How do we know when an impact is likely to occur? These are questions which are addressed in this paper.

A discussion of the concept of impact opens the paper. Most attention will be paid to the socio-political character of impacts, but other aspects will be considered. Throughout the paper, reference will be made to EIA methods which are able to deal with some of the complex problems posed by the nature of impacts. A fuller, more descriptive discussion of many of these methods can be found in references. This paper is essentially wide-ranging and discursive in attempting to deal with some basic issues related to impacts and EIA methods. It may provide a useful framework for other papers in this volume.

The Aim of EIA

EIA is a planning aid and is concerned with identifying, predicting and assessing impacts arising from proposed activities such as policies, programmes, plans and development projects which may affect the environment. The main aim of EIA is to improve decisions on development by increasing the quality and scope of information on likely impacts presented to decision-makers and members of the public. Although concerned with different levels of developmentactivities (policies, plans and projects), EIAs are implemented mostly for major development projects such as highways, nuclear power stations and water-related projects. The importance of assessing policies and plans is increasingly recognized, but is proving difficult to implement because of institutional, political and technical problems. Consequently, most EIAs are undertaken for major projects in which a number of alternatives may be assessed.

In previous papers presented to this seminar, considerable attention has focused on the variety of institutional procedures which exist for the organization of national EIA systems. These procedures are generally administrative in nature and provide for the production of EISs to be used in decision-making. Usually, also, they attempt to ensure that EISs are produced to a uniform standard in accordance with established mechanisms. Procedures by themselves, however, cannot ensure that EISs contain structured information, produced in a "scientific" manner, likely to be of most use in decision-making. For this reason, methods to aid identification and assessment of impacts have been developed. Of course, the provision of information on impacts is no guarantee that it will be used by decision-makers, but that is a separate issue.

This part deals simply with outlining EIA methods developed to ensure that the best possible information from EIA procedures is available to decision-makers and the public. All methods involve impacts. It is vital to consider the nature of impacts as they are the central concern of EIA methods.

The Political Dimension of Impacts

An important reason for considering the meaning and nature of the impact concept is that it may influence to some extent, the nature and operational characteristics of EIA methods. The use of these methods will subsequently have an effect on the content of EISs produced. Such EISs will form at least part of the information base upon which decision-makers and members of the public will come to conclusions concerning proposals. It is also useful to examine the impact concept to be aware of the complexity of the concept which forms most of the contents of EISs.

Reading the academic literature on EIA and a random selection of EISs leads one to the view that an impact is an event which results as a direct consequence of a prior event. However, not all such events would be included in EISs even if they were to arise from a development. A process of selection is constantly in operation to decide which events will be termed impacts and included in EISs and which will be excluded. For example, the use of pile-driving to prepare a development site gives rise to increased ambient noise levels. These increased noise levels are a direct result of the development activity. Are they an impact? As far as the air molecules are concerned circumstances have changed. Yet a description of the increased excitation rate of air molecules is unlikely to figure in many EISs. For decision-makers and the public it is the further indirect or second-order "impact" of these increased noise levels which are of concern. Will these levels disrupt the sleep of local residents at night? Will they annoy them during the day by making normal conversation more difficult? If these events were likely then they would in most cases be included in EISs as impacts.

Those likely to experience these events would consider them to be impacts as would most EIS writers and other people not likely to experience the noise levels. However, if we continue with this noise example, such unanimity may break down. For example, interrupted sleep patterns may affect work performance and increase family tensions. Thus, a second-order event may result in further events, some of which may not be perceived by those experiencing them. Individuals may blame increased noise levels for disrupted sleep, but not realize that increased family tension is due to this factor. For the operators of a development the impacts of increased noise levels may be thought of only in terms of the number of complaints received. This example shows some of the complexities involved in the concept of impact.

This example of the series of changes which can result from an initial increase in ambient noise levels indicates a major point about EIA. Not all environmental changes will find their way into EISs nor will all individuals or social groups agree on which changes should be classed as impacts for inclusion in EISs. It matters little what terminology is used-whether one distinguishes between effects and impacts, or changes in physical and environmental parameters and impacts-the fact remains that not all changes induced by a development will be classed as impacts. Trying to devise a suitable nomenclature, although laudable and perhaps necessary, cannot disguise the nature of the social and political choices involved in selecting and giving meaning to those likely changes resulting from a proposed project which will be termed impacts and included in EISs.

The social or political choices involved in EIA work are not made by society as a whole. It would be much easier if they were, as much disagreement in project planning would vanish. However, society is not a monolithic structure exhibiting a unitary structure. Societies are made up of groups and individuals with differing interests and values. Therefore, the allocation of significance to environmental changes depends on who is aware of the changes involved. For example, a particular change in the composition of a wildlife habitat may be seen as an ecologically significant local impact by local conservation groups. Other interest groups, for example, the unemployed looking for new work opportunities from a project, might consider the likely habitat change to be of no significance or not even recognize it as an impact at all.

Not only can non-expert social groups and individuals assign different values to environmental changes, but those preparing EISs, also, may act in this manner. For example, the scope of an EIS prepared by the regional office of the US Environmental Protection Agency (EPA) for a proposed wastewater treatment plant in Wyoming, reflected the values of EPA staff. A number of alternative sites were examined and, of these, one site was favoured by EPA whereas a different site was favoured by the council of the local community requiring the treatment plant.

The EIS was, therefore, a component of a political process revolving round the proposed plant. It was subject to the value choices of EPA staff, who ensured that the EIS considered the impacts of induced urbanization, thought to be harmful by EPA staff in many different ways. The impact of the expected urbanization was divided into a number of separate impacts; for example, loss of agricultural productivity, induced development cost, growth inducement and adherence to existing planning policies. These separate impacts were all different forms of expressing the same general impact. A scoring system was used to compare these impacts for the alternative sites. Each individual form of the general impact scored worst for the site favoured by the local council. Thus the format of impact identification was used in the EIS to further political ends. This EIS, like many others, one suspects, was not a purely technical, scientifically "objective" document.

Another characteristic of this social aspect of "impacts" is the time element. As techniques develop and more is understood of the relationships between environmental, social and health factors and their combined influence on individual well-being, "new" impact may be recognized. Therefore, improved knowledge of the causal relations between development activity and environmental/social change leads to public recognition of impacts which were previously unsuspected.

This attempt to show the socio-political dimension of EIA is not new. Many working in EIA will consider the argument unoriginal both theoretically and practically. However, the significance of this aspect of EIA is often ignored or not considered by those producing EISs and developing EIA methods. For example, many EISs describe, in considerable technical detail, the likely concentrations of differing air or water pollutants at various locations relative to the sources of emission, but fail to relate these environmental changes, which are, in themselves, neither good or bad, to the interests and concerns of those groups which may use (economically, aesthetically, symbolically or by recreation) the resource being affected. Environmental changes are not converted into impacts and described in terms of their consequences for people. This cannot be done for all changes. Some, for example ecological changes, may affect only specialized social groups such as scientists and may only be considered impacts by such groups. An attempt should be made to trace all changes to social interests and how they will be affected.

Many environmental assessment methods leave out this social dimension. They are concerned solely with measuring environmental changes and ascribing a significance to them. For example, a changes in dissolve O_2 in a river may be predicted and some concept of the importance of this change applied to the prediction. However, we have seen that such direct changes lead on, in many cases, to indirect impacts on social and individual interests. Methods which compartmentalize the environment into discrete units such as BOD levels, SO_2 concentrations and noise contours do not show the indirect effects of changes in these parameters. They stop short of attempting to assess the effect of the changes on people. Data on changes in these discrete units are frequently manipulated arithmetically and aggregated, often with weighting, to produce composite impact scores. Such scores are also of limited relevance because they do not express the ways in which social groups will be affected differentially by a proposal.

The importance of always tracing environmental changes to people is reinforced when it is realized that impacts (beneficial or harmful) from projects are not evenly and homogeneously distributed among social groups or individuals. Everyone does not benefit or suffer equally. Some people may only benefit, suffering none of the deleterious impacts borne by others. In addition, the mix of beneficial and harmful impacts is likely to be differentially distributed. It is important for decision-makers and members of the public to appreciate that certain social groups may be subject to more harmful and fewer beneficial impacts, while other groups may be in a more favourable position. This type of information is essential for public debate on the future of a proposed project.

Selected from "*Ronald Bisset, PADC Environmental Impact Assessment and Planning Unit, Martinus Nighoff Publishers, The Hague Printed in the Netherlands, UK, 1983*"

Unit 29

Text: Impact of Wastewater Effluents on Water Quality of River

Predicting Effects of Point Source Discharges

It is common to[①] use mathematical models to estimate how point source discharges influence concentrations within receiving waters. Although the residuals have been modeled in streams, lakes and estuaries, this introduction considers only streams. Discharges and stream-flows are considered to be constant and steady-state conditions are assumed. It is also assumed throughout that only a single point source is involved.

Models for predicting how the discharge of conservative substance affects stream quality are invariably based on the law of conservation of mass, also referred to as[②] the "mass balance equation." Chlorides and total dissolved solids are the indicators modeled most frequently. As shown in Figure 1, a mass balance analysis for conservative substances uses two equations, one for the conservation of flow and one for the conservation of residuals. Solving these equations

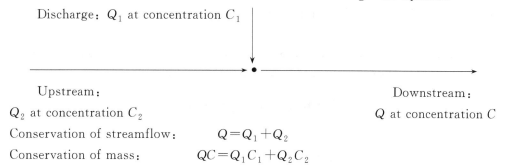

Discharge: Q_1 at concentration C_1

Upstream: Downstream:
Q_2 at concentration C_2 Q at concentration C
Conservation of streamflow: $Q = Q_1 + Q_2$
Conservation of mass: $QC = Q_1 C_1 + Q_2 C_2$

FIGURE1 steady-state mass balance analysis for a conservative substance discharged to a stream.

simultaneously yields C, the concentration of residuals downstream of the discharge point:

$$C = \frac{Q_1 C_1 + Q_2 C_2}{Q_1 + Q_2} \tag{1}$$

where Q_1 = flowrate upstream of the discharge (cfs)

Q_2 = rate of wastewater flow (cfs)

C_1 = residuals concentration upstream of the discharge (mg/L)

C_2 = residuals concentration of the wastewater (mg/L)

The units shown (cfs and mg/L) are illustrative; other flowrate and concentration units can also be used. Because the substance is conservative and streamflow is constant, the increase in concentration caused by the discharge is independent of the distance downstream. This analysis assumes the effluent is completely mixed once it enters the stream and the concentration of residuals is uniform throughout any stream cross section[3]. Equation (1) is used to determine how chloride reduction by the Margarita Salt Company influences the concentration of chlorides in the Cedro River.

The next level of modeling complexity is for a substance that decreases in concentration in accordance with "first-order reaction kinetics[4]." Underlying physical or biochemical cause of the decrease is not treated explicitly. At any instant, the rate of decay is assumed proportional to the amount of substance present; the proportionality factor is called the rate constant[5]. As indicated in the discussion of equation (2), coliform bacteria[6] are predicted using a first-order reaction model. More generally, a first-order has the form

$$C(x) = C_0 e^{-k(x/\mu)} \tag{2}$$

where x = distance downstream of the discharge (miles)

$C(x)$ = residuals concentration at location x (mg/L)

C_0 = residuals concentration at $x = 0$ (mg/L)

μ = velocity of streamflow (miles/day)

k = empirically determined rate constant (per day)

The concentration at $x = 0$ is calculated using the mass balance concepts yielding equation (1). The value of C_0 is the total mass of residuals per time just below the discharge point divided by the volume of streamflow per time at that point. Because stream velocity is assumed constant, (x/μ) represents the "time of travel" below the discharge location.

Selected from "*Environmental Planning and Decision Making, writted by Leonard Ortolano in Stanford University. Copyright 1984, by John Wiley & Sons, Inc. New York*"

Words and Expressions

flowrate *n*. 流率

illustrative *a*. 用作说明的，解说性的

estuary *n.* 江口，河口
kinetics *n.* 动力学
steady-state *a.* 稳态的
underlying *a.* 潜在的，在下的
streamflow *n.* 溪流量，溪流速
biochemical *a.* 生物化学的
conservative *a.* 守恒的
explicitly *ad.* 明晰地，清楚地
invariably *ad.* 不变地，总是
proportionality *n.* 比例（性），相称
empirically *ad.* 以经验为根据地，经验地
upstream *ad.* 在上游，向上游

Notes

① It is common to do…… 做……是普通的，干……是平常的
② refer to sth. as   称某物为
③ stream cross section   溪流断面
④ first-order reaction kinetics   一级反应动力学
⑤ rate constant   速率常数
⑥ coliform bacteria   大肠杆菌

Exercises

1. Put the following into Chinese.

   point source discharge, receiving water, constant and stead-state condition, mass balance equation, dynamic mode, tributaries, chemical decay processes, cummlative.

2. Put the following into English.

   标准偏差　残留物浓度　溪流断面　一级反应速率常数
   统计数据　　模型应用的范围　　影响预测　溶解氧

3. Translate the following sentences into Chinese.

（1）It is common to use mathematical models to estimate how point source discharges influence concentrations within receiving water.

（2）Models for predicting how the discharge of conservative substance effects stream quality are variably based on the law of conservation of mass, also referred to as the "mass balance equation".

Reading Material: The Aims and Objectives of Environmental Impact Assessment

Since the 1950s growing environmental awareness is increasingly focussing attention upon the interactions between development actions and their environmental consequences. In developed countries this has led to the public demanding that environmental factors be explicitly considered in the decision-making process and a similar situation is now occurring in developing countries.

Early attempts at project assessment were crude and often based upon Technical Feasibility Studies and Cost Benefit Analysis (CBA). CBA was developed as a means of expressing all impacts in terms of resource costs valued in monetary terms. Up to the 1960s, a number of major developments such as the proposed third airport at London and the Aswan Dam, which were assessed using CBA techniques, caused considerable public disquiet. Flaws in CBA became more apparent and one consequence was the development of a new evaluation approach which came to be known as Environmental Impact Assessment (EIA). It has evolved as a comprehensive approach to evaluation, in which environmental considerations, as well as economic and technical considerations, are given their proper weight in the decision-making process.

When EIA was first conceived, it was regarded as an "add on" component to CBA, and was designed to incorporate all those potential impacts that had proved troublesome in CBA. As a consequence, early practitioners used the system as a means of collecting information but often failed to comprehend the policy environment in which the development was proposed. Today, more fundamental questions are being posed such as:

(1) Is the development required?

(2) What are the relevant alternatives which could provide the same benefits and how do respective EIAs compare?

(3) What is the appropriate level of public safety in relation to hazardous technologies?

(4) What degree of environmental protection should be guaranteed for areas of significant ecological and landscape value?

EIA can, therefore, be thought of as a basic tool for the sound assessment of development proposals. Whilst EIA can play an important role in formulating environmentally sound policies and plans, and in their evaluation, this paper will concentrate on the application of EIA at a project level. Aspects of EIA such as procedural mechanisms, including selection of projects for EIA (screening), terms of reference (scoping), EIA methods and EIA reviews, have also been excluded from the discussion.

The Rationale for EIA

Projects until recently were often formulated and assessed according to technical, economic and political criteria, and the potential environmental, health and social impacts of projects were rarely considered in a vigorous manner. Even when considered, such assessment usually took the form of cost-benefit analysis which crudely attempted to place a monetary value upon non-economic variables such as the destruction of marine ecosystems or the social and health impacts of air pollution. As a consequence of such restricted assessment, many developments resulted in unforeseen harmful impacts which reduced predicted benefits. An example is the Aswan Dam which created deleterious secondary effects such as a loss of agricultural productivity and a reduction in the importance of the Mediterranean fishing industry.

The purpose of an EIA is to determine the potential environmental, social and health effects of a proposed development. It attempts to define and assess the physical, biological and socioeconomic effects in a form that permits a logical and rational decision to be made. Attempts can be made to reduce potential adverse impacts through the identification of possible alternative sites and/or processes.

There is no general and universally accepted definition of EIA. The following examples illustrate the great diversity of definitions:

(1) "… an activity designed to identify and predict the impact on the biogeophysical environment and on man's health and well-being of legislative proposals, policies programmes, projects and operational procedures, and to interpret and communicate information about the impacts".

(2) "… to identify, predict and to describe in appropriate terms the pros and cons (penalties and benefits) of a proposed development. To be useful, the assessment needs to be communicated in terms understandable by the community and decision-makers and the pros and cons should be identified on the basis of criteria relevant to the countries affected".

(3) "… an assessment of all relevant environmental and resulting social effects which would result from a project".

(4) "… assessment consists in establishing quantitative values for selected parameters which indicate the quality of the environment before, during and after the action".

Such definitions provide a broad indication of the objectives of EIA, but illustrate differing concepts of EIA. The scope of EIA is clearly defined in only the first definition. Three definitions include socio-economic impacts, but in the last definition only the environment is mentioned. The definition of the United Nations Environment Programme in (2) above, implies that decision-making on the relative importance, based on local conditions, of beneficial and adverse impacts, should be part of EIA. The other definitions merely indicate that EIA is an "objective", technical and predictive exercise, with no decision-making component. For the purposes of this paper it is assumed that the main objective of EIA is to provide decision-makers with an account of the implications of proposed courses of action before a decision is made. Decision-makers are not usually involved in the assessment process, as this is normally considered to be a technical exercise. The results of the assessment are assembled into a document called an Environmental Impact Statement (EIS) which contains a discussion of beneficial and adverse impacts considered to be relevant to the project, plan or policy. The completed report, or EIS, is one component of the information upon which the decision-maker ultimately makes a choice. At this stage, other factors such as unemployment, energy requirements or national policies may influence the outcome of the decision. A final decision can be made with due regard being paid to the likely consequences of adopting a particular course of action, and where necessary introducing appropriate monitoring and auditing programmes to reduce any deleterious impacts that may have been identified in the assessment process.

EIA is a mechanism which aids the efficient use of natural and human resources and which has proved valuable to both those promoting developments and those responsible for their authorization. EIA may reduce costs and the time taken to reach a decision by ensuring that subjectivity and duplication of effort are minimized, as well as identifying and attempting to quantify the primary and secondary consequences which might necessitate the introduction of expensive pollution control equipment, compensation or other costs at a later date.

There are a number of ways in which EIA can improve the efficiency of decision-making, but to be effective EIA should be implemented at an early stage of project planning and design. It must be an integral component in the design of projects, rather than something utilized after the design phase is complete. Preferably, EIAs should be part of an incremental decision-making process which has a number of decision-points in the project planning procedure. This means that there can be a continuous feedback between EIA findings, project design and locations. EIAs can be implemented to test alternative project designs at an early stage, to help choose the project design which emphasizes benefits and minimizes harmful effects. EIA therefore can be used not only to investigate and avoid harmful impacts but also to increase likely benefits.

The emergence of an optimum alternative in terms of the objectives or goals relevant to a proposed project means that EIAs may have significant long-term financial advantages. If a potential problem is identified early in project planning it may allow considerable financial savings to be achieved. At the crudest level the abandonment of a project may be required if all alternative designs or locations are considered unsuitable in terms of likely detrimental effects. This would save capital costs. It is more likely, however, that design modifications may obviate the need for expensive ameliorating action once a project becomes operational. If a development is not assessed for its likely impacts, it may cause serious social and health problems. For example, a proposed dam and reservoir may have health effects which may require an expensive health care programme. The wrong location for a resettled population may result in agricultural failure and the need for food supplies to be sent to the relocated people from other areas.

The incorporation of EIA into decision-making may create a number of benefits. If a forecast of the likely impacts of developments is available, allowances can be made and the infrastructure can be provided in a manner whereby impacts are minimized. Where uncertainty exists as to future development, EIA can identify those areas most susceptible to adverse impacts and so guide site selection. To be effective EIAs can only be used when the alternative sites are few in number; otherwise EIA can be time-consuming and very expensive. EIA can, nevertheless, aid the identification of the most suitable site in terms of benefit maximization and reduction of harmful effects. Should no site be considered suitable, then the results of EIA aid the determination of broad environmental, social or health criteria to be used when a large number of sites are screened for their suitability. The relevance and importance of EIA for site selection has been recognized in a report entitled Draft Guidelines for Assessing Industrial Environmental Impact and the Siting of Industry.

The Scope of EIA

In principle, an EIA procedure should apply to all actions likely to have a significant environmental effect. The potential scope of a comprehensive EIA system is, therefore, considerable and could include the appraisal of policies, plans, programmes and projects. The potential advantages of a tiered approach over a procedure which is restricted to development projects are:

(1) At the development project stage, the available options are often severely limited by earlier decisions made at a higher level. Mis-specification of a project assessment (e.g. a road scheme) may occur if the higher levels (e.g. transport policies) were not subject to such evaluation. The issue may be where environmental impacts will occur, and not whether they ought to.

(2) Assessment of individual projects can only be conducted once proposals have been made. It cannot guarantee optimum site selection, and a thorough assessment of all alternative actions may be prohibitively expensive and time consuming.

(3) The scope of viable alternatives decreases at the project level; also the willingness to contemplate alternatives may decline.

(4) The available time for collection and analysis of environmental data will become increasingly restricted at the lower tier unless a programme of establishing environmental baseline data is undertaken independently of individual project EIAs.

(5) When projects are individually small in size, but collectively large in number (e.g. housing), an EIA at the plan and programme stages may lead to a reduction in the time required for evaluation.

If plans are based on a sound environmental assessment, specific project proposals which fail to conform are likely to require detailed environmental assessment. The preparation of plans must be based on adequate data relating to the existing environment and the implications of possible changes. The systematic collection, analysis, storage and regular updating of such data must, therefore, greatly improve the quality of subsequent project EIAs as well as reducing time and costs. EIA and landuse planning ought, therefore, to be seen as complementary to each other.

The Relationship of EIA to the Decision-making Process

The purpose of EIA is to generate and provide information on the environmental consequences of a proposal. By itself it does not assure that adverse impacts are averted or minimized and beneficial impacts increased. If EIA is to be effective, it must be related in form and timing to the decision-making process in order that optimum strategies for averting or reducing adverse consequences can be formulated and evaluated.

Whilst the overall objective of an EIA is to safeguard the environment, it has often been regarded as a tool of advocacy by environmental interest groups. Such interests are, however, rarely the only or dominant criteria governing decision. The conclusions of an EIA must be presented in such a manner that they can be taken into account alongside other relevant economic and social factors and be as scientifically objective as possible.

A frequent objection to EIA in countries with well developed decision-making procedures, is that it imposes an additional stage in the process and may duplicate the existing system. Evidence exists, however, to suggest that EIA may be beneficial even in countries with a well developed decision-making procedure. By the systematic examination of environmental constraints and the development of measures designed to protect or conserve the environment, EIA prevents such considerations becoming a late addition to the decision-making process, thereby saving financial resources and time and also ensuring a more environmentally-sound project.

Public Information and Participation in EIA

Public involvement should be an integral part of any EIA system. Efforts should be made to obtain the views of, and to inform, the public and other interests who may be directly or indirectly affected by the project. The authorizing agencies may not always identify the environmental issues which the public perceives to be important and they may also lack the detailed local knowledge that the public possesses. Advantages of participation may lead to the provision of information about local environmental, economic and social systems; the possible identification of alternative actions; an increase in the acceptability of the project as the public will better understand the reasons for the project; and a minimization of conflict and delay. Problems may nevertheless arise. Public participation may, in the short-term, be time-consuming, and increase costs, and participants may be unrepresentative of the community. In spite of these potential problems, many countries are actively encouraging public involvement in EIA.

Selected from "*Brian D Clark, DADC Environmental Impact Assessment and Planning Unit, Martinus Nighoff Publishers, The Hague Printed in the Netherlands, UK, 1983*"

Unit 30

Text: Environmental Impact Assessment of Air Quality

Air Quality Impact Assessment Process

A comprehensive assessment of a proposed project's air quality impacts involves the following steps:

1. Establish background air quality levels.
2. Identify applicable air quality criteria and standard.
3. Forecast future air pollutant emissions with and without the project.
4. Forecast future ambient air pollutant concentrations with and without the project.
5. Compare predicted air quality with applicable standards.
6. Modify plans, if necessary, to deal with potential air quality problems.

The first step, establishing background levels of air quality is carried out only for air quality indicators likely to be influenced by the proposal. For example, if the proposed project were a highway, the main indicators of interest would be hydrocarbons, nitrogen oxide①, carbon monoxide②, and photochemical oxidants③. In the United States, data on background air quality levels are available from local, regional, and state air quality management agencies, various state implementation plans proposed in response to federal air quality laws, and EPA's computerized data retrieval systems. The EPA data include measurements from the federal Continues Air Monitoring Program (CAMP) stations established in the early 1960s.

Applicable air quality criteria and standards are determined in the second step of the assessment process. The EPA has issued criteria relating levels of air pollutants to human health and welfare. In the United States, standards used in impact assessments include the national ambient air quality standards and pertinent state or local standards.

The NAAQS, as illustrated by carbon monoxide standards in 1980, stipulate both concentration and time of exposure. The CO standards require that the average concentration for any 1-hr period be less than 9 parts of CO per million parts of ambient air. These CO measurements are said to have an "averaging time" of 1-hr. The NAAQS also require average CO concentrations during any 8-hr period to be less than 35 parts per million (ppm). Limits on CO are not to be exceeded more than once per year. As of 1980, the NAAQS applied to particulates, SO_x, CO, NO_2, hydrocarbons, ozone, and lead. The national standards are revised periodically and up-to-date versions are obtainable from EPA.

In the third step of the assessment process, the proposed project's emissions are estimated in units of④ weight (or mass) per time period. Procedures for estimating emissions are reviewed in the next section.

The fourth step, predicting changes in the ambient concentrations of air quality indicators due to a new discharge, is often complex. In fact, sometimes it is not carried out and the assessment considers only the increased emissions from the proposed project. Reasons for not estimating concentrations include (1) inadequate understanding of the underlying physical and chemical processes and (2) unwillingness to commit the time and money needed to utilize existing forecasting procedures.

The final steps in a comprehensive impact assessment are to compare forecasted concentrations with applicable standards and to modify the proposed project if expected air quality degradation is unacceptable. Air-borne residuals are commonly reduced by changing combustion processes and using emission control devices such as scrubbers and filters. However, there are many options for mitigating adverse air quality effects that do not involve control devices. Examples include reducing the scale of a facility or changing the locations of discharges.

Forecasts of air pollutant emissions and concentrations are carried out at various levels of sophistication. Some impact assessments are limited to quick and simple estimates of increases in emissions. Others use elaborate computer-based mathematical models to translate increases in emissions into changes in concentrations at various times and places.

Selected from "*Environmental Planning and Decision Making, written by Leonard Ortolano in Stanford University. Copyright © 1984, by John Wiley & Sons, Inc. New York*"

Words and Expressions

emission *n.* 散发；排放物
inadequate *a.* 不充足的；不适当的
ambient *a.* 周围的，包围着的
implementation *n.* 补充
retrieval *n.* （可）重新获得，（可）收回
stipulate *v.* 约定，规定
mitigate *v.* 减缓，使缓和
sophistication *n.* 改进；复杂化
version *n.* 版本
elaborate *a.* 精巧的，精细的

Notes

① nitrogen oxide   氮氧化物
② carbon monoxide   一氧化碳
③ photochemical oxidants   光化学氧化剂
④ in unit of   以……为单位

Exercises

1. Put the following into English.
   大气质量背景值   光化学氧化剂   综合评价   标准的大气监测方法
   气象资料   环境状况   污染物排放速度
2. Put the following into Chinese.
   air-borne residuals, elaborate computer-based mathematical models, ecosystem, unacceptable reduction, human health and welfare, carbon monoxide
3. Translate the following sentences and paragraphs into Chinese.
(1) In the United States, data on background air quality levels are available from local, regional, and state air quality management agencies, various state implementation plans proposed in response to federal air quality laws, and EPA's computerized data retrieval systems.

(2) The identification of potential changes of air quality can only be made if data both on the project activities and on the existing ambient air quality are available. The combined evaluation of these data will result in the identification of potential air quality changes. Project data have to be collected and analyzed for sources of impacts to be identified. It is essential to distinguish sources of potential impacts by analyzing the activity involved including secondary activities, i. e. those generated as a result of the proposed activity. Furthermore, assessment should include construction, operational, and post-operational phases. A description of the environmental conditions, and of data collection and analysis, must define what variables constitute the back ground of ambient concentrations. The comparison of identified sources of potential impacts with the description of the background concentrations will constitute the assessment of potential changes.

Reading Material: Risk Assessment

In the late 1980s, Ulrich Beck's book, *Risikogesellschaft*, was already famous in Germany. In 1992, right after the Chernobyl nuclear plant disaster in Ukraine and Belorus, his book was published in English as *Risk Society: Towards a New Modernity* (1992) and became instantly famous in the English-speaking world. In that book he coined the term "risk society" which referred to the dangers that globalization and modern technology pose to the planet:

In advanced modernity the social production of wealth is systematically accompanied by social production of risks. Accordingly, the problems and conflicts relating to distribution in a society of scarcity overlap with the problems and conflicts that arise from production, definition and distribution of techno-scientifically produced risks. (*Beck* 1992, P. 19)

Put another way, industrial society generates new forms of risk:

Today's risks, such as global warming, the hole in the ozone layer, ionizing radiation, or the contamination of foodstuffs by pesticides tend to be invisible, and can only be assessed by scientific methods and culturally represented by scientific and media knowledge systems. Moreover, Beck argues, these risks often reflect the dangers posed by science itself. (*Tulloch* 2008, P. 146)

Modern society's scientific advances and capitalistic activities have generated risks that are unevenly distributed and which are often invisible. Indeed, unlike tornadoes or mountains of trash, toxins are often imperceptible, and their effects are not necessarily obvious. their production, distribution, and use tends to be controlled by agencies that have access to the best possible scientific understanding of toxic materials and the political power to shape, implement, enforce, and possibly ignore policy.

The danger of toxins is usually determined through a process known as risk assessment which involves a systematic evaluation of the "likelihood of an adverse effect resulting from a given exposure". Objectives of risk assessment include:

Balancing risks and benefits (e. g. , drugs and pesticides)

Setting target levels of risk (e. g. , food contaminants and water pollutants)

Setting priorities for program activities (e. g. , regulatory agencies, manufacturers, and environmental and consumer organizations)

Estimating residual risks and success of steps taken to reduce risk.

In the U. S. , the Environmental Protection Agency (EPA) follows a widely accepted, four-step procedure to assess risk:

1. Hazard assessment

This initial step identifies sources of risk and hazards that might result in harmful exposure. We are not always aware of the source of risks. As already mentioned, the books *Silent Spring and Our Stolen Future* drew public attention to chemicals and exposures that had previously not been thought to be dangerous. Lead in paint and gasoline, asbestos, and Bisphenol-A plastics are other toxic hazards that have been at the forefront of public and scientific attention at different times.

Although the hazards of many toxins are universal, some chemicals pose more of a hazard to some people than to others. For example, anyone who reads labels has probably noticed the warning on beverages containing aspartame that the product contains phenylketonurics. Most people cannot pronounce the name of that chemical never mind know why they might be concerned. People with phenylketonuria, a metabolic disorder involving a deficiency of an enzyme required to metabolize an amino acid called phenylalanine, however, know to avoid that chemical as much as possible. For most people, the presence of phenylalanine in their diet pop does not pose a problem, but it is a case for special labeling since, for some people, it poses a hazard.

2. Dose-response assessment

Paracelsus, a Renaissance alchemist, physician, and botanist, said "In all things there is a poison, and there is nothing without a poison. It depends only upon the dose whether a poison is poison or not". How much exposure or how great a dose will lead to a toxic effect? Two main ways to assess exposure and effect are through animal studies and through epidemiological studies. Animal studies allow for controlled experimentation, the results of which are extrapolated to humans. Animal studies may be critiqued in terms of biological extrapolation and in terms of numerical extrapolation: Can the system of a mouse, guinea pig, or rabbit realistically stand in for a human system? Is it realistic to study the effects of extremely large doses given to animals over a short time frame and assume that smaller doses in humans over a longer time period will have similar effects? Animal studies do provide a way to conduct research on toxins without harming humans, but they raise the question of the ethics of inflicting harm on animals for human benefit. Animal studies cannot test real world human exposures to toxins, but epidemiological studies can. These studies are done after the fact of exposure. For example, if a known site is contaminated, an epidemiological study would identify people known to have lived near that site and assess symptoms or reactions that those people present from that exposure. One problem with epidemiological studies is that the exact level of exposure to a suspected toxin is unknown. Additionally, it is unknown what other substances a person has been exposed to that might exacerbate or ameliorate the effects of the chemical in question. In both animal studies and epidemiological studies, it is nearly impossible to study with any accuracy the latent effects of a suspected toxin in humans since latent effects can show up years after exposure.

3. Exposure Assessment

Controlled laboratory experiments do not tell us about real world exposures. This step of risk assessment aims to identify and measure our actual interactions with toxins or other hazards. Who is likely to be exposed? Under what circumstances? Calculations of exposure are made based on standard human body mass, skin surface area, and resting respiration rate to estimate exposure through drinking, inhalation, or contact with contaminated air, water, or food.

4. Risk Characterization

Assessing dosage and exposure provide valuable measures for understanding risk, but the scale or scope at which risk is characterized and communicated is also important. Is risk described in terms of "individual lifetime risk" or as "societal risk"? Measuring risk in terms of the "loss of life expectancy" recognizes a difference between risk for older people and risk for younger people who might be expected to live many more years.

An ideal outcome in risk assessment is the identification of a bright line or threshold that divides safe exposure from dangerous exposure. A bright line provides a sense of certainty and brings clarity to policy. Speed limits, for example, are a type of bright line that makes it possible to set a policy on which to base decisions: How fast is it safe to drive on this road? Driving how much faster than the speed limit is likely to result in a speeding ticket? Yet a bright line is also arbitrary depending on how safety or danger is defined and for whom, for how long, and under what conditions the bright line is established. How was it determined that fewer than four rodent hairs is an acceptable number to have in 100 grams of apple butter but more than that is not?

Another perspective on risk and uncertainty, a psychometric approach, focuses on risk perception and what kinds of risks people worry about. This kind of approach is demonstrated in comparison of government approaches to the nuclear industry in the US and in France. In both countries, public perception of risk associated with nuclear power is high. In France, government control over the nuclear industry is centralized and public has less input but more trust in experts and in the government. The system in France allows people to exercise power on the basis of merit: scientific training and certification. In the US, the public also has a high degree of risk perception associated with nuclear power, but there is general distrust of government, science, and industry. In the US, the public tends to believe it has some control over risks. The system allows for the public to intervene, to question expert opinion, and to litigate. Public risk perception, in this example, shapes and reinforces how the nuclear industry is managed in both countries. This nuanced understanding of risk would not be captured by a realist approach that would more likely focus on exposures to nuclear radiation and other tangible hazards.

Selected from "*Shannon O'Lear, Environmental Politics, Scale and Power, Cambridge University Press, USA, 2010*".

Unit 31

Text: The Role of Environmental Monitoring in Pollution Science

Environmental monitoring is based on scientific observations of changes that occur in our environment. Scientists need to observe changes to study the dynamics of not only natural cycles, but also anthropogenic-based impacts. The effects of pollution in the environment, as in humans, can be slow (chronic) and may require multiple observations over time, or they can be fast-acting (acute) and be assessed with simultaneous observations[①]. The effects of pollution occur at all spatial scales; therefore, observations are also made at multiple scales of space. However, since the environment is a continuum, observations must be made using physical, chemical, and biological methods. Only science-based observations, standards-based data processing, and objective interpretations can produce the knowledge and level of understanding required to accurately evaluate and solve environmental problems. There are numerous examples of knowledge-based regulations that benefit modern society and protect the environment from pollution. These include waste management regulations involving disposal, treatment, or reuse; regulations governing the protection of water resources including natural and public water supplies; and regulations protecting endangered species.

There are numerous agencies and world institutions involved in environmental monitoring. In the U.S., pollution monitoring and prevention is the primary focus of the Environmental Protection Agency (EPA). This agency is mandated to develop and enforce laws and regulations that are protective of our health and the environment.

Sampling and Monitoring Basics

Developing a program to monitor the extent or effects of pollution in a particular environment requires careful consideration of following: (1) The purpose of the monitoring program-this includes determining, for example, pollution level changes in space and time; (2) the objectives, which may include specific chemical, physical, or biological analyses; and (3) the approach, which will assist in defining the number and types of measurements. To achieve our objectives, we must also consider the environmental characteristics or uniqueness of each environment. Finally, we must consider sampling methods, including locations, timing, and type. There are several types of sampling methods. Random sampling, for example, assumes that all units from an environment have an equal chance of being selected. In contrast, systematic sampling selects sampling locations at predefined intervals in space or time.

Specific sampling plans are needed to insure that all aspects of monitoring are detailed and described. The U. S. EPA defines several critical elements of a sampling plan to insure that all the data quality objectives are met; these are described in Table 1[②]. In the development of sampling plans, environmental scientists must be thoroughly familiar with types of sampling such as destructive and nondestructive methods. They should also know the accepted methods of analyses by consulting the latest scientific and regulatory reference manuals and books. Environmental scientists should also be familiar with the basic concepts and applications of analysis and measurement principles.

Table 1   Data Quality Objectives of a Sampling Plan

1 Data quality control and objectives	They are needed in a sampling plan. Although the following requirements are borrowed from U. S. EPA pollution monitoring guidelines, these are generic enough that they should be included in any type of environmental sampling plan.
2 Quality	It discusses statistical measurement.
3 Accuracy (bias)	It determines how data will be compared to reference values when known. Estimates over bias of the project based on criteria and assumptions made.
4 Precision	It discusses the specific (sampling methods, instruments, measurements) variances and overall variances of the data or data sets when possible using relative standard deviations (%CV).
5 Defensible	It insures that sufficient documentation is available after the project is complete to trace the origins of all data.
6 Reproducible	It insures that the data can be duplicated by following accepted sampling protocols, methods of analyses, sound statistical evaluations, and so on.
7 Representative	It discusses the statistical principles used to insure that the data collected represents the environment targeted in the study.
8 Useful	It insures that the data generated meets regulatory criteria and sound scientific principles.
9 Comparable	It shows similarities or differences between this and other data sets, if any.
10 Complete	It addresses any incomplete data and how this might affect decisions derived from these data.

Statistics and Geostatistics

Statistical methods are necessary in pollution science because it is impossible to characterize all properties of an environment everywhere all of the time. Statistics are used to select samples from a population in an unbiased manner. They also help to interpret the data with the appropriate degree of confidence. Descriptive statistics are very useful in environmental science, because they provide a summary of the properties or characteristics of an environment. Descriptive statistics include sample means, standard deviations, and coefficients of variation.

Data samples should be collected randomly or systematically from an environment. Biased sampling (usually based on convenience) will produce biased data. The range of values that can be expressed from sampling an environment varies randomly, but values have a likelihood of occurrence that is defined by a probability distribution.

A reoccurring question in pollution science is how many samples are needed to define the extent of pollution with some statistical certainty. This is a nearly impossible task, since pollution distributions are not usually known beforehand. However, we can estimate the number of samples needed, assuming that the sample standard deviation from previous studies is equivalent to the population standard deviation, and that the sample mean is equivalent to the population mean. Also we must assume a normal distribution of values and be willing to accept that our sample mean value will be within a specific confidence interval.

Sampling and Monitoring Tools

Modern data collection uses automated data acquisition methods to monitor environmental variables related to pollution. For example, urban air quality is monitored using a network of automated stations that measure pollutants such as carbon monoxide (CO) and ozone (O_3) at close intervals. Coupled with weather stations, these gas monitors produce near-realtime environmental information, which is in turn used to predict future pollution-related events[3].

A critical component of the system is the sensor, which responds to an environmental stimulus such as temperature. Thermistor[4], for example, is able to respond to temperature changes by changing their internal resistance. The data acquisition system (DAS) usually includes an analog-to-digital converter used to convert analog (continuous) signals from the sensor to a digital (discrete) value that can be stored and manipulated by computer processors. Data storage and transmission systems are also needed in modern DAS to collect raw and processed data, and send it to the user.

Environmental scientists that use modern data collection systems must have a rudimentary working knowledge of electricity, computer processing, and computer programming, and be familiar with basic laws of environmental physics and chemistry.

Selected from "Ian L. Pepper, Charles P. Gerba, Mark L. Brusseau. Environmental Pollution Science, Elserier Inc. USA, 2006"

Words and Expressions

multiple observations　多重观察
multiple scales of space　多尺度空间
objective interpretations　客观解释，客观说明
endangered species　濒临危险物种
uniqueness　*n.* 独特性，唯一性
random sampling　随机采样
predefined intervals　预定的间隔
reproducible　*a.* 重复的
representative　*a.* (有) 代表性的
geostatistics　*n.* 地球统计学

characterize *vt.* 表征
reoccurring question 再发生问题
standard deviation 标准偏差
confidence interval 可靠的区间,可靠的间隔
rudimentary *a.* 基本的,初步的,早期的
thermistor *n.* 热敏电阻器
data acquisition system 数据获取系统
analog-to-digital converter 模拟数字转换器

Notes

① The effects of pollution in the environment, as in humans, can be slow (chronic) and may require multiple observations over time, or they can be fast-acting (acute) and be assessed with simultaneous observations. 可译为:对环境或人类,污染产生的影响可以是缓慢的,需要长时间的多方面观察;或者这些影响是急剧的,需用同步观察来评价。

② The U.S. EPA defines several critical elements of a sampling plan to insure that all the data quality objectives are met; these are described in Table 1. 可译为:美国环境保护署定义了几种重要的采样规划中需分析的元素,以确保所有的数据真实客观,如表1所示。

③ Coupled with weather stations, these gas monitors produce near-realtime environmental information, which is in turn used to predict future pollution-related events. 可译为:与气象站一起,这些气体监测结果几乎产生实时环境信息,依次被用于预测未来污染相关的事件。

④ Thermistor 热敏电阻器,是一种因温度变化导致原件内阻改变从而可以测量温度变化的仪器。

Exercises

1. Put the following into Chinese.
   previous study, data storage and transmission systems, sample mean, population mean, public water supplies, all spatial scales, descriptive statistics, regulatory criteria
2. Put the following into English.
   模拟数字转换器 随机和系统采样 独特性 濒临危险物种 客观说明
   采样与监控工具 精确性 准确性 监控指南 水源保护
3. Translate the part of "Sampling and Monitoring Tools" into Chinese.
4. Give a brief summary of Data Quality Objectives of a Sampling Plan.
5. What is the difference between systematic sampling and random sampling?

Reading Material: Environmental Chemical Processes and Chemicals

Environmental Chemical Processes

The chemical reactions affecting trace gases in the atmosphere generally have quite significant activation energies and thus occur on a timescale of minutes, days, weeks, or years. Consequently, the change to such chemicals is determined by the rates of their reactions and atmospheric chemistry is intimately concerned with the study of reactions kinetics. On the other hand, some processes in aquatic systems have very low activation energies and reactions occur extremely rapidly. In such circumstances, provided there is good mixing, the chemical state of matter may be determined far more by the thermodynamic properties of the system than by the rates of chemical processes and therefore chemical kinetics.

The environment contains many trace substances at a wide range of concentrations and under different temperature and pressure conditions. At very high temperatures such as can occur at depth in the solid earth, thermodynamics may also prove important in determining, for example, the release of trace gases from volcanic magma. Thus, the study of environmental chemistry requires a basic knowledge of both chemical thermodynamics and chemical kinetics and an appreciation of why one or other is important under particular circumstances. As a broad generalization it may be seen that much of the chapter on atmospheric chemistry is dependent on knowledge of reaction rates and underpinned by chemical kinetics, whereas the chapters on freshwater and ocean chemistry and aqueous aspects of the soils are very much concerned with equilibrium processes and hence chemical thermodynamics. It should not however be assumed that these generalizations are universally true. For example, the breakdown of persistent organic pollutants in the aquatic environment is determined largely by chemical kinetics, although the partitioning of such substances between different environmental media (air, water, soil) is determined primarily by their thermodynamic properties and to a lesser degree by their rates of transfer.

Environmental Chemicals

This part is not concerned explicitly with chemicals as pollutants. This is a topic covered by a companion volume on *Pollution Science*. This section, however, is nonetheless highly relevant to the understanding of chemical pollution phenomena. The major areas of coverage are as follows:

(1) **The chemistry of freshwaters.** Freshwaters comprise three different major components. The first is the water itself, which inevitably contains dissolved substances, both inorganic and organic. Its properties are to a very significant degree determined by the inorganic solutes, and particularly those which determine its hardness and alkalinity. The second component is suspended sediment, also referred to as suspended solids. These are particles, which are sufficiently small to remain suspended with the water column for significant periods of time where they provide a surface onto which dissolved substances may deposit or from which material may dissolve. The third major component of the system is the bottom sediment. This is an accumulation of particles and associated pore water, which has deposited out of the water column onto the bed of the stream, river, or lake. The size of the sediment grains is determined by the speed and turbulence of the water above.

(2) **Salt waters.** The waters of seas and oceans differ substantially from freshwaters by

virtue of their very high content of dissolved inorganic material and their very great depth at some points on the globe. These facets confer properties, which although overlapping with those of freshwaters, can be quite distinct. Some inorganic components will behave quite differently in a very high salinity environment than in a low ionic strength freshwater. Historically, therefore, the properties of seawater have traditionally been studied separately from those of freshwaters and are presented separately, although the important overlaps such as in the area of carbonate equilibriums are highlighted.

(3) **The chemistry of soils and rocks.** There are very significant overlaps with freshwater chemistry but the main differences arise from the very large quantities of solid matter providing very large surfaces and often restricting access of oxygen so that conditions readily become anoxic. However, many of the basic issues such as carbonate equilibriums and redox properties overlap very strongly with the field of freshwater chemistry. Soils can, however, vary very greatly according to their location and the physical and chemical processes which have affected them during and since their formation.

(4) **Environmental organic chemistry.** Much of the traditional study of the aquatic and soil environment has been concerned with its inorganic constituents. Increasingly, however, it is recognized that organic matter plays a very important role both in terms of the contribution of natural organic substances to the properties of waters and soils, but also that specific organic compounds, many of them deriving from human activity, show properties in environment which are not easily understood from traditional approaches and therefore these have become a rather distinct area of study.

(5) **Atmospheric chemistry.** The atmosphere contains both gas phase and particulate material. The study of both is important and the two interact very substantially. However, as outlined previously, chemical processes in the atmosphere tend to be very strongly influenced by kinetic factors, and to a large extent are concerned with rather small molecules, which play only a minor part in the chemistry of the aquatic environment or solid earth. Inevitably, there are important processes at the interface between the atmosphere and the land surface or oceans, but these are dealt with more substantially in the companion volume on *Pollution Science*.

Selected from "*Roy M Harrison. Principles of Environmental Chemistry, The Royal Society of Chemistry. UK, 2007*"

Unit 32

Text: Toxins

Introduction

There is renewed energy to ban aspartame[①], commercially known as Equal, Nutrasweet, and Canderel, on the basis that it causes cancer. Aspartame is the second most widely used artificial sweetener in the world after saccharin. It was initially approved by the U. S. Food and Drug Administration in 1974, but its approval process is a dramatic story of high-level politics and corporate involvement in the policy-making process. When faced with the ubiquitous pink (saccharin), blue (aspartame), and yellow (Splenda or Sucralose) packets of sweeteners, many people already have a sense that these artificial sweeteners are not good for human consumption…are they right? How would we know?

Another similar but less widely known case of possible toxins intentionally included in our food supply is diacetyl. Just because you have never heard of it does not mean you have not eaten it. It is not required that this chemical be listed on any packaging by name, and even if it were, few people would know about the serious lung disease attributed to inhaling this chemical. Diacetyl safely (or ominously, depending on your view-point) is suggested only by the ingredient "butter flavor" in microwave popcorn. Inhaling diacetyl, either in the process manufacturing microwave popcorn or simply in opening a freshly popped bag of popcorn, is the most dangerous form of exposure to this toxin.

It is easy to assume that if artificial sweeteners and other food additives, as well as ingredients in drugs, cosmetics, or other items such as building materials and paint, have been approved by a government, then surely these chemicals are safe. Aspartame and diacetyl highlight two particular features of toxins. First, they are usually invisible to the public. People are often unaware of chemicals they are consuming or using in a wide array of products. Unless they are avid readers of product labels and have educated themselves about chemicals that may be considered risky, they probably have no idea that they are exposed to a wide range of toxins on a daily basis. Second, since most toxins are invisible to the public, the public tends to trust that appropriate research has been done and that measures have been taken to ensure a safe, non-toxin environment. Toxins raise the issue of risk assessment and the question what is safe or at least what kind of risk is acceptable. It would be impossible to create a risk-free world due to economic, technical, and logistical realities. At some point in a decision-making process, someone has determined what will be considered a risk and what level of that risk is acceptable. There are clear power implications in any risk assessment process: Who is assessing the risk? Who decides how much and what kinds of risks will be tolerated or faced and by whom? What if something goes wrong?

Toxins are a concern at multiple spatial scales that represent different kinds of interactions and exposures. For example, this chapter will discuss persistent organic pollutants (POPs)[2] which literally travel the globe as cross-media pollutants that move through air, water, soil, and organisms. Yet due to their chemical nature, these pollutants tend to concentrate in specific regions. The example of DDT was discussed in the next section. That pesticide was banned in the US but is still produced there and shipped to other places such as Ecuador[3] for use on crops to be imported back into the US. DDT is widely recognized as being toxic to humans (as well as an excellent means to fight mosquito-borne malaria[4]), and the commodity chain or lifecycle path of DDT illustrates the spatial scale of that particular toxin: Where is it produced? Do its handling and transportation affect people and places along the way? Where is it used and for what purpose? Who is affected (for better or worse and in the long run) by the use of this chemical? Different state policies for toxins generate spatial patterns of toxins that are banned, controlled, traded, and tracked. For example, the U.S.-Mexico border has come to be referred to as "the Chemical Border" since the production, transport, and disposal of various toxic and hazardous materials near the border exacerbate existing, uneven patterns of environmental health and degradation. Borderlands can be vague frontiers where lax policies encourage dumping or other forms of mishandling of toxic materials. Other types of borders differentiate distinctive spaces with very different regulations for toxins. Differences between toxins regulations in the U.S. and in the European Union will be discussed as creating a spatial scale of toxins sloping towards the U.S. where less stringent regulation enables international dumping of toxic products that do not meet Western European standards. Even though the U.S. and the E.U. do not share an actual border, toxins regulations between these regions has established a clear borderline in the geography of toxins.

Toxins also have important spatial dimensions at other scales such as the scale of the body. The body is a spatial scale that reflects culture, economy, and politics. For example, the wearing of religious symbols, women wearing a chador, tattoos, piercing, and scarring, male and female circumcision, the cultural significance in Western societies of a diamond ring—all of these examples are ways that we use our bodies to reflect our identity, status, and beliefs. Toxins draw our attention to how these chemicals affect the health and well-being of individual bodies of actual people as they interact with a particular environment or lifestyle. As we think about how toxins may be affecting us in the air we breathe, in the water we drink, in the foods we consume, in the products we use, or in the home and workplace, it becomes clear that the spatial scale or administrative level at which decisions are made about toxin regulations is often quite removed from the spatial scale of toxin effects.

Selected from "*Shannon O' Lear, Environmental Politics: Scale and Power, Cambridge University Press, USA, 2010*".

Words and Expressions

toxin　　*n.* 毒素
aspartame　　*n.* 天门冬酰苯丙氨酸甲酯
nutrasweet　　*n.* 天冬甜素
canderel　　*n.* 阿斯巴甜
saccharin　　*n.* 糖精
ubiquitous　　*a.* 无所不在的，普遍存在的
splenda　　*n.* 三氯蔗糖

sweetener　*n.* 甜味剂
diacetyl　*n.* 二乙酰
butter flavor　牛油香精
popcorn　*n.* 爆米花
lax　*a.* 松懈的，不严格的，松弛的
dump　*v.n.* 倾销，倾倒
chador　*n.* 黑布
tattoos　*n.* 文身，刺青
scarring　*n.* 伤疤

Notes

① aspartame　一种食品添加剂，是比蔗糖甜 200 倍的甜味剂。
② persistent organic pollutants（POPs）　持久性有机污染物，该类污染物很难降解，可以长期存在于自然环境中，如二噁英等芳香烃类有机物。
③ Ecuador 厄多尔共和国，位于南美洲西北部的国家。
④ fight mosquito-borne malaria。治疗由蚊媒虫害引起的疟疾，民间把疟疾称之打摆子。

Exercises

1. Put the following into Chinese.
   （1）Toxins are a concern at multiple spatial scales that represent different kinds of inter-actions and exposures.
   （2）Borderlands can be vague frontiers where lax policies encourage dumping or other forms of mishandling of toxic materials.
   （3）For example, the wearing of religious symbols, women wearing a chador, tattoos, piercing, and scarring, male and female circumcision, the cultural significance in Western societies of a diamond ring—all of these examples are ways that we use our bodies to reflect our identity, status, and beliefs.
2. Put the following into English.
   无所不在的　甜味剂　松懈的　文身　伤疤　倾销
3. Based on the first paragraph, how many sweeteners are mentioned?.

Reading Material: Toxins Basics

People in Western societies did not always know about germs largely because germs are invisible to the naked eye. Physicians did not always wash their hands prior to performing surgery. Similarly, people were not always aware of toxins. In 1962, Rachel Carson published her book Silent Spring, which drew public attention to the dangers of using DDT as a pesticide and its effects on wildlife and human health. Her book is often cited as the starting point of a widespread realization of toxins in the environment and risks associated with long-term exposure. We have a better understanding now of how many chemicals are in use. A study in

1980 identified "1500 active ingredients of pesticides, 4000 active ingredients of therapeutic drugs, 2000 drug additives to improve stability, 2500 food additives with nutritional value, 3000 food additives to promote product life, and 50000 additional chemicals in common use". That was 30 years ago. The volume, not to mention the number, of chemicals in use is no doubt greater now. With this many chemicals in use, it would be a phenomenal task to understand the toxicity of each one as well as possible combinations of chemicals that are toxic. There is still a lot we do not know about toxins.

Of particular concern among this boundless selection of chemicals is a group of toxins known as persistent organic pollutants, known by the shorthand POPs. The pesticides DDT and toxophene are POPs as are industrial chemicals such as dioxins, furans, and polychlorinated biphenyls (PCBs). POPs represent hundreds if not thousands more toxins since these chemicals bread down into other chemicals, many of which are unknown or untested. The Green Revolution in the 1960s contributed to the diffusion of the use of agricultural POPs since it encouraged the use of genetically modified crops in less economically developed countries. Those crops required the use of chemical pesticides and fertilizers, many of which contained POPs, for maximum production. Although some countries such as Canada and the U.S. have banned agricultural POPs, other countries such as Russia, China, and India continue to use them. Other POPs, such as dioxins, are released when heavy metals, such as those found in computer components, are burned for disposal.

Individual country bans on POPs help to reduce the overall level of POPs in the environment, yet it matters that POPs are still in use in many places. One of the reasons these chemicals have earned their name is that not only do they persist for a long time in the environment, but they also are a cross-media pollutant. That means they can travel long distances in air, water, soil, and in living organisms. POPs are lipophilic which means they concentrate in fatty tissue within living organisms. This trait allows them to move easily up the food chain in a process known as bioaccumulation. At each trophic level, for example as small fish are eaten by large fish which are then eaten by seals, POPs reach higher concentrations in an animal's fatty tissue. The higher up the food chain an organism is, the greater the concentration of POPs in their system.

In 2001, The Stockholm Convention on Persistent Organic Pollutants was adopted and signed, although not yet ratified and implemented, by 150 countries. The motivation behind this international agreement was the recognition that no state working alone could protect its population from persistent organic pollutants. The Stockholm Convention recognizes 12 POPs hazards of which far outweigh their industrial or agricultural benefits, shown in Table 1. The Convention requires an end to the production and use of nine of these chemicals. DDT, however, is permitted for use in some places for controlling disease vectors for malaria until suitable substitutes are found. The convention requires governments to develop plans for the appropriate transport and disposal of unintentionally produced POPs such as dioxins and furans.

Table 1 The dirty dozen POPs of the Stockholm Convention

Aldrin	A pesticide applied to soils to kill termites, grasshoppers, corn rootworm, and other insect pests.
Chlordane	Use extensively to control termites and as a broad-spectrum insecticide on a range of agricultural crops
DDT	Perhaps the best known of the POPs, DDT was widely used during the World War II to protect soldiers and civilians from malaria, typhus, and other diseases spread by insects. It continues to be applied against mosquitoes in several countries to control malaria
Dieldrin	Used principally to control termites and textile pests, dieldrin has also been used to control insect-borne diseases and insects living in agricultural soils
Dioxins	These chemicals are produced unintentionally due to incomplete combustion, as well as during the manufacture of certain pesticides and other chemicals. In addition, certain kinds of metal recycling and pulp and paper bleaching can release dioxins. Dioxins have also been found in automobile exhaust, tobacco smoke, and wood and coal smoke
Endrin	This insecticide is sprayed on the leaves of crops such as cotton and grains. It is also used to control mice, voles, and other rodents
Furans	These compounds are produced unintentionally from the same processes that release dioxins, and they are also found in commercial mixtures of PCBs
Heptachlor	Primarily employed to kill soil insects and termites, heptachlor has also been used more widely to kill cotton insects, grasshoppers, other crop pests, and malaria-carrying mosquitoes
Hexachlorobenzene (HCB)	HCB kills fungi that affect food crops. It is also released as a by-product during the manufacture of certain chemicals and as a result of the processes that give rise to dioxins and furans
Mirex	This insecticide is applied mainly to combat fire ants and other types of ants and termites. It has also been used as a fire retardant in plastics, rubber, and electrical goods
Polychlorinated Biphenyls(PCBs)	These compounds are employed in industry as heat exchange fluids, in electric transformers and capacitors, and as additives in paint, carbonless copy paper, sealants, and plastics
Toxaphene	This insecticide, also called camphechlor, is applied to cotton, cereal grains, fruits, nuts, and vegetables. It has also been used to control ticks and mites in livestock

Source: Stockholm Convention on Persistent Organic Pollutants, http://www.pops.int/documents/pops/default.htm, accessed 30 September 2009.

One way to access how toxins bio-accumulate in humans is to study breast milk. In the 1980s, researchers wanted to look at the accumulation of toxins in the breast milk of women living in cities. The researchers needed a control group and sought women who were unlikely to be exposed to toxins on a daily basis. The researchers analyzed samples of breast milk from indigenous women living in Arctic communities thinking that these women would be far removed from toxins. To their surprise, the researchers found that women in Arctic had among the highest recorded levels of POPs in the world. This seemingly unlikely concentration of toxins is explained by another feature of POPs.

Because POPs are semi-volatile compounds, they can travel from lower to higher latitudes through a series of volatilizations and condensations—volatilizing in warmer climates, condensing in colder climates, and repeating the process with the seasonal cycles. This process, known as the "grasshopper effect", tends to move atmospheric POPs pole-ward.

Entire Arctic ecosystems and food chains are contaminated with toxic POPs. One researcher commented that "The Arctic is contaminated to the degree that wildlife re-productivity and survival is compromised, and humans are consuming food so toxic that it would be prohibited from placement in modern landfill".

Selected from *"Shannon O' Lear, Environmental Politics: Scale and Power, Cambridge University Press, USA, 2010".*

Unit 33

Text: Sampling Sediment and Soil

Sampling Natural Environments

Individually, molecular and classical microbial ecology techniques are powerful tools in microbial ecology, but both are limited with respect to relating the presence and/or diversity of microorganisms to their function/activity in that environment. By combining the two approaches this obstacle can be overcome. Paramount in this respect are in situ sampling procedures that produce samples in a form representative of that environment or habitat[①].

To determine the roles played by microorganisms in a particular habitat, some form of procedure has to be undertaken to obtain representative samples upon which representative measurements can be made. There are usually three options available. First, a sample can be removed from an environment and returned to the laboratory for analysis. This approach is often synonymous with "destructive sampling," which renders the sample nonrepresentative of the environment from which it is removed (e.g., grab sample from benthic environment). The nonrepersentative nature of the sample is owing to complete or partial loss of functional integrity. This is particularly apparent when studying geochemical processes that rely on redox gradients or those directly affected by the ingress of oxygen. Consequently, any process measurements made are no longer representative of that environment. Second, a sample can be removed from the environment while attempting to maintain "*in situ*" conditions during transportation and subsequent laboratory analysis. During the laboratory analysis, the sample can be maintained as close as possible to "in situ" conditions, with one or two parameters being varied for experimental purposes. Further development of this principle led to systems that are often termed "microcosms." These can vary from the simple two-phase systems (lake water/air or soil/air) to the complex (three-phase flow-through sediment/lake water systems). Microcosms have provided model systems with which to study survival, movement, transport, gene transfer, and microbial interactions. A third option is to perform the experiments in the field, with a minimum of disturbance to the habitat. This is the least flexible of the options, and only a limited range of parameters can be measured,

e. g. , methane flux from upland soils. These limitations arise owing to logistical constraints particularly when the transportation of delicate equipment to remote locations is required.

Sampling Aquatic Environments

1. Grab Samplers

There are a considerable number of grab samplers available. All have their advantages and disadvantages and none suit all environments. For in situ sampling, all grabs are not appropriate because the sediment obtained is mixed. In general, grab samplers penetrate the sediment by approx 10cm and cover an area of 0. 1~0. 12m^2. Grab samplers include the Petersen grab, van Veen grab, Shipek grab, Ekman grab, Okean grab, and the Smith-McIntyre grab, all of which can be operated from a boat; however, they differ in size and complexity of operation, with the latter suitable for sampling continental shelf sediments. Larger amounts of sediments can be obtained using the Reineck box sampler in which the increase in sampler size is compromised by its relatively large size and cumbersome nature[2].

2. Core Samplers

The most crucial aspect of corer design is the retention of the core on removal from the sediment environment. This is particularly important for corers penetrating the sediments to depths >10~20cm and those in which access to the device occurs at the surface after retrieval. Other important features that are characteristic of all corers are the compression of the sediment and disturbance of the fine upper layers. Both are inevitable consequences of the coring operation. The simplest corer consists of a perspex tube (e. g. , 5cm diameter, 30 cm length), that can be driven into sediments by hand in shallow waters (or intertidal zones) or manipulated by divers. After removal from sediment, the ends of the tube can be sealed with bungs, preventing loss of the core material in transit. Several larger devices that are remotely operated are available for sampling at depth. The gravity corers, akin to small missiles attached to a rope, penetrate the sediment and the core is collected in the central Perspex core tube. The most notable disadvantage of this type of device is that it causes both disturbance and compression of the core material. The Emery and Dietz corer, the gravity corer with external retaining devices, and the Sholkovitz corer produce cores of increasing length in the marine environment (up to 3m). The Makereth corer retrieves long intact freshwater sediments cores. This corer permits the removal of undisturbed sediment of up to 6m in length from lake-water environments. Operation of the apparatus is pneumatic, whereby hydrostatic pressure acts on a cylindrical anchor chamber that embeds into the sediment on the lake bed. The anchor chamber holds the apparatus firmly in place while the corer tube is driven downward into the sediment by means of compressed air. Once coring is completed, the anchor chamber is automatically filled with air, the coring tube is removed from the sediment, and the whole apparatus is recovered to the surface by buoyancy lift and returned to the laboratory by boat. The core is extruded back on dry land and carried to the laboratory as four covered sections.

Frame-mounted corers minimize both disturbance and compression. The best example of this type is the "Jenkin surface-mud sampler" which retrieves 30~40cm cores with overlying water sealed at both ends. The number of manipulations that can be performed on Jenkin core tubes samples demonstrates their versatility. Extrusion of the core, sectioning followed by processing the sample is the most common manipulation performed. However, the intact core and overlying water can be set up as a microcosm, and processes measured after incubating under varying conditions or after addition of substrates, e. g., acetate. Some tubes have been adapted with a spiral of sampling ports that allow substrate additions or sample removal at a variety of depths. This versatility is shared by others in this group, which includes the Craib sampler and simultaneous multicore samplers, both of which work on the same principle.

3. Sampling Soil Environments

Despite the heterogeneity of the soil environment, there are few methods for obtaining soil cores. The basic principle is to use either a large rubber mallet to hammer the core tube into the soil or a cutting device applied with a downward pressure to ensure penetration. Core tubes can comprise a variety of lengths and diameters. Obstructions in the soil such as stones, rocks, and branches and fibrous material, in general, can prevent successful penetration and affect the integrity of the core after removal. Once removed, the core can be transported to the laboratory. Maintenance of anaerobic conditions is possible using the appropriate apparatus and will be discussed in the next section, which focuses on sampling peat soils and the maintenance of anaerobic conditions. However, the apparatus and procedure are directly applicable to sampling of soil types.

4. Peat Sampling

The fibrous nature of undisturbed surface peat prevents intact cores from being sampled without distortion of the vertical profile. The transition between aerobic conditions in the surface layers and the water-logged anaerobic horizons is characterized by steep oxygen concentration gradients[3]. Such gradients play an important role in the vertical distribution of microbial populations. To accurately determine the vertical distribution of microbial activities within the peat, any compaction of the profile or disturbance of the redox conditions should be avoided. Moreover, the exposure of some obligate anaerobic bacteria[4] to air, e. g., the methanogens, even for short periods, could affect the rate of methane production and, therefore, reduce the activity of the population relative to the undisturbed condition.

Two strategies have been used to avoid compaction and oxygen contamination when sampling peat cores. One approach is to cut the peat, using a long knife or other cutting device, to the shape and depth of the sample tube, which is then inserted into the preformed space. The initial cutting of the peat could introduce oxygen to the deep anaerobic layers, and, thus, the shape of the cut should be accurate to avoid compaction of the peat on insertion of the sample tube. In the second approach, a cutting device is attached to the base of the sample tube. The tube (and cutter) can be rotated on the surface of the peat, cutting the fibrous peat deposits while enclosing the peat core in the sample tube, thereby reducing the potential for oxygen contamination. The core sample tubes must be excavated from the peat deposits because attempts to remove these directly would result in the core sample remaining in place or, at best, breaking along its length.

Methods used to section the core, and obtain subsamples from depth, should also avoid distortion of the profile and exposure of anaerobic layers to oxygen. Many reports describe the gas flushing of incubation chambers to establish anaerobic conditions prior to the determination of methane production. This would imply that the sample material had been exposed to oxygen at some stage during the sectioning procedure. Moreover, many samples are slurried, which removes the spatial relationship among different populations of organisms⑤. These associations are known to be important for the activity of different physiological groups of bacteria and, if possible, the structure of the sample should be maintained. Methods that sample and section cores of peat and avoid exposure of the anaerobic layers to oxidizing conditions must benefit the interpretation of activity measurements.

Selected from "*Clive Edwards, Environmental monitoring of Bacteria, Humana press Inc. USA, 1999*"

Words and Expressions

microbial *a.* 微生物的，由细菌引起的
grab sample 攫取样品
integrity *n.* 完整，完整性，完全
microcosms *n.* 微观世界
gene transfer 基因转移
corer *n.* (苹果，梨等) 去心器，挖核器
crucial *a.* 至关紧要的
penetrate *v.* 穿透，渗透，看穿，洞察；刺入，看穿，渗透，弥漫
perspex *n.* 塑胶玻璃，透明塑胶；[化] 聚合的 2-甲基丙烯酸甲酯
bung *v.* 堵；*n.* 桶等的塞子，桶孔
hydrostatic *a.* 静水力学的，流体静力学的
extrude *v.* 挤压出，挤压成，凸出，伸出，逐出
incubate *v.* 孵卵
heterogeneity *n.* 异种，异质，不同成分
fibrus *a.* 含纤维的，纤维性的
anaerobic *a.* [微] 没有空气而能生活的，厌氧性的
intact *a.* 完整无缺的
obligate *v.* 使担义务；*a.* 有责任的
methanogen *n.* [微] 产甲烷生物
excavate *v.* 挖掘，开凿，挖出，挖空
incubation *n.* 孵卵，抱蛋；[医学] 潜伏期

slurried *a.* 泥浆的；浆化的

physiological *a.* 生理学的，生理学上的

Notes

① Paramount in this respect are in situ sampling procedures that produce samples in a form representative of that environment or habitat. 可译为：在此方面，那些能够体现栖息地或环境得到标本方式的现场采样程序才是极其重要的。

② Larger amounts of sediments can be obtained using the Reineck box sampler in which the increase in sampler size is compromised by its relatively large size and cumbersome nature. 可译为：使用雷恩奈克盒式取样器可以获得大量的沉积物，而它本身尺寸比较大以及笨重的特点不利于取样器的尺寸增大。

③ The transition between aerobic conditions in the surface layers and the water-logged anaerobic horizons is characterized by steep oxygen concentration gradients. 可译为：表层好氧环境过渡到充满水的厌氧层的重要特征就是氧浓度的急剧变化。

④ obligate anaerobic bacteria 专性厌氧菌，只能在专一厌氧环境下活动的细菌。

⑤ Moreover, many samples are slurried, which removes the spatial relationship among different populations of organisms. 可译为：更有甚者，许多标本是泥浆状的，这会破坏不同微生物种群间的空间生存关系。

Exercises

1. Put the following into Chinese.

   The most crucial aspect of corer design is the retention of the core on removal from the sediment environment. This is particularly important for corers penetrating the sediments to depths $>10 \sim 20$cm and those in which access to the device occurs at the surface after retrieval. Other important features that are characteristic of all corers are the compression of the sediment and disturbance of the fine upper layers. Both are inevitable consequences of the coring operation. The simplest corer consists of a perspex tube (e.g., 5cm diameter, 30cm length), that can be driven into sediments by hand in shallow waters (or intertidal zones) or manipulated by divers. After removal from sediment, the ends of the tube can be sealed with bungs, preventing loss of the core material in transit. Several larger devices that are remotely operated are available for sampling at depth. The gravity corers, akin to small missiles attached to a rope, penetrate the sediment and the core is collected in the central Perspex core tube. The most notable disadvantage of this type of device is that it causes both disturbance and compression of the core material.

2. Put the following into English.

   微生物生态学　　大陆架沉积物　　核芯采样　　厌氧/好氧条件
   垂直剖面　　　　氧气污染　　　　培养室

3. Answer the following questions.

(1) Why is the sample nonrepresentative of the environment from which it is removed?

(2) Why are all grabs not appropriate for in situ sampling?

(3) What are the notable disadvantages of core samplers?

(4) Which approach is used to avoid compaction and oxygen contamination when sampling peat cores?

Reading Material: EMAP Overview—Objectives, Approaches, and Achievements

The Environmental Protection Agency's Office of Research and Development (ORD) has developed an Ecosystem Protection Strategy that is based on a risk assessment Paradigm. This multi-disciplinary approach focuses on four major research areas: monitoring, process modeling, regional assessments methods, and restoration. One key to the successful operation of this strategy is accurate, representative monitoring information. Monitoring provides the critical baseline information that develops the problem formulation for these other research areas. Monitoring through time can be used to detect changes, and eventually trends, in the condition of the environmental resources in relation to agency policies. The Environmental Monitoring and Assessment Program (EMAP) is the primary ORD program devoted to advancing the science of statistically based ecosystem monitoring for establishing baselines and trends in the condition of regional and national aquatic resources. This provides EPA with critical, measurable performance standards for its programs and policies in reaching its environmental goals at the state, regional and national level.

The key research areas at the heart of EMAP have been indicators and multiple-scale, statistically-based monitoring designs. These indicators measure some aspect of the biological condition of a characteristic component of the resource. In some cases, it might be physiologic constraints, such as dissolved oxygen or temperature, that can limit biological occurrence. In other cases, a more complex suite of individual biological indicators may be developed and combined into an index of biotic integrity that can represent the biological condition of an entire component of a resource. Indicators must be valid and interpretable and should apply at different levels within a resource (e.g., from 1st order streams to 4 + order streams) and across ecoregions. Because EMAP data may be used in formulating policy and management decisions, it is important that indicators used are thoroughly evaluated. All indicators proposed for use in EMAP must meet 15 guidelines. In general, these guidelines are evaluated in four phases: conceptual foundation, feasibility of implementation, response variability, and interpretation and utility. These phases follow a logical progression for indicator development, but in practice they may be iterative and not necessarily sequential.

The EMAP designs provide sampling templates that provide unbiased estimates of resource condition with quantifiable confidence limits. EMAP's statistically random sampling design was developed to allow representative characterization of a resource's condition over large geographical areas and to provide the comparability necessary to detect trends over time. The design is flexible enough to be intensified to adequately characterize subpopulations spanning a wide range of densities and distributions (e. g., oligotrophic lakes rather than all lakes). Because of EMAP's probability-based design, inference can be made from the sampling results to the entire population from which the sample was taken. Thus, rather than censussing the population, EMAP assessments can be made with relatively small sample sizes, leading to faster, more responsive determinations of environmental condition.

EMAP researchers have been leaders in the development and use of landscape indicators and landscape ecology to produce broad-scale assessments of the condition of the environment. This approach accounts for the spatial arrangements of different types of land use (e. g., forest, agriculture, urban) and for the changes in the relationships between ecological patterns and processes with changes in the scale of observation. In this way, humans and their activities can be incorporated as integral parts of the environment. Our landscape approach links land use and its changes (e. g., forest fragmentation, riparian buffer zone width) at multiple scales to the water quality and biological integrity of streams and estuaries. Water links numerous resources in a watershed, such that major changes to the watershed, either natural or human-induced, will likely result in a change in water quality. By developing new landscape indicators, we hope to use remote sensing imagery to broadly assess conditions relevant to water quality.

EMAP researchers have used geographic studies as a scientific testing and development area for indicators, design and remotely sensed data collection. In recent years the major geographic study area has been the mid-Atlantic region of the United States. This study, known as the Mid-Atlantic Integrated Assessment (MAIA), was EMAP's first region-wide determination of environmental condition. The development and implementation of MAIA required the partnering of EPA's ORD, other federal agencies, the states and tribes, nongovernmental organizations (NGOs), and the public stakeholders in the area. MAIA has been a successful "proof-of-concept" for large-scale monitoring. We have been able to demonstrate the utility of a number of indicators at different scales, the feasibility of probabilistic sampling over several different ecoregions, and a scientific basis for selecting reference sites for use in determining the condition of a resource. Because of the success of our monitoring effort, we are now shifting our emphasis in MAIA to assessment.

The MAIA assessment phase includes interpreting and evaluating EMAP monitoring results for multiple ecological resources for use in policy decisions. This includes determining the fraction of each resource population in "good" or "impaired" condition, and the stressors associated with impaired conditions. Results from streams in the mid-Atlantic highlands area of MAIA have already provided insights into the major stressors in these systems. These results suggest that point source chemical contamination may still be an important local stressor, but it is not the most widely distributed stressor. The most widely distributed stressors appear to be due to non-point source pollution, typically due to human alteration of the landscape. To reduce these widespread land use impacts on ecosystems will require dealing with human socio-economic behaviors in relation to land use; this is not typically included in current ecological management practices.

EMAP's future monitoring research will be focused primarily in the western United States. This Western EMAP geographic pilot study (known as the Western Pilot), encompasses ecosystems which are very different from MAIA. The Western Pilot includes 12 contiguous western states in EPA's Regions, an area roughly seven times the size of the MAIA region and including approximately 18 ecoregions. This region of the country was selected for a pilot study because it was the most geographically and ecologically distinct from MAIA. How variable and/or applicable the MAIA-tested sampling designs and indicators will be in the West is unknown. To aid in testing the EMAP design and developing indicators for the West, we will initiate a stream survey to examine the health of streams, including ephemeral streams. Because estuaries are important ecological resources and are the final integrators of upstream water quality, we will begin assessing the condition of western coastal estuaries. As part of the overall evaluation of the West, EMAP researchers will develop a landscape atlas that documents the landscape characteristics and the land-use patterns for this region.

As suggested above, new landscape and aquatic health indicators will likely be necessary to establish the condition of the western ecosystems. Thus, EMAP researchers will continue their development of indicators for measurement of aquatic health, and subsequent assessment of federal and state policies and practices on maintaining or improving the condition of the environment. The next generation of monitoring indicators will be needed as we progress from cataloging problems to maintaining and restoring ecosystems. Possibly, new EMAP indicators will address watershed sustainability given current practices, but they will have to include a better understanding and incorporation of human/ecosystem linkages.

EMAP researchers will also begin a nationwide study of the health of coastal estuaries. As part of the proposed Coastal 2000 Monitoring Initiative (beginning in the summer of 2000), EMAP and its federal and state partners are planning the first nationwide measurement of estuarine condition. The work on West Coast estuaries will begin through the Western Pilot and will be linked with the EMAP design and sampling of the Gulf and Atlantic Coast estuaries. This will provide a snapshot of the ecological health of U.S. estuaries and allow for future assessments of trends in their condition.

Regional applications of EMAP allow testing of EMAP indicators and designs in different EPA regions, and serve to transfer our technology to the states in these regions. Because measures of biological quality are central to implementing the Clean Water Act (to restore and maintain the chemical, physical, and biological integrity of the Nation's waters) and States are required to report on the conditions of their waters, our R-EMAP studies will focus on developing biological criteria for aquatic Systems. If quantifiable, scientifically defensible biological criteria can be determined for the states and regions, then the amount of the aquatic resource which meets or exceeds these conditions can be determined. Similarly, the amount of the resource which does not meet these conditions, and is in need of restoration, can also be determined.

EMAP continues to tackle problems relevant to EPA and the nation. Our researchers are developing solutions to the problems associated with measuring the health of aquatic ecosystems. The science done within EMAP has been carefully scrutinized. The program has undergone extensive scientific peer-review and its scientists have produced over 1000 peer-reviewed journal and EPA publications. But the real proof of EMAP's success is in the increasing acceptance and use of the EMAP approach. Fifteen states are currently using the EMAP design in at least one component of their monitoring programs, and 22 others are considering or have requested our design support. Lastly, EMAP has been able to partner with EPA's Regional and Program Offices, other federal agencies, the states, tribes, NGOs, and the public stakeholders. These partnerships have helped make EMAP responsive and relevant at a number of levels. The systematic monitoring of natural resources on a national basis is more than any one program can accomplish. To be successful, it is crucial that monitoring efforts be coordinated and the environmental data shared. In this way, true partnerships will emerge that range from the individual at the local level to the policy-makers at the national level, and will lead to a general improvement in the condition of the nation's aquatic resources for future generations.

Selected from "*Shabeg S Sandhu, Brian D Melzian, Edward T Long, et al, Monitoring Ecological Condition in the Western Unite States, Kluwer Academic Publishers, USA, 2000*"

Unit 34

Text: Pollution Control Strategies [I]

Improvement in the quality of our environment will come about[①] only if action is taken by someone to correct an existing pollution problem or to prevent a new problem from occurring. Knowledge of the cause of the pollution problem and the means to correct it are, of course, necessary for a successful resolution, but unless the polluter takes action, voluntarily or through enforcement, nothing will change. There are many examples of pollution problems where the cause and the remedy are well known, but no action has been taken. Why is this so and what can be done about it? Who should do something about it? These questions, as well as the economic and legal aspects involved in dealing with them, are considered next.

Economic Aspects

Homeowners wishing to improve the physical environment of their surroundings can do so by purchasing goods (trees, shrubs, furniture, roofing material, etc.) and labor (a gardener, carpenter, plumber, etc.) to accomplish this. The person who owns the land and building provides or borrows the money to make the improvements, enjoys the benefits of the improvements, and suffers the consequences of not making the improvements (e.g., water damage caused by a leaking roof). In extreme cases of neglect of the home or the yard, the local municipality may take action under a section of the Public Health Act or Weed Control Bylaw[②]. Otherwise, there is no need for governmental interference. But what about the situation where the homeowner is concerned about the quality of the water in the creek at the rear of the property, or the quality of the surrounding air? What actions can be undertaken to correct these problems? The owner is powerless to do anything independently and will most likely complain to the local government, which may or may not be in a position to take action. The concept of ownership of resources is important in such a situation: the air and water are not owned by any individual, but are public goods[③], as compared to the lot, house, furniture, etc., which are private goods.

The economic system of capitalist countries, and to a lesser extent that of state-controlled socialist countries, has been remarkably efficient in producing consumer goods of all kinds, priced so that there is a balance between supply and demand. Neither of these economic systems has been as successful in providing public goods, such as clean air and water. Since no individual or identifiable group of individuals owns these resources, no one has taken the responsibility to maintain their quality. An individual or an industry which pollutes the air or water reduces the quality of that resource for everyone in the area. Everyone suffers the consequence for the activities or negligence of the polluter, although not always equally.

The costs of pollution have been generally external to the polluter. That is, the costs do not appear on the polluter's balance sheet[④], nor do they make up a part of the cost of production. This is true for a steel mill discharging untreated wastes or for a town discharging untreated sewage. These external costs are paid by society in two ways. One of these is the cost of pollution damage: the reduction in the value of available resources. For example, the value of a public beach is reduced if it is too polluted for swimming. And a fishing lodge operator[⑤] loses business if the fish in the nearby stream or lake are killed by pollution.

The other component of external cost[⑥] is the expense of repairing or reducing the damage caused by pollution. The cost of building and operating a municipal water treatment plant[⑦] is an example: the water which has been "damaged" by pollution must be restored in order for it to be fit to drink.

Because the rights to public goods are not owned by anyone, they cannot be bought and sold, and therefore there is no marketplace in which values can be placed on them. Sometimes compensation for these external costs can be determined relatively easily. For example, suppose that the fish in a certain pond had been grown by a fishing lodge operator. If the effluent from an upstream pulp mill were to kill the fish, the lodge owner might have the chance to collect the value of the lost fishery operation from the pulp mill—often, however, not without a court case. To avoid this heavy toll, the pulp mill, by applying normal business management principles, would determine how much it needed to spend on pollution abatement to avoid killing the fish and paying the consequent damages. In this simple example, the external costs of pollution would be transferred to the polluter. Of course, in the long run, these costs would be passed on to the consumers of pulp and paper and, therefore, to society as a whole.

It is much more difficult to establish a value for the loss of enjoyment of the recreational and aesthetic pleasures of a river or lake by a group of cottagers once the river or lake is polluted by the same pulp mill. It is also more difficult for the cottagers, as individuals or as a group, to initiate a possibly expensive court action against the pulp mill. Accordingly, we look to our elected representatives in government, at the local, state, or national level, for protection of our public resources—for clean air and clean water. Governments are more likely to respond if they perceive that a cause is very important to the electorate. Many citizens' groups concerned about environmental protection have therefore made it their business to use every available means to pressure governments to enact legislation and to introduce incentives which will make the polluter implement pollution control measures. In one way or another, these methods "internalize" the external costs of pollution as much as possible. When it is not possible to achieve this, such external costs will have to be borne by society as a whole, as a social cost of pollution, just as we share other social costs, such as those created by unemployment, welfare, and occupational health requirements.

Selected from "*J Glynn Henry, Gary W. Heinke, Environmental Sci. and Eng., Prentiu-Hall Inter. Edition, USA, 1989*"

Words and Expressions

municipality　*n.* 市政当局；自治市
lodge　*n./v.* 容纳，寄存，存放，(向有关当局) 提出 (声明)
pulp and paper　纸浆
aesthetic　*a.* 美学的，审美的，有审美感的
cottager　*n.* 住在农舍 (别墅) 者，佃农
electorate　*n.* 选民，选区，有选举权者
enact　*v.* 制定法律，颁布，扮演

Notes

① come about　出现，到来
② Weed Control Bylaw　草科植物监控法
③ public goods　公共物品
④ balance sheet　资产负债表
⑤ fishing lodge operator　钓鱼场经营者
⑥ external cost　外部成本
⑦ municipal water treatment plant　市政水处理厂

Exercises

1. Translate the last paragraph into Chinese.
2. Answer the following questions.
   (1) In which ways are the external costs paid by society?

(2) Why was it said that in the long run, these costs would be passed on to the consumers of pulp and paper and, therefore, to society as a whole?

(3) What is their business when many citizens' groups are concerned about environmental protection?

Reading Material: Pollution Control Strategies [Ⅱ]

Legal Aspects and the Role of Government

Legislation to control the quality of water and air and the disposal of solid and hazardous wastes was introduced, with particular emphasis on the United States and Canada. It is the responsibility of government, at the national, state, provincial, or local level, to enact and update environmental control legislation. This legislation is generally written to provide the broad goals and objectives for environmental quality. It does not provide the means and methods by which these goals are to be achieved. Nor need it provide the details which are necessary to monitor and control the performance of pollution control facilities. It is, therefore, necessary for governments at all levels to establish regulatory strategies, in order to implement the broadly stated objectives of general legislation.

The goal of environmental management strategies is to maintain or improve the quality of the ambient or surrounding environment. Ambient standards are determined for a number of different characteristics or pollutants within a medium such as air or water. These standards are designed to minimize risks to the health of humans, animals, or the environment in general. The components for which these ambient standards are set must be quantifiable and scientifically measurable. In water and air, criteria are set for allowable concentrations of a variety of pollutants. Furthermore, the pollutants for which ambient standards are set must be related to their sources. A regulatory agency can set ambient standards and monitor ambient conditions, but it cannot control or manage conditions except by controlling the sources of the pollutants which affect the ambient conditions. For example, in the air, it is desirable to maintain the concentration of particulates below a certain level. To do this, we must determine the possible sources of the particulates. Some of these sources may be identifiable, such as a smokestack or a burning garbage dump. But much of the particulate matter may come from unidentifiable or nonpoint sources, such as open fields, highways, or a forest fire many miles away. After the sources have been identified, it is necessary to relate the rate at which the pollutants are being released from the sources to the ambient concentrations. When this is done, it is possible to set allowable limits on the discharge of pollutants at the sources. This forms the basis for effluent standards.

Effluent standards are more useful for environmental management than ambient standards because they can be monitored and controlled in many cases. Even though the ambient quality is what we are interested in preserving, we normally try to achieve this by controlling effluent quality and quantity.

Three main instruments are available to government for environmental control: direct regulation, polluter subsidies, and polluter charges. They are all a means of controlling effluents or discharges of pollutants, and they all work to internalize pollution costs to the polluter. They can be applied independently, but are usually applied in combination. Each of these instruments appears in a variety of forms. We shall consider some of the more common forms in which they are applied, as well as other interesting possibilities for controlling pollution.

Direct regulation. The government can use its legislative powers to regulate the actions of individuals, corporations, and lower levels of government. Therefore, through direct legislative action, the quantity, quality, and location of discharges of pollutants can be regulated. The main forms of direct regulation are zoning; prohibition, or zero discharge; and effluent standards.

Zoning. Zoning regulations are one of the simplest and oldest forms of pollution control and are still a part of almost every pollution control strategy. The objective is to separate the polluter from the rest of society by either space or time. A result of the so-called sanitary awakening in mid-nineteenth century Britain was the realization that open garbage dumps had to be removed from areas of dense population and kept away from public water supplies. Local bylaws were enacted to ensure that this was done so that the benefits to public health were realized. The prohibition on the burning of coal in nineteenth-century London while parliament was in session is another example of this type of zoning. More recent examples of zoning to separate polluters from the public are the location of airports, the use of curfews on airport operations, and the construction of tall chimneys or long marine sewage outfalls.

Prohibition or Zero Discharge. Another form of direct regulation of pollution is prohibition, also known as zero discharge. The advantages of such a concept are obvious. First and foremost, there would be no change in environmental quality. Moreover, all resources would have to be completely converted into useful products or stored indefinitely. And the legislation would appear to be equitable, since the same regulation would apply to everyone. Such a concept, however, is normally impossible to realize. A simple materials balance shows that any resource taken from the environment, including energy, must be returned in some form. Even if it were conceivable to recycle all wastes into new products, there would still be a large energy requirement to achieve this. For most activities, zero discharge would be expensive if not impossible to achieve. At present, producers of extremely hazardous wastes, for which no treatment is available, are the only ones subjected to zero discharge requirements. They must store their wastes until a means of safe disposal is found.

Effluent Standards. Effluent discharge standards are the most common and the most useful form of direct regulation. They can be in the form of across-the-board standards which require that effluents of all polluters meet the same criteria, or they may be individually developed for each polluter. The advantages of an across-the-board type of approach are that it is easy to administer, it appears fair to all polluters, and it provides the most rigid control over environmental quality. The disadvantages are that it may be uneconomical, and therefore impractical, to insist that all polluters meet the same effluent standards. Some polluters may easily meet standards that others will be unable to meet at all, or only at a very high cost. The different assimilative capacities of the environment in different locations can be taken into account only on a case-by-case basis. For example, a large, fast-moving river can accept a much larger amount of organic pollution than a small creek, and therefore pollutant concentrations from point-source discharges could be much higher before river quality is seriously affected. Nevertheless, most jurisdictions prefer to set common effluent discharge guidelines, which must be met unless the contributor is specifically exempted.

Subsidies. One method of encouraging polluters to comply with regulations is to provide money to help cover their costs. These subsidies may be in the form of direct payments or grants based on a percentage of the cost of pollution abatement or on a percentage reduction in effluent quantity or strength. They may also take the form of low-interest loans for the capital costs of improved treatment facilities. Alternatively, governments can reduce or defer taxes or relax other government requirements to encourage spending on pollution control.

The main advantage of subsidies is that they reduce the costs of pollution abatement to the polluter and limit the associated increase in production costs. Government grants can be used to cover capital costs, and tax incentives can be used to relieve operation and maintenance costs. Subsidies (the carrot) combined with regulations (the stick) can be used by government to reduce stress on the environment and at the same time encourage research and development by industry in pollution abatement technology. The main disadvantage of polluter subsidies is that the government will have to increase taxes or direct money from other programs in order to pay the subsidies. This is partially offset by decreased expenditures needed to correct the effects of damage due to pollution (i.e., expenditures on water treatment plants or public health care). However, these returns may be small compared to the costs involved. A general tax increase may seem fair when everyone benefits from an increase in environmental quality. In fact, however, people benefit to varying degrees, and some may balk at paying money for what appears to be someone else's problem.

Another serious drawback to the subsidy system is that it can be easily abused. The idea of paying someone to stop damaging the environment sounds suspiciously like a criminal protection racket. All potential polluters will want to be paid for not polluting. Companies may find that the subsidy available for waste reduction exceeds their actual costs of making the change. They may then increase their production above normal simply to receive a subsidy and go on to dump the extra goods at a lower price. In this situation, a polluting industry has been rewarded while its competitors who already treat their wastes adequately get no benefit.

Service charges. Service or user charges are similar to subsidies in that monetary means are used to encourage a polluter to comply with effluent requirements. Charges are the most direct way of internalizing the costs of pollution to a polluter. There are numerous types of service charges, but in general, money is paid to the local government or agency in proportion to the amount of pollution. The government or agency may then use the money to pay for and operate central pollution control facilities.

The obvious advantage of a service charge is that it is the polluter who pays for the costs of polluting. The system rewards those industries that are clean and efficiently run and penalizes those that are dirty and wasteful. Also, it does not encourage increases in polluting activity, as a subsidy system might. Finally, the administration of such a system is relatively easy, requiring only the monitoring of discharges.

The disadvantages are that production and operating costs for the industries connected may rise. If the service charges are nominal, industries may find it less expensive to simply continue polluting. If the charges are high enough to force an industry to stop or severely restrict its effluent discharges, the industry may close down. In any event, the charges will be passed on in the form of increased prices for the industry's products. Since each industry has different capabilities and costs related to controlling its wastes, a uniform service charge could upset the economic balance between competing industries. However, to customize effluent charges for each polluter would be an administrative burden and appear to be unfair.

We are all familiar with charges for municipal services. In urban areas, we pay through property taxes or special levies to have refuse and sewage removed from our homes. In the same way, industries may find it more convenient to pay to have their untreated wastes removed and disposed of at a central treatment facility. In some of the heavily industrialized areas of Europe, this has been found to be an attractive and efficient way to dispose of industrial wastes. In many cases, the extra cost of waste collection is offset by the economy of scale of large, specialized treatment plants.

In general, all wastes which do not harm the system or affect the operation of the treatment plant should be accepted without pretreatment. If the wastes are stronger than "normal" sewage, then a charge, or more correctly, a surcharge, should be assessed against the industry for the extra cost of sewage treatment. For this approach, a surcharge formula setting out the charges for accepting wastes stronger than normal would have to be included in the industrial waste bylaw. Ideally, charges for sewage treatment should be related to the cost of providing the facilities and the benefits received. The practical application of this method is difficult, however, and various methods of charging for industrial wastes have evolved.

Selected from "*J Glynn Henry, Gary W. Heinke, Environmental Sci. and Eng. , Prentiu-Hall Inter. Edition, USA, 1989*"

Unit 35

Text: Food Security

Introduction

When we think of security, we might think first of international relations and conflict, military concerns, or systems that monitor borders, personal movements, and computer networks. We might not initially think of farming, food processing, or access to food. This section considers food security at multiple spatial scales and ways in which food supply is connected to power. Three key themes are relevant to a consideration of food security: stability, access, and quality. Beginning at the global scale, we will examine climate change and implications for the world food supply. Yet food is not evenly distributed around the globe, so we must then consider issues of access to food. We will look at how food became an international policy issue, how the definition of food security has changed over time and how it has also been challenged. Examining issues pertaining to access to food draws our attention to the question "Who controls the food supply?" and leads us to investigate aspects of food quality and content. Asking the question "What are we eating?" brings the discussion full circle back to the overarching themes of the book, namely, power and spatial scale.

Climate Change and Food Supply

If the earth's atmosphere, precipitation, and temperature patterns are changing, what kinds of effects might there be global food supply? Projections of climate imply impacts on soil quality, water resources, temperature patterns, duration of growing seasons, and net primary productivity in different places. Changing concentrations of soil carbon and atmospheric CO_2 could influence agricultural productivity in specific places while the global demand for food increases or shifts. Carbon trading schemes [1] could further influence agricultural practices and patterns of food production, for example, if arable land is shifted from supporting food production to supporting carbon sink [2] forests. These physical and social processes are likely to result in new patterns of food production and concerns for food security. Climate change will affect agriculture positively in some places and negatively in others depending in par on latitude. Climate change can exacerbate processes that degrade soil quality such as hydrolysis (the leaching of silica), cheluviation (the removal of aluminum and iron), ferrolysis (oxidation of clay), dissolution of minerals by increased acids, and reverse weathering altering clay formation. The decline in soil quality resulting from such processes is likely to be steeper in the tropics where soils tend to have low productivity and do not respond well to inputs such as fertilizers.

Food Security: Definition and Policy Approaches

The definition of food security has shifted from availability of food to food individual access. Recent work on food security reflects this trend and has considered, for example, gender dimensions of food security such as ways that societal differences between men and women contribute to different levels of access to food. Research has considered gender inequities in different societies, such as women's limited entitlement, education, and expenditure as well as their role as care givers, as factors of food security. Women's status influences children's food security. Other work has considered urban-rural differences in individuals' access to food. For example, supermarkets are the retail end of the food commodity chain. They influence the food security of both rural food producers who supply urban markets and urban consumers who rely on supermarkets for their food supply. Just looking at these few examples, we began to appreciate that food security involves power: which groups in society have control over their own food supply or that of other people? Who is in a position to make decisions about who gets to eat and what they are eating?

Improving food security would not seem to be an insurmountable challenge compared to other public health issues, and a first step would be to identify who is most food insecure.

Who is vulnerable to food insecurity? Poor people are the most vulnerable. In an early contribution to the literature on food security, some scientists (Amartya Sen) examined how poverty and famine are linked. They discussed poverty as a complex category that requires us to think about how we identify "the poor" (Is poverty absolute? Is poverty relative?) and the fact that there can be different groups of poor in a given society. They also differentiated between starvation and famine. In their work they describe starvation as an ongoing condition of people living on inadequate food. They argue that "Famines imply starvation", but not vice versa. And starvation implies poverty, but not vice versa.

As we consider the issue of food security, it is useful to examine the idea of vulnerability more closely. We can think of vulnerability as a continuum of uncertainty. Since people move in and out of poverty, especially in conflict-prone areas, it is important to assess vulnerability over time rather than at one point in time. This kind of approach exposes a longer term dimension of food insecurity than does a focus on poverty alone. Existing food insecurity requires ex post efforts to deal with an already existing situation. Examining vulnerability, on the other hand, is a forward-looking approach that encourages us to consider who faces the most uncertainty or who is most likely to fall and remain below a threshold of food security. A focus on vulnerability motivates ex ante strategies to identify who is likely to be food insecurity in the future, why they are likely to be food insecure, and what policy options or mitigation efforts could address these risks before they are manifested.

Selected from "*Shannon O'Lear, Environmental Politics: Scale and Power, Cambridge University Press, USA, 2010*".

Words and Expressions

overarching   *a.* 首要的
exacerbate   *v.* 激怒，使恶化
cheluviation   *n.* 螯合淋溶作用
ferrolysis   *n.* 铁解作用
care giver   关怀者，照顾者，护理者
not vice versa   反过来就不一样
ex ante   事前

Notes

① Carbon trading schemes   碳交易（贸易）计划。是指《京东议定书》的温室气体 CO_2 排

放权交易。

② carbon sink 碳汇，是指森林树木吸收并储存 CO_2 多少的能力。可通过植树造林，森林管理，植被恢复等措施，利用植物光合作用吸收大气中的 CO_2，并将其固定在植被和土壤中，从而减少温室气体在大气中的浓度过程和活动或机制。与其相反的词是 carbon source，是指自然界向大气释放 CO_2 的母体。

Exercises

1. Put the following into Chinese.

（1）This section considers food security at multiple spatial scales and ways in which food supply is connected to power.

（2）Asking the question "What are we eating?" brings the discussion full circle back to the overarching themes of the book, namely, power and spatial scale.

（3）Climate change can exacerbate processes that degrade soil quality such as hydrolysis (the leaching of silica), cheluviation (the removal of aluminum and iron), ferrolysis (oxidation of clay), dissolution of minerals by increased acids, and reverse weathering altering clay formation.

（4）They argue that "Famines imply starvation", but not vice versa. And starvation implies poverty, but not vice versa.

（5）A focus on vulnerability motivates ex ante strategies to identify who is likely to be food insecurity in the future, why they are likely to be food insecure, and what policy options or mitigation efforts could address these risks before they are manifested.

2. Put the following into English.

反之就不一样　事前　事后　碳汇　使恶化　铁解作用

Reading Material: Challenging Food Security

How we define food security will shape how we go about identifying solutions and policy approaches. Two critiques of concept of food security suggest that food security, as generally understood, is not the best guide for policy. Food sovereignty and food justice perspectives argue that the dominant definition of food security does not lead to useful discussions or to sound policy outcomes. In the policy world, food security is about maximizing supply or availability. It is also about increasing access to food at multiple scales beyond the state to include the region, the community, the household, and the individual. Critiques offered by food sovereignty and food justice perspectives, however, point to the limited usefulness of a food security approach: "Food security's focus on accessibility and hunger is further problematized by viewing individuals, or even households, as mere containers for calories". A focus on availability of and access to food miss other, important dimensions of human well-being. We have already seen this idea in the promotion of good nutrition, not just adequate calories. Food justice perspectives go further and view food not as a commodity but as a human right. Food, like other human rights, is often at the center of political struggle. Fundamental ideas of a food justice movement include:

• The rights of consumers must be fought over, maintained or expanded, not merely assumed.

• Human health and environmental health are tightly inter-wined.

• There is no such thing as an average consumer.

• It is important to look beyond what is eaten to consider how it is produced and distributed.

• It is worthwhile to work to change policy, but policy is often influenced through creative activism outside of predetermined political channels.

Food justice movements aim to create political spaces or networks through which people can be supported in their resistance of corporatized agriculture and globalization. Food justice movements seek to reconnect people with their food supply.

In Toronto, for example, activists have created a new political space—beyond and in between existing venues for input into the policy process to address concerns about community food security. This food justice movement involves networks of various groups and individuals rather than being formally institutionalized within the established political infrastructure. The movement engages citizens in a "bottom up", citizen-initiated planning process to address local hunger, poverty, and unban agriculture. Different elements and stages of Toronto's food justice movement have focused on emergency food provision, community gardens, and local economic development paying attention to ways in which these features are influenced by policy decisions and are linked to other special scales:

Working together, Food Share and the Toronto Food Policy Council, and the coalitions and networks that have formed in Toronto, have created a new political space that operates at a multiplicity of scales—the global, national, and regional scales of networks and flows of information that incorporate the Global Food Summit in Rome and community gardens and back ovens at the neighborhood level.

Food justice movements are a type of social movement aiming to generate social change and support everyday acts of resistance in relation to food supply: what and how much food is available, who controls the supply, and who benefits from it? Even if food justice movements do not alleviate all food insecurity, they can foster the development of skills and experience in participatory citizenship and strengthen bonds of local community. As noted earlier, Michael Mann (a scientist) has observed that a significant obstacle to public dissent tends to be the inability of the public to organize collectively. Yet food justice movements, like other types of social movements, encourage participatory citizenship and provide groups and individuals with experience in collective organization and critical thinking about political processes. These movements can help public groups to build their power and capacity to bring about change. Although food justice movements may be primarily focused on the community or local spatial scale, they often "jump scales" to connect their efforts with processes at other scales such as regional distribution systems, federal farm policies, local, small farms, individual consumption choices, etc. Groups and organizations may be empowered by their ability "to jump scale to organize themselves at the scale with the most promising opportunities to achieve their goals". When food justice movements are able to link their efforts to different spatial scales, they can enhance their own knowledge base and their capacity to bring about change in the system.

Food justice movements may be described as "bottom up" projects since they tend to focus on empowering local food security initiatives by connecting them to other spatial scales. Food sovereignty perspectives may be described as "top to bottom" views since they tend to look at ways in which larger scale international connections have impacts on food security in specific places. Food sovereignty, which also critiques the focus on food security, sees the push for increasing the global food supply as inflicting negative consequences on less economically developed countries. For example, when Haiti obliged World trade Organization regulations and allowed the import of rice grown in the U.S., the result was "food dumping": US-grown rice dominated the domestic market and made it impossible for Haitian rice farmers to compete. Local rice farms were abandoned, and the Haitian economy suffered downward spiral effects. Rather than enhancing Haiti's food self-sufficiency and ability to meet local needs. World Trade Organization regulations, like international trade agreements often do, strengthen global markets while causing detrimental, local effects in particular places. A food sovereignty perspective views this process as backwards and would instead encourage us to consider how well poorer countries can feed themselves, to what extent they have control over their food supply, and how that food supply might serve to enhance their domestic economy and culture.

In other words, food justice and food sovereignty expand our understanding of food security by bring our attention to different spatial scales. More than the availability or quantity of food, these views emphasize access to food at household, community, and the state level. These views encourage us to think about the influence of power at different spatial scales and ways in which policy decisions in one place or at one scale can affect other places or create other spatial scales of connection and activity.

Consider this observation on policy and food insecurity:
The main reason why there are more food insecure people today than…20 years ago is not that the required action is unknown or that the world does not possess the resources to assure food security for all, but rather that food security for all is not an important priority among most policy-markets.

If food security is not a top priority concern, does that mean that a society's political leaders and institutions have failed? Or is the lack of attention to food security a temporary condition that will be overcome as economic development proceeds? The next section considers this question about the relationship between a society's wealth and its food security.

Selected from "*Shannon O'Lear, Environmental Politics: Scale and Power, Cambridge University Press, USA, 2010*".

PART 7 THE BIOSPHERE: ECOSYSTEMS AND BIOLOGICAL COMMUNITIES

Unit 36

Text: Life and the Biosphere

The water-rich boundary region at the interface of Earth' surface with the atmosphere, a paper-thin skin compared to the dimensions of Earth or its atmosphere, is the biosphere where life exists. The biosphere includes soil on which plants grow, a small bit of the atmosphere into which trees extend and in which birds fly, the oceans, and various other bodies of water. Although the numbers and kinds of organisms decrease very rapidly with distance above Earth' surface, the atmosphere as a whole, extending many kilometers upward, is essential for life as a source of oxygen, medium for water transport, banket to retain heat by absorbing outgoing infrared radiation, and protective filter for high-energy ultraviolet radiation. Indeed, were it not for the ultraviolet-absorbing layer of ozone in the stratosphere, life on Earth could not exist in its present form.

This part deals with life on Earth. It considers the highly varied locations where life exists and the vastly different conditions of moisture, temperature, sunlight, nutrients, and other factors to which various life-forms adapt. Such conditions may be those of the tropics, with abundant moisture, intense sunlight, high temperatures, and relatively little variations in these and other factors. Or they may be characteristics of inland deserts that are hot during the daytime and cold at night, generally very dry, but subject to occasional torrential rainstorms and flash flooding. Life thrives on land surfaces, in bodies of water, and in sediments in water. The extreme variability of environments in which life exists is matched by the remarkable variety, versatility, and adaptability of the communities of organisms that populate these environments. These range from tropical rain forest communities containing thousands of plant, animal, and microbial species in a small area to austere, exposed mountain rocks subjected to extremes of weather and populated by a thin coating of tenacious lichen, a symbiotic combination of fungi and algae that clings as a thin layer to the rock surface. In addition to dealing with organisms and their environment, this part also discusses the intricate relationships among organisms that enable them to coexist with each other and their surroundings.

Understanding life requires defining what it really is[①]. Living organisms are constituted of cells that are bound by a membrane, contain nucleic acid genetic material (DNA), and possess specialized structures that enable the cell to perform its functions. A living organism may consist of only one cell or of billions of cells of many specialized types. All living organisms have two characteristics (1) they process matter and energy through metabolic processes, and (2) they reproduce. The ability of an organism to process matter and energy is callled metabolism. Another important characteristic of living organisms is their ability to maintain an internal environment that is favorable to metabolic processes and that may be quite different from the external environment. Warm-blooded animals, for example, maintain internal temperatures that may be much warmer or even cooler than their surroundings. Finally, through succeeding generations, living organisms can undergo fundamental changes in their genetic composition that enable them to adapt better to their environment.

Living species are present in the biosphere because they have evolved with the capability to survive and to reproduce. Every single species in the biosphere has become an expert in these two things; otherwise it would not be here[②]. The key factors for existing—at least long enough to reproduce—are the ability to process energy and process matter. In so doing, life systems and processes are governed by the principles of thermodynamics and the law of conservation of matter. Organisms handle energy and matter in various ways. Plants, for example, process solar energy by photosynthesis and utilize atmospheric carbon dioxide and other simple inorganic nutrients to make their biomass. Herbivores are animals that eat the matter produced by plants, deriving energy and matter for their own bodies from it. Carnivores in turn feed upon the herbivores.

Life-forms require several things to exist. The appropriate chemical elements must be present and available. Energy for photosynthesis is required in the form of adequate sunlight. Temperatures must stay within a suitable range and preferably should not be subjected to large, sudden fluctuations. Liquid water must be available. And, as noted above, a sheltering atmosphere is required. The atmosphere should be relatively free of toxic substances. This is an area in which human influence can be quite damaging, through release of air pollutants that are directly toxic or which react to form toxic products, such as life-damaging ozone produced through the photochemical smog-forming process.

Individual organisms and groups of organisms must maintain a high degree of stability (homeostasis, meaning "same status") through a dynamic balance involving inputs of energy and matter and interaction with other organisms and with their surroundings[③]. This requires a high degree of organization and the ability to make continuous compensation adjustments in response to external conditions. For an individual organism, homeostasis means maintaining temperature, levels of water, inputs of nutrients, and other crucial factors at suitable levels. Arguably, the most advantageous evolutionary trait of mammals is their ability to keep their body temperatures within the very narrow limits that are optimum for their biochemical processes. The concept of homeostasis applied to whole ecosystems consisting of groups of organisms and their surrounding is termed ecosystem stability[④]. Indeed, homeostasis applies to the entire biosphere. To a large extent, environmental science addresses the homeostasis of the biosphere, and how the critical factors involved in maintaining the dynamic equilibrium of the biosphere are affected by human activities.

The nature of life is determined by the surroundings in which the life-forms must exist. Much of the environment in which organisms live is described by physical factors, including whether or not the surroundings are primarily aquatic or terrestrial. For a terrestrial environment, important physical factors are the nature of accessible soil, availability of water, and availability of nutrients. These are abiotic factors. There are also important biotic factors relating to the life-forms present, their wastes and decomposition products, their availability as food sources, and their tendencies to be predatory or parasitic.

Selected from "*Stanley E Manahan, Environmental Science and Technology, A Sustainable Approach to Green Science and Technology, Taylor & Franicis Group, USA, 2007*"

Words and Expressions

banket   *n*. 护坡堤，填土
biosphere   *n*. 生物圈
outgoing   *a*. 外逸的，外出的
infrared   *n*./*a*. 红外线/红外线的
torrential rainstorm   暴雨
flash flooding   暴洪，猝发洪水
austere   *a*. 苛刻的，恶劣的
tenacious lichen   坚韧的青苔
symbiotic   *a*. 共生的
fungi   *n*. 真菌
algae   *n*. 海藻
herbivore   *n*. 食草动物（植物）
carnivore   *n*. 食肉动物
homeostasis   *n*. 动态平衡
arguably   *ad*. 可论证地
aquatic   *a*. 水生的
terrestrial   *a*. 陆生的
predatory   *a*. 食肉（捕食）的
parasitic   *a*. 寄生的
abiotic   *a*. 无生命的，无生物的
biotic   *a*. 生命的，生物的

Notes

① Understanding life requires defining what it really is. 可译为：了解生命，需要定义生命实质上是什么。

② Every single species in the biosphere has become an expert in these two things; otherwise it would not be here.：可译为：生物圈中的每一物种在两方面成为专家，否则它就不

存在。

③ Individual organisms and groups of organisms must maintain a high degree of stability (homeostasis, meaning "same status") through a dynamic balance involving inputs of energy and matter and interaction with other organisms and with their surroundings. 可译为：单个生物和生物群必须通过能量与质量输入与其他生物及与它们的环境相互作用的动力平衡维持高度稳定性（动态平衡，指的是相同状态）。

④ The concept of homeostasis applied to whole ecosystems consisting of groups of organisms and their surrounding is termed ecosystem stability. 可译为：适用于由生物群及它们的环境组成的整个生态系统动力平衡概念称之为生态稳定。

Exercises

1. Put the following into Chinese.
   symbiotic combination of fungi and algae, feed upon the herbivores, ultraviolet radiation tenacious lichen, homeostasis, nucleic acid genetic material, warm-blooded animal, predatory

2. Put the following into English according to the text.

   生命因子　　相同状态　　寄生的　　共生的　　动力学平衡　　水丰富的边界区
   红外辐射　　湿条件　　　光合作用　暴雨　　　热力学　　　　质量守恒定律

3. Translate the following passage into Chinese.

   Materials and energy balances are key tools in achieving a quantitative understanding of the behavior of environmental systems. They sever as a method of accounting for the flow of energy and materials into and out of environmental systems. Mass balances provide us with a tool for modeling the production, transport, and fate of pollutants in the environment. Energy balances likewise provide us with a tool for modeling the production, transport, and fate of energy in the environment. Examples of mass balances include prediction of rainwater runoff, determination of the solid waste production from mining operations, oxygen balance in streams, and audits of hazardous waste production. Energy balances allow us to estimate the efficiency of thermal processes, predict the temperature rise in a stream from the discharge of cooling water from a power plant, and study climate change.

4. Answer the following questions.

   （1）What is the difference between abiotic and biotic factors?

   （2）What is the distinction between herbivores and carnivores?

Reading Material: Nutrient Cycles for Ecosystem

Materials and energy balances are key tools in achieving a quantitative understanding of the behavior of environmental systems. They sever as a method of accounting for the flow of energy and materials into and out of environmental systems. Mass balances provide us with a tool for modeling the production, transport, and fate of pollutants in the environment. Energy balances likewise provide us with a tool for modeling the production, transport, and fate of energy in the environment. Examples of mass balances include prediction of rainwater runoff, determination of the solid waste production from mining operations, oxygen balance in streams, and audits of hazardous waste production. Energy balances allow us to estimate the efficiency of thermal processes, predict the temperature rise in a stream from the discharge of cooling water from a power plant, and study climate change.

These same balances, although very complex, hold true globally. The basic elements of which all organisms are composed are carbon, nitrogen, phosphorus, sulfur, oxygen, and hydrogen. The first four of these elements are much more limited in mass and easier to trace than are oxygen and hydrogen. Because these elements are conserved, they can be recycled indefinitely (or cycled through the environment). Because the pathways used to describe the movement of these elements in the environment are cyclic, they are referred to as the carbon, nitrogen, phosphorus, and sulfur cycles.

Carbon cycle

Although carbon is only the 14th by weight in abundance on earth, it is by far, one of the most important elements on earth as it is the building block of all organic substances and thus, of life, itself. Carbon is found in all living organisms, in the atmosphere (predominately as carbon dioxide and bicarbonate), in soil humus, in fossil fuels, and in rock and soils (predominately as carbonate minerals in limestone or dolomite or in shale). Although it was once thought that the largest reservoir of carbon is terrestrial (plants, geological formations, etc.), the ocean actually serves as the greatest reservoir of carbon. Approximately 85% of world's carbon is found in the oceans.

Photosynthesis is the major driving force for the carbon cycle. Plants take up carbon dioxide and convert it to organic matter. Even the organic carbon compounds in fossil fuels had their beginnings in photosynthesis. The "bound", or stored, CO_2 in fossil fuels is released by combustion processes. The cycling of carbon also involves the release of carbon dioxide by animal respiration, fires diffusion from the oceans, weathering of rocks, and precipitation of carbonate minerals.

The ocean is a major sink of carbon, much of which is found in the form of dissolved carbon dioxide gas, and carbonate and bicarbonate ions. Primary productivity is responsible for the assimilation of inorganic carbon into organic forms. Productivity is limited by the concentrations of nitrogen, phosphorus, silicon, and other essential trace nutrients. The concentrations of CO_2 vary with depth. In shallow waters, photosynthesis is active and there is a net consumption of CO_2. In deeper waters, there is a net production of CO_2 due to respiration and decay processes. Because ocean circulation occurs over such a long time scale, the oceans take up CO_2 at a slower rate than the rate at which CO_2 from anthropogenic sources is accumulating in the atmosphere. In addition, as the amount of CO_2 dissolved in the ocean increases, the chemical capacity to take up more CO_2 decreases. The rate of uptake of CO_2 is driven by two main cycles: the solubility and biological pumps.

The solubility pump, as it is known, is the net driving force for dissolution of CO_2 into waters. Polar waters are colder at the surface than in deeper regions. As a result of the cold temperatures, CO_2 dissolution is enhanced in colder waters, driving dissolution of CO_2 from the atmosphere into the waters. Because these colder waters are denser than the warmer waters below, the colder waters tend to sink, taking with them CO_2. Because ocean circulation is slow, much of this CO_2 is "lost" to deep waters, keeping surface waters lower in CO_2 and driving dissolution from the atmosphere.

Phytoplankton, zooplankton and their predators, and bacterial make up the biological pump. These organisms take up carbon, resulting in a cycling of much of the carbon and nutrients found in the surface ocean waters. However, as these organisms die, they settle into deeper regions of ocean, taking with them bound CO_2. Additionally, as the dead organisms settle, some of this bound CO_2 finds its way into the ocean depths with the fecal matter of these organisms. Some is carried by currents to deeper regions. Thus, the depths of the ocean become a CO_2 sink, releasing carbon mainly through "upwelling" of water, diffusion across the thermocline, and seasonal, wind-driven mixing, which brings the deep water to the surface. This mixing of the deeper waters returns nutrients and carbon to the ocean surface, continuing the cycle of photosynthesis and respiration.

Nitrogen cycle

Nitrogen in lakes is usually in the form of nitrate (NO_3^-) and comes from external sources by way of inflowing streams or groundwater. When taken up by algae and other phytoplankton, the nitrogen is chemically reduced to amino compounds (NO_2-R) and incorporated into organic compounds. When dead algae undergo decomposition, the organic nitrogen is released to the water as ammonia (NH_3). At normal pH values, this ammonia occurs in the form of ammonium (NH_4^+). The ammonia released from the organic compounds plus that from other sources such as industrial wastes and agricultural runoff (e. g., fertilizers and manure) is oxidized to nitrate (NO_3^-) by a special group of nitrifying bacterial in a two-step process called nitrification:

$$4NH_4^+ + 6O_2 \Longleftrightarrow 4NO_2^- + 8H^+ + 4H_2O \tag{1}$$

$$4NO_2^- + 2O_2 \Longleftrightarrow 4NO_3^- \tag{2}$$

The first reaction is mediated by the organism nitrosomonas sp., the second by nitrobacter sp.

The overall reaction is

$$NH_4^+ + 2O_2 \Longleftrightarrow NO_3^- + 2H^+ + H_2O \tag{3}$$

Nitrogen cycles from nitrate to organic nitrogen, to ammonia, and back to nitrate as long as the water remains aerobic. However, under anoxic conditions, for example, in anaerobic sediments, when algal decomposition has depleted the oxygen supply, nitrate is reduced by bacterial to nitrogen gas (N_2) and lost from the system in a process called denitrification. Denitrification reduces the average time nitrogen remains in the lake. Denitrification can also result in the formation of N_2O, (nitrous oxide). The denitrification reaction is

$$2NO_3^- + \text{organic carbon} \longrightarrow N_2 + 2CO_2 + 2H_2O \tag{4}$$

Some photosynthetic microorganisms can also fix nitrogen gas from the atmosphere by converting it to organic nitrogen and are, therefore, called nitrogen-fixing microorganisms. In lakes the most important nitrogen-fixing microorganisms are photosynthetic bacterial called cyanobacteria, also known as blue-green algae because of the pigments they contain. Because of their nitrogen-fixing ability, cyanobacteria have a competitive advantage over green algae when nitrate and ammonium concentrations are low but other nutrients are sufficiently abundant. Nitrogen fixation also occurs in the soil. The aquatic fern Azolla is the only fern that can fix nitrogen. It does so by virtue of a symbiotic association with a cyanobacterium (Znabaena azollae). Azolla is found worldwide and is sometimes used as a valuable source of nitrogen for agriculture. Similarly, lichens can also contribute to nitrogen fixation. For example, Lobaria pulmonaria, a common N-fixing lichen in Pacific northwest forests, fixes nitrogen by a symbiotic relationship with the cyanobacterium Nostoc. Lichens such as this are a major source of nitrogen in old growth forests.

Nitrogen fixation is also accomplished in the soils by almost all legumes. Nitrogen fixation occurs in the root nodules that contain bacterial (Bradyrhizobium for soybean, Rhizobium for most other legumes). The legume family (Leguminosae or Fabaceae) includes many important crop species such as pea, alfalfa, clover, common bean, peanut, and lentil. The reaction describing nitrogen fixation is

$$N_2 + 8e^- + 8H^+ + ATP \longrightarrow 2NH_3 + H_2 + ADP + P_I \tag{5}$$

where P_I = inorganic phosphate.

Phosphorus Cycle

Phosphorus in unpolluted waters is imported through dust in precipitation or via the weathering of rock. Phosphorus is normal present in watersheds in extremely small amounts, usually existing dissolved as inorganic orthophosphate, suspended as organic colloids, adsorbed onto particulate organic and inorganic sediment, or contained in organic water. In polluted waters, the major source of phosphorus is from human activities. The only significant form of phosphorus available to plants and algae is the soluble reactive inorganic orthophosphate species (HPO_4^{2-}, PO_4^{3-}, etc.) that are incorporated into organic compounds. During algal decomposition, phosphorus is returned to the inorganic form. The release of phosphorus from dead algal cells is so rapid that only a small fraction of it leaves the upper zone of a stratified lake (the epilimnion) with the settling algal cells. However, little by little, phosphorus is transferred to the sediments; some of it in under-composed organic matter; some of it in precipitates of iron, aluminum, and calcium; and some bound to clay particles. To a large extent, the permanent removal of phosphorus from the overlying waters to the sediments depends on the amount of iron, aluminum, calcium, and clay entering the lake along with phosphorus.

Human activities have led to a release of phosphorus from the disposal of municipal sewage and from concentrated livestock operations. The application of phosphorus fertilizers has also resulted in perturbations in the phosphorus cycle, although these changes are thought to be more localized than the perturbations in the other cycles. Phosphorus releases can have a significant effect on lake and stream ecosystems.

Sulfur Cycle

Until the Industrial Revolution the effect of sulfur on environmental systems was quite small. However, with the Industrial Revolution, our use of sulfur-containing compounds as fertilizers and the release of sulfur dioxide during the combustion of fossil fuels and in metal processing has increased significantly. Mining operations have also resulted in the release of large quantities of sulfur in acid mine drainage. Like the nitrate ion, sulfate is negatively charged and is not adsorbed onto clay particles. Dissolved sulfates thus can be leached from the soil profile by excess rainfall or irrigation. In the environment, sulfur is found predominantly as sulfides (S^{2-}), sulfates (SO_4^{2-}), and in organic forms.

As with the nitrogen cycle, microorganisms play an important role in the cycling of sulfur. Bacterial are involved in the oxidation of pyrite-containing minerals, releasing large quantities of sulfate. In anaerobic environment, sulfate-reducing bacterial reduce sulfate to release hydrogen sulfide. In marine waters, the biological production of dimethylsulfide may occur.

Selected from *"Mackenzie L Davis, Susan J Masten. Principles of Environmental Engineering and Science, The McGraw-Hill Companies, Inc. USA 2004"*

Unit 37

Text: Ecology

Ecology, the study of the interrelationships between plants and animals that live in a particular physical environment, requires that we first define the physical environment, that is, the ecosystem that is to be studied. Ecosystems, short for ecological systems, are communities of organisms that interact with one another and with their physical environment, including sunlight, rainfall, and soil nutrients. Organisms within an ecosystem tend to interact with one another to a greater extent than do the organisms between ecosystems. Ecosystems can vary greatly in size. For example, a tidal pool of only about 2 m in diameter could be considered an ecosystem because the plants and animals living in this environment depend on one another and are unique to this type of system. On a large scale, a tropical rainforest is also an ecosystem[①]. Even larger is our global biosphere, which could be considered the "ultimate" Earthbound ecosystem[②]. Within each ecosystem are habitats, which are defined as the place where a population[③] of organisms lives.

Ecosystems can be further defined as systems into which matter flows. Matter also leaves ecosystems; however, this flow of matter into and out of the ecosystem is small compared with the amount of matter that cycles within the ecosystem. If we think of a lake as an ecosystem, matter flows into the lake in the form of carbon dioxide that dissolves into the water, nutrients that runoff from the land, and chemicals that flow with any streams or rivers feeding the lake. Within the lake, matter flows from one organism to the next in the form of food, excreted material, or respired gases (either oxygen or carbon dioxide). This flow of matter is critical for the existence of an ecosystem.

Another characteristic of an ecosystem is that it can change with time. Later we will discuss how lakes change (naturally or anthropogenically) over time, from a system that has very clear water, low levels of nutrients, and low numbers of a large variety of species to one that contains highly turbid water, high levels of nutrients, and large numbers of a few species. Both of these systems (the same lake at different times) are very different ecosystems. Similarly, severe flooding or droughts and extreme temperature changes or other extreme environmental conditions (e.g., volcanic activity or forest fires) can cause significant changes in an ecosystem.

Ecosystems can be natural or artificial. The lake, the tidal pool, or the forest is usually natural (although lakes can be human-made and forests can be cultivated). Constructed wetlands are increasingly being used for the treatment of storm runoff, mining wastes (acid mine drainage), or municipal sewage. Agricultural land is another example of an artificial (or human-made) ecosystem. The criteria explained earlier are met in all ecosystems, whether they are natural or artificial, large or small, long-lasting or temporary.

A biological community consists of an assembly of organisms that occupy a defined space in the environment. The nature of such a community depends on the physical and chemical characteristics that influence the life-forms in it and on the interactions of organisms in the community. A biological community is the biological component of an ecosystem, which includes the organisms and their physical environment. The community functions in a manner such that it tends to utilize and convert energy and materials in the most efficient manner that will enable the organisms in it to reproduce and thrive. Therefore, the exchange of matter and energy among the organisms and with their physical environment is a key aspect of an ecosystem. The study of biological communities is called community ecology.

Human influences on ecosystems

As environmental engineers and scientists we have a responsibility to protect ecosystems and the life that resides within them. Although ecosystems change naturally, human activity can speed up natural processes by several orders of magnitude (in terms of time).

Seemingly harmless or beneficial activities can wreak havoc on the environment. For example, large-scale agricultural operations, although producing inexpensive food to feed millions, can result in the release of pesticides, fertilizers, and carbon dioxide and other greenhouse gases to the environment. Hydroelectric power is seen as a clean, renewable energy source. However, dam construction can have detrimental effects on river ecosystems, drastically reducing fish populations, as well as causing erosion of soil and vegetation during powerful water surges.

Human activity can also change ecosystems through the destruction of species. The loss of habitat can threaten the existence of individual species within an ecosystem. For example, the destruction of the rainforest in Mexico threatens the very existence of the monarch butterfly. If these forests are destroyed to the extent that the monarch loses its winter roosting grounds, global extinction of this butterfly could result—the complete and permanent loss of this animal species across the entire planet. However, the localized destruction of the milkweed plant deprives the butterfly of its nesting environment, resulting in local extinction.

The destruction of an ecosystem is not the only way humans can affect animal populations. The release of toxic chemicals can also threaten wildlife. Dichlorodiphenyltrichloroethane (DDT) is a classical example of such a chemical. The release of DDT from the 1940s to the 1960s threatened the very existence of the American bald eagle. DDT weakened the eggs laid by the female birds, so much so that they were often crushed in the nest by the mother bird[④]. In some areas, DDT uptake from contaminated prey by the adult eagles resulted in a nearly complete failure to reproduce.

A third way species can be threatened is by the introduction of nonnative (exotic) species into ecosystems. The introduction of the rabbit onto Norfolk Island, the zebra mussel into the Great Lakes or the Asian fungus causing Dutch elm disease onto the U. S. east coast has had a significant impact on ecosystems. The introduction of rabbits onto Norfolk Island in 1830 resulted in the loss of 13 species of vascular plants by 1967. Scientists believe that the zebra mussel (Dreissena polymorph) was introduced into the Great Lakes from the ballast water of a transatlantic freighter that previously visited a port in eastern Europe, where the mussel is common. The zebra mussel can now be found in inland waters in 21 states and the Province of Ontario and is thought to be responsible for the reduction of some 80% of the mass of phytoplankton in Lake Erie. Because the zebra mussel is an efficient filterer of water its presence significantly increases water clarity, allowing light to penetrate deeper into the water column, increasing the density of rooted aquatic vegetation, benthic forms of algae, and some forms of insect-like benthic organisms. The mussel has also cause the near extinction of many types of native unionid clams in Lake St. Clair and the western basin of Lake Erie. The zebra mussel attaches itself to the native clams, eventually killing them.

The last method by which species can become extinct is through excessive hunting, some legal, others illegal. The manatee whose habitat is the Everglades is threatened by poaching, along with harm due to boat propellers, loss of habitat and vandalism. The rhinoceros is threatened by poaching, mainly for its horns.

Selected from *"Mackenzie L Davis, Susan J Masten. Principles of Environmental Engineering and Science, The McGraw-Hill Companies, Inc. USA 2004"*

Words and Expressions

tidal *a.* 潮汐的,潮水的
habitats *n.* 动植物
excreted material 排泄物
respired gas 呼吸气体
anthropogenical *a.* 人为的
wreak *vt.* 造成,报复
havoc *n.* 浩劫,大破坏
criteria *n.* 判据
monarch butterfly 君主蝴蝶
roosting ground 栖息地,居住地
milkweed *n.* 牧草
nesting environment 居住环境
nonnative (exotic) 外来的,非本土的
mussel *n.* 河蚌
Dutch elm 荷兰榆
vascular *a.* 充满活力的
transatlantic freighter 横过大西洋货轮
province of Ontario 安大略省
phytoplankton *n.* 浮游植物群落
Lake Erie 伊利湖
benthic *a.* 底栖生物的,水底植物的
clams *n.* 蚌
everglades *n.* 湿地,沼泽地
poaching *n.* 偷捕鱼
manatee *n.* 海牛科
vandalism *n.* 故意破坏
rhinoceros *n.* 犀牛

Notes

① On a large scale, a tropical rainforest is also an ecosystem. 可译为:从大的范围来看,热带雨林也是一种生态系统。a tropical rainforest 热带雨林(是生物群落的例子,生物群落指的是某一地区和某一气候环境下动植物区域)。

② Even larger is our global biosphere, which could be considered the "ultimate" Earth-bound ecosystem. 可译为:更大的是地球生物圈,它可看为总的地球生态系统。biosphere 生态系统(由大气,水,构成地球土壤组分的岩石,矿物质及泥土组成)。

③ population 这里指的是同一时间,同一方式生活相同种类的生物群体。

④ DDT weakened the eggs laid by the female birds, so much so that they were often crushed in the nest by the mother bird. 可译为:DDT会导致雌鸟产的蛋壳强度不够,因此这些蛋

通常在鸟巢中会被雌鸟压破。

Exercises

1. Put the following into Chinese according to the text.
   exotic species, local extinction, Asian fungus, zebra mussel, roosting ground, harmless or beneficial activities, tidal pool, volcanic activity, flooding and drought
2. Put the following into English according to the text.
   故意破坏　浮游植物群落　　居住环境　　君主蝴蝶　　　生态系统　　生态学
   生物圈　　天然或人造生物系统　湿地　　　暂时或持久的　水力发电
3. Translate "the second paragraph" into Chinese.
4. What is the third way species can be threatened?

Reading Material: Benefits of Biodiversity

Biodiversity is a concept as multifaceted as life itself, and biogogists employ different working difinitions according to their own aims and philosophies. Yet scientists agree that the concept applies acrose the major levels in the organization of life. The level that is easiest to visuslize and most commonly used is species diversity.

Biodiversity loss matters from and an ethical perspective, because many people feel that organisms have an intrinsic right to exist. However, losing biodiversity is also a problem for human society because of the many tangible, pragmatic ways that biodiversity benefits people and supports our society.

Biodiversity Provides Ecosystem Services

Contray to popular opinion, some things in life can indeed be free—as long as we protect the ecological systems that provide them. Intact forests provide clean air and water, and they buffer hydrologic systems against flooding and drought. Native crop varieties provide insurance against disease and drought. Wild-life can attract tourism and boost economies. Intact ecosytems provide these and other valuable processes, known as ecosystem services, for all of us, free of charge.

Maintaining these ecosystem services is one clear benefit of protecting biodiversity. According to UNEP, biodiversity:

- Provides food, fuel, fiber, and shelter
- Purifies air and water
- Detoxifies and decomposes wastes
- Stabilizes Earth's climate
- Moderates floods, droughts, and temperature extremes
- Cycles nutrients and renews soil fertility
- Pollinates plants, including many crops
- Controls pests and diseases

- Maintains genetic resources for crop varieties, livestock breeds, and medicines
- Provides cultural and aesthetic benefits
- Gives us the means to adapt to change

Biodiversity Helps Maintain Ecosystem Function

Ecological research demonstrates that biodiversity tends to enhance the stability of communities and ecosystems. Research has also found that biodiversity tends to increase the resilience of ecological systems—their ability to weather disturbance, bounce back from stress, or adapt to change. Thus, when we lose biodiversity, this can diminish a natural system's ability to function and to provide services to our society.

Will the loss of a few species really make much difference in an ecosystem's ability function? Consider a metaphor first offered by Paul and Anne Ehrlich: The loss of one rivet from an airplane's wing—or two, or three—may not cause the plane to crash. But at the some point as rivets are removed the structure will be compromised, and eventually loss of just one more rivet will cause it to fail. Keeping this metaphor in mind, we would be wise to preserve as many components of our ecosystems as possible to make sure these systems continue to function.

That said, some species are more vital to an ecosystem than others. Removing a species that can be functionally replaced by others may make little difference in how the system functions. However, as with the keystone that holds together an arch, removing a keystone species can significantly alter an ecological system. If a keystone species is lost, other species may disappear in response.

Biodiversity Enhances Food Security

Biodiversity provides the food we eat. Throughout our history, human beings have used 7000 plant species and several thousand animal species for food. Today many experts are concerned because industrial agriculture has narrowed our diet. Globally, we now get 90% of our food from just 15 crop species and eight livestock species, and this lack of diversity leaves us vulnerable to failures of particular crops. In a world where nearly 1 billion people go hungry, we can improve food security (the guarantee of an adequate, safe, nutritious, and reliable food supply to all people at all times) by finding sustainable ways to capitalize on the nutritional opportunities offered by wild species and rare crop varieties.

Many new or underutilized food sources could be harvested or farmed in sustainabl ways. Table 1 shows a selection of promising food resources from just one region of the world—Central and South America. Plenty more exist elsewhere worldwide.

Biodiversity Boosts Economies through Tourism and Recreation

Besides providing for our food and health, biodiversity can generate income through tourism, particularly for developing countries in the tropics that boast impressive species diversity. Many people like to travel to experience protected natural areas, and in so doing they

Table 1　Potential new food sources

Amaranths (three species of Amaranthus) Grain and leafy vegetable, livestock feed; rapid growth, drought resistant	Capybara (hydrochoeris hydrochaeris) World's largest rodent; meat esteemed; easily ranched in open habitats near water
Buriti palm (Mauritia flexuosa) "tree of life" to Amerindians; vitamin-rich fruit; pith as sources for bread; palm heart from shoots	Vicuna (Lama Vicugna) Threatened species related to llama; source of meat, fur, and hides, can be profitably ranched
Maca (Lepidium meyenli) Cold-resistant root vegetable resembling radish, with distinctive flavor; near extinction	Chachalacas (Ortalis, many species) Tropical birds; adaptable to human habitations; fast-growing

create economic opportunities for residents living near those areas. Visitors spend money at local businesses, hire local people as guides, and support parks that employ local residents. Ecotourism can thereby bring jobs and income to areas that otherwise might suffer poverty.

The parts and wildlife of Kenya and Tanzania are prime examples. Ecotourism brings in fully a quarter of all foreign money entering Tanzania's economy each year. Leaders and citizens in both nations recognize biodiversity's economic benefits, and as a result they have managed their parks and reserves diligently. Ecotourism is a vital source of income for nations such as Costa Rica, with its rainforests; Australia, with its Great Barrier Reef; and Belize, with its reefs, caves, and rainforests. The United States, too, benefits from ecotourism; its national parks draw millions of visitors from around the world.

Ecotourism can serve as a powerful financial incentive for nations, states, and local communities to preserve natural areas and reduce impacts on the landscape and on native species. Yet as ecotourism increases, an overabundance of visitors to natural areas can degrade the outdoor experience and disturb wildlife. Anyone who has been to Yosemite, the Grand Canyon, or the Great Smoky Mountains on a crowded summer weekend can attest to this. As ecotourism continues to grow, so will debate over its costs and benefits for local communities and for biodiversity.

Selected from " *Jay Withgott, Matthew Laposata. , Environment, the Science Behind the Stories*, 5th Ed., Pearson Education Inc., US. 2014"

Unit 38

Text: Ecology and Life Systems

To consider the biosphere and its ecology in their entirety, it is necessary to look at several levels in which life exists. The unimaginably huge numbers of individual organisms in the biosphere belong to species (kinds of organisms). Groups of organisms of the same species living together and occupying a specified area over a particular period of time constitute a population; and that part of Earth on which they dwell is their habitat. In turn, various populations coexist in biological communities. Members of a biological community interact with each other and with their atmospheric, aquatic, and terrestrial environments to constitute an ecosystem. An ecosystem describes the complex manner in which energy and matter are taken in, cycled, and utilized; the foundation on which an ecosystem rests is the production of organic matter by photosynthesis. Assemblies of organisms living in generally similar surroundings over a large geographic area constitute a biome. Each biome may contain many ecosystems. The following are examples of important kinds of biomes:

(1) Tropical rain forests characterized by warm temperatures throughout the year and having most of their nutrients contained in the organisms populating the forest

(2) Warm-climate evergreen forests found in the southeastern U. S.

(3) Coniferous forests in temperate climates that have distinct summer and winter seasons and that are polulated by cone-bearing trees with needles, such as cedar, hemlock, and pine

(4) Temperate deciduous forests growing in regions with hot, wet summers and cold winters populated by trees that grow new leaves and shed them annually

(5) Grasslands in which grass is anchored in a tough, dense mass of grass roots, and soil called sod

(6) Hot deserts populated by cacti, creosote bush, yucca, and other species adapted to high temperatures and sparse moisture

(7) Cold deserts in which tough perennial plants, such as sagebrush and some grasses, survive under cool, dry conditions

(8) Tundra found in arctic regions or at high altitudes. Tundra regions have no trees, cold winters with frost possible even in the summer, permanently frozen subsurface soil called permafrost, low productivity, relatively few species, and high vulnerability to environmental insult.

In order to sustain life, an ecosystem must provide energy and nutrients. Energy enters an ecosystem as sunlight. Part of the solar energy is captured by photosynthesis, and part is abosorbed to keep organisms warm, which enables their metabolic processes to occur faster. In addition to capturing energy, an ecosystem must provide for recycling essential nutrients, including carbon, oxygen, phosphorus, sulfur, and trace-level metal nutrients, such as iron.

Much of the organization of ecosystems has to do with the acquisition of food by the organisms in it. Virtually all food on which organisms depend is produced by the fixation of carbon from carbon dioxide and energy from light in the form of energy-rich, carbon-rich biomass through the process of photosynthesis[①]. Photosynthesis can be represented by

$$CO_2 + H_2O + h\nu \longrightarrow \{CH_2O\} + O_2$$

Where $h\nu$ represents light energy absorbed in photosynthesis and $\{CH_2O\}$ represents biomass. Thus the photosynthetic plants in the biosphere are the basic producers on which all other members of the community depend for food and for their existence. The rate of biomass production is called productivity. It is conventionally expressed as energy or quantity of biomass per unit area per unit time. The food manufactured by producers is utilized by other organisms generally classified as consumers.

The sequence of food utilization, starting with biomass synthesized by photosynthetic producers is called the food chain. Numerous food chains exist in ecosystems, and there is crossover and overlap between them. Therefore, food chains are interconnected to form intricate relationships called food webs. An example of a food chain would be one in which biomass is produced by unicellular algae in a lake and consumed by small aquatic organisms (copepods), which are eaten by small fish which are eaten, in turn, by large fish. Finally, a bald eagle atop the food chain may consume the large fish. As shown by this example, there are several levels of consumption in a food chain called trophic levels. In going up the chain, the first through fourth trophic levels are (1) producers, (2) primary consumers (herbivores), (3) secondary consumers (carnivores), and (4) tertiary consumers, sometimes called "top carnivores". In addition to herbivores and carnivores, there are several other classifications of consumers. Omnivores eat both plant matter and flesh. Parasites draw their nourishment from a living host. Scavengers, such as beetles, flies, and vultures, feed on dead animals and plants. Detritovores, such as crabs, earthworms, and some kinds of beetles, feed on detritus composed of fragments of dead organisms and undigested wastes in feces[2]. Ultimately, fungi and microorganisms complete the degradation of food matter to simple inorganic forms that can be recycled through the ecosystem, a process called mineralization[3].

Food webs can be divided into two main categories broadly based upon whether the food is harvested from living populations or from the remains of dead organisms. In a grazing food web food, along with the energy and nutrient minerals that it contains, is transferred from plants to herbivores and on to carnivores. Dead organisms become part of a detritus food web in which various levels of scavengers degrade the organic matter from the dead organisms.

The energy content of food passing through each trophic level may be consumed by respiratory processes to maintain the metabolic activity and movement of the organisms at that level, eventually to be dissipated as heat. Energy incorporated into the bodies of organisms or excreted from them as waste products can become part of detritus food web. A relatively small, but very important fraction of the energy is passed on to the next trophic level when the organisms are consumed by predators.

It is instructive to consider the flow of energy utilization through various trophic levels. Typically, based upon 100% of the energy from primary producers, the percentages utilized by various trophic levels are decomposers 25%, herbivores 15%, primary carnivores 1.5%, and top carnivores less than 0.1%.

Selected from "*Stanley E Manahan, Environmental Science and Technology, A Sustainable Approach to Green Science and Technology, Taylor & Franicis Group, USA, 2007*"

Words and Expressions

evergreen forests　常青林
coniferous　针叶树松柏类植物
temperate climates　温和气候
cedar　*n.* 红松杉
hemlock　*n.* 铁杉
pine　*n.* 松树
deciduous　*a.* 落叶性
sod　*n.* 草地
cacti　*n.* 仙人掌
creosote bush　杂酚油灌木
yucca　*n.* 丝兰植物
perennial　*adj.* 四季不断的，常年的
sagebrush　*n.* 灌木蒿丛，蒿属植物
tundra　*n.* 冻原带，苔原，冻土带
arctic regions　北极区，寒冷区
vulnerability　*a.* 脆弱
insult　*vt./n.* 损害，刺激
food web　食物网
unicellular　*a.* 单细胞的
trophic　*a.* 营养的
omnivore　*n.* 杂食动物
detritus　*n.* 碎岩，泥河

Notes

① Virtually all food on which organisms depend is produced by the fixation of carbon from carbon dioxide and energy from light in the form of energy-rich, carbon-rich biomass through the process of photosynthesis. 可译为：实际上，生物所依赖的所有食物是通过光合作用来自二氧化碳固碳以及富含能量和富含碳的生物质产生的光能。

② Detritovores, such as crabs, earthworms, and some kinds of beetles, feed on detritus composed of fragments of dead organisms and undigested wastes in feces. 可译为：食废弃物动物，如螃蟹、蚯蚓和某些甲虫是以组成动物尸体碎片和排泄物中未消化废物的腐殖质为食物。

③ Ultimately, fungi and microorganisms complete the degradation of food matter to simple inorganic forms that can be recycled through the ecosystem, a process called mineralization. 翻译为：最终，真菌和微生物通过称为矿化过程的生态系统循环实现食物降解为简单的无机物形式。

Exercises

1. Put the following into Chinese.

   fungi, microorganism, mineralization, detritus food web, trophic level, herbivore parasite, permafrost, food chain, productivity, intricate relationship, algae, desert

2. Put the following into English according to the text.

   草地　生物群落区　常青林　温和气候　环境损害　金属营养
   生产者　消费者杂食动物　新陈代谢活动　呼吸过程　寒冷区

3. Translate " the first paragraph" into Chinese.

4. How many kinds of biomes are there according to the text? What are they?

Reading Material: The Five Major Components of an Industrial Ecosystem

Industrial ecosystems can be broadly defined to include all types of production, processing, and consumption. It is useful to define five major components of an industrial ecosystem. These are (1) a primary materials producer, (2) a source or sources of energy, (3) a materials-processing and manufacturing sector, (4) a waste-processing sector, and (5) a consumer sector. In such an idealized system operating with the best practice of industrial ecology, the flow of materials among the major hubs is very high. Each constituent of the system evolves in a manner that maximizes the efficiency with which the system utilizes materials and energy.

It is convenient to consider the primary materials producers and the energy generators together because both materials and energy are required for the industrial ecosystem to operate. The primary materials producer or producers may consist of one or several enterprises devoted to providing the basic materials that sustain the industrial ecosystem. Most generally, in any realistic industrial ecosystem, a significant fraction of the material processed by the system consists of virgin materials. In a number of cases, and increasingly so as pressures build to recycle materials, significant amounts of the materials come from recycling sources.

The processes that virgin materials enter the system are subjected to vary with the kind of material, but can generally be divided into several major steps. Typically, the first step is extraction, designed to remove the desired substance as completely as possible from the other substances with which it occurs. This stage of materials processing can produce large quantities of waste material requiring disposal, as is the case with some metal ores in which the metal makes up a small percentage of the ore that is mined. In other cases, such as corn grain providing the basis of a corn products industry, the "waste" — in this specific example, the cornstalks associated with the grain — can be left on the soil to add humus and improve soil quality. A concentration step may follow extraction to put the desired material into a purer form. After concentration, the material may be put through additional refining steps that may involve separations. Following these steps, the material is usually subjected to additional processing and preparation leading to the finished materials. Throughout the various steps of extraction, concentration, separation, refining, processing, preparation, and finishing, various physical and chemical operations are used, and wastes requiring disposal may be produced. Recycled materials may be introduced at various parts of the process, although they are usually introduced into the system following the concentration step.

The extraction and preparation of energy sources can follow many of the steps outlined above for the extraction and preparation of materials. For example, the processes involved in extracting uranium from ore, enriching it in the fissionable uranium -235 isotope, and casting it into fuel rods for nuclear fission power production include all of those outlined above for materials. On the other hand, some rich sources of coal are essentially scooped from a coal seam and sent to a power plant for power generation with only minimal processing, such as sorting and grinding.

Recycled materials added to the system at the primary materials, and energy production phase may be from both pre- and postconsumer sources. As examples, recycled paper may be macerated and added at the pulping stage of paper manufacture. Recycled aluminum may be added at the molten metal stage of aluminum metal production.

Finished materials from primary materials producers are fabricated to make products in the goods fabrication and manufacturing sector, which is often a very complex system. For example, the manufacture of an automobile requires steel for the frame, plastic for various components, rubber in tires, lead in the battery, and copper in the wiring, along with a large number of other materials. Typically, the first step in materials manufacturing and processing is a forming operation. For example, sheet steel suitable for making automobile frames may be cut, pressed, and welded into the configuration needed to make a frame. At this step some wastes may be produced that require disposal. An example of such wastes consists of carbon fiber/epoxy composites left over from the foming parts such as jet aircraft engine housings. Finished components from the forming step are fabricated into finished products that are ready for the consumer market.

The materials processing and manufacturing sector presents several opportunities for recycling. At this point it may be useful to define two different streams of recycled materials:

(1) Process recycle streams consisting of materials recycled in the manufacturing operation itself.

(2) External recycle streams consisting of materials recycled from other manufacturers or from postconsumer products.

Materials suitable for recycling can vary significantly. Generally, materials from the process recycle streams are quite recyclable because they are the same materials used in the manufacturing operation. Recycled materials from the outside, especially those from postconsumer sources, may be quite variable in their characteristics because of the lack of effective controls over recycled postconsumer materials. Therefore, manufacturers may be reluctant to use such substances.

In the consumer sector, products are sold or leased to the consumers who use them. The duration and intensity of use vary widely with the product; paper towels are used only once, whereas an automobile may be used thousands of times over many years. In all cases, however, the end of the useful lifetime of the product is reached and it is either (1) discarded or (2) recycled. The success of a total industrial ecology system may be measured largely by the degree to which recycling predominates over disposal.

Recycling has become so widely practiced that an entirely separate waste-processing sector of an economic system may now be defined. This sector consists of enterprises that deal specifically with the collection, separation, and processing of recyclable materials and their distribution to end users. Such operations may be entirely private or they may involve cooperative efforts with governmental sectors. They are often driven by laws and regulations that provide penalties against simply discarding used items and materials, as well as positive economic and regulatory incentives for their recycle.

Selected from "*Stanley E Manahan, Environmental Science and Technology, A Sustainable Approach to Green Science and Technology, Taylor & Franicis Group, USA, 2007*"

PART 8 ENVIRONMENTAL SUSTAINABLE DEVELOPMENT AND GREEN SCIENCE AND TECHNOLOGY

Unit 39

Text: The Dilemma of Sustainability

Sustainable Development

In October 2007, the Nobel Peace Prize was awarded to the Intergovernmental Panel on Climate Change (IPCC) and the former US presidential candidate, Al Gore, for "their efforts to build up and disseminate greater knowledge about man-made climate change, and to lay the foundations for the measures that are needed to counteract such change" [1] (www. nobelprize. Org, 12 October 2007). The award marked their work, in the IPCC's series of monumental reports and Gore's tireless lectures and successful documentary film *An Inconvenient Truth*, to identify and build awareness of the connection between human activities and climate change. Extensive climate changes were likely to alter and threaten the living conditions of much of humankind, placing particularly heavy burdens on the world's most vulnerable countries.

These connections between environment and human welfare are uncomfortable for world leaders. In April 2007, release of the second volume of the IPCC's Fourth Assessment Report Impacts, *Adaptation and Vulnerability* was held up by last-minute political wrangling[2]. As the *New Scientist* headline put it, "as polluters quibble, the poor learn their fate". The report showed the significance of climate change for the world's poor. Storms, drought, heat waves, early flowering seasons, changes in insect migrations and dwindling water supplies from mountain regions were global problems the world's poorest countries and people were least well equipped to deal with. Greenhouse gas emissions in industrial and rapidly industrializing economics were directly linked to the day-to-day problems of the poor. The United Nations Development Program (UNDP) regards climate change as "proven scientific fact", and comments we now know enough to recognize that there are large risks, potentially catastrophic ones". The connections between wealth and wealth creation, environmental change and poverty are laid bare by scientific understanding of planetary carbon metabolism. No wonder the diplomats squabbled over the small print[3].

In the first decade of the new century, the issue of human impacts on global climate change has mostly been framed within a broader debate about sustainability. The challenge of doing something about this and other global issues (such as biodiversity depletion and pollution), while simultaneously tackling global inequality and poverty and not letting the wheels come off the world economy, is labelled as sustainable development. In the aftermath of two global conferences, the United Nations Conference on Environment and Development (UNCED) in Rio de Janeiro in 1992 (the Rio Conference or the "Earth Summit") and the World Summit on Sustainable Development in Johannesburg in 2002, these concepts have become staples in any debate about environment and development. The classic oxymoron "sustainable development" (combining two seemingly contradictory concepts) had "come of age". But where had it come from, and what did it mean?

The idea of development attracts new concepts at a ferocious rate. New terms are coined and adapted faster than old ones are discarded. This is an important process, for words change the way we think and what we do, modifying mindsets, legitimating actions and stimulating research and learning. The last twenty years of research in development studies, influenced by postmodernism and post-structuralism, leave no double of the enormous power of language and discourse to structure the way we think about—and therefore take action about—development. Development action is driven forwards by texts ranging from humanitarian tracts to national development plans. The way these texts portray the world, often in a crisis of some kind, determines what knowledge (and whose knowledge) provides a frame for problems and solutions, constitutes the basis for action and determines who has the authority to act.

Sustainable development gained its salience largely as a result of the United Nations Conference on Environment and Development in Rio in 1992[④]. The 170 governments represented made public proclamations of support for the idea of environmentally sensitive economic development, egged on by a vast array of non-governmental organizations, meeting nearby in the parallel Global Forum. The media danced attendance, and the conference was promoted as a global event, although many a journalist pointed out the stark contrasts between the lifestyles and life chances of delegates and poverty of people in Rio de Janeiro's favelas. The media had built up hopes that UNCED would bring about a new environmental world order, and, once the razzmatazz had died down, many commentators reported that the chance had been blown. A series of international agreements had been signed, but had anything really changed? Over the next decade, many commentators pointed out that the economy was carrying on much as before, rich and poor, polluter and polluted. The words had changed, but it was said that deeds had not: it was "business as usual" at Earth plc, despite the calls from its shareholders and the high-profile statements for chief executives.

Sustainable development as panacea

The range of context in which the phrase "sustainable development" is now employed is very wide. In research, it seems to offer the potential to unlock the doors separating academic disciplines, and to break down the barriers between academic knowledge and policy action. It does this because it seems to draw together ideas in ecology, ethics, economics, development studies, sociology and many other disciplines. Yet it looks forward to action and practical projects of social and environmental improvement. The term is beguilingly simple, yet at the same time capable of carrying a wide range of meanings. It can be used by political actors with divergent interests, a convenient rhetorical flag under which favored projects can be launched. It has become a powerful term in the lexicon of development studies, but also a theoretical maze of remarkable complexity.

It has been recognized for decades that sustainable development can be defined in many ways. Many definitions are rhetorical and vague[5]. The most commonly quoted is that from the Brundtland Report, in Our common Future: "development that meets the needs of the present without compromising the ability of future generations to meet their own needs". The longevity of this formulation stems from simultaneous appeal both to those concerned about poverty and development and to those concerned about the state of the environment, and the preservation of biodiversity. Moreover, it demands that attention be focused on both intra-generational equity (between rich and poor now) and intergenerational equity (between present and future generations). The appealing, moralistic but slightly vague form of words of the Brundtland Report allowed sustainable development to become the "new jargon phrase in the development business". It also became a vital element in the discourse of researchers trying to explain the relations between economy, society and environment.

Environmentalists speak of "sustainable development" in trying to demonstrate the relevance to development planners of their ideas about proper management of natural ecosystems. The conviction behind works such as the world conservation strategy was that sustainable development is a concept that could truly integrate environmental issues into development planning. In using terminology of this sort since that time, environmentalists have attempted to capture some of the vision and to exert influence in development debates. Sadly, they often have no understanding of their context or complexity. Environmentalist prescriptions for development, short of any explicit treatment of political economy, can have a disturbing naivety.

Politicians and governments have been enthusiastic in their incorporation of the language of sustainable development. In the U.K., for example, the 1997 U.K. government White Paper on international development made a specific commitment to the elimination of poverty in poorer countries through sustainable development; specific objectives include the promotion of "sustainable livelihoods", a Sustainable Development Strategy was published in 2005. An independent Sustainable Development Commission was established in 2000, building on the work of the U.K. Round Table on Sustainable Development and Panel on Sustainable development. In 2005 the U.K. government charged the Commission with the role of "watchdog for sustainable development". The U.K. has identified four priority areas of immediate action, shared across the U.K.: sustainable consumption and production (working towards achieving less with more); natural-resources protection and environmental enhancement (protecting the natural resources on which we depend); "from local to global" (building sustainable communities) and climate change and energy.

Selected from "*W. M. Adams, Green Development, Environment and sustainability in developing world, third edition Published by Routledge 2009*".

Words and Expressions

dilemma　　*n.* 进退两难，困境，二难推理
disseminate　　*vt.* 传播，宣传，扩散
quibble　　*n. v.* 诡辩，支吾，双关语
wrangle　　*v.* 争论，口角
flowering seasons　　开花季节，花粉季节
dwindle　　*v.* 缩小，减小，变坏
catastrophic　　*adj.* 大事故的，毁灭的，灾祸的
planetary　　*adj.* 行星的，漂泊不定的
metabolism　　*n.* 新陈代谢
diplomat　　*n.* 外交官，有手腕者
squabble　　*v.* 争吵，搞乱
panacea　　*n.* 万能药，治百病的灵药
beguilingly　　*ad.* 欺骗地
maze　　*n.* 迷宫，混乱，错综复杂（事物）
longevity　　*n.* 长寿，耐久性
ferocious rate　　极大速率，惊人速率
salience　　*n.* 显著，特征

Notes

① In October 2007, the Nobel Peace Prize was awarded to the Intergovernmental Panel on Climate Change (IPCC) and the former US presidential candidate, Al Gore, for "their efforts to build up and disseminate greater knowledge about man-made climate change, and to lay the foundations for the measures that are needed to counteract such change". 可译为：2007年10月，诺贝尔和平奖授予气候变化国际政府专门小组成员，美国前总统候选人戈尔（戈尔，1993—2000年为美国总统克林顿的副手，副总统，2000年作为民主党候选人竞选美国总统失败），奖励该小组成员，在努力构建和广泛宣传有关人造气候变化知识，以及为需要抵制这类变化采取的措施奠定的基础所做出的贡献。

② In April 2007, release of the second volume of the IPCC's Fourth Assessment Report Impacts, Adaptation and Vulnerability was held up by last-minute political wrangling. 可译为：2007年4月，气候变化国际政府专门小组成员第四次评估报告（影响冲击，适应与脆弱性）在最后一分钟政治争论后公布。

③ No wonder the diplomats squabbled over the small print. 可译为：难怪有手腕者（外交家）搞乱排好的印刷（出版）物。

④ Sustainable development gained its salience largely as a result of the United Nations Conference on Environment and Development in Rio in 1992. 可译为：可持续发展引起极大关注主要因为1992年在里约召开的联合国环境与发展会议的结果。

⑤ Many definitions are rhetorical and vague. 可译为：许多定义都是花言巧语，含糊不清的。

Exercises

1. Put the following phrases into English.

迷宫　　大事故的　　减小（变坏）　　特征（显著）　　极大速率　　万能药
诡辩　　争吵　　传播（宣传）　　新陈代谢　　进退两难　　欺骗地

2. Put the following sentences into Chinese.

(1) Extensive climate changes were likely to alter and threaten the living conditions of much of humankind, placing particularly heavy burdens on the world's most vulnerable countries.

(2) The United Nations Development Program (UNDP) regards climate change as "proven scientific fact", and comments "we now know enough to recognize that there are large risks, potentially catastrophic ones".

(3) Environmentalists speak of "sustainable development" in trying to demonstrate the relevance to development planners of their ideas about proper management of natural ecosystems.

(4) Politicians and governments have been enthusiastic in their incorporation of the language of sustainable development.

3. What are the main reasons of sustainable development dilemma?

Reading Material: The Origins of Sustainable Development

Environmentalism and Emergence of Sustainable Development

The phrase "sustainable development" has become the focus of debate about environment and development. It is not only the best-known and most commonly cited idea linking environment and development; it is also the best documented, in a series of publications beginning with the World Conservation Strategy (WCS) and the Brundtland Report, Our Common Future, and leading to the documents arising out of the Rio Conference in 1992 and the Johannesburg Summit in 2002. These mainstream documents are the subject of the new two chapters, which discuss their arguments and assess the nature of the ideology that shapes their ideas about development. However, the concept of sustainable development cannot be understood in a historical vacuum. It has many antecedents, and over time has taken on board many accretions and influences. These are the subjects of this chapter.

The history of thinking about sustainable development is closely linked to the history of environmental concern and of the conservation of nature in Western Europe and North America. An understanding of the evolution of sustainable development thinking must embrace the way essentially metropolitan ideas about nature and its conservation were expressed on the periphery in the twentieth century, initially on the colonial periphery and latterly within in the countries of the independent developing world. This focuses attention in particular on the rise of international environmentalism in the second half of the twentieth century. The phenomenon of the emergence is well described elsewhere. Its intellectual roots lie a great deal further back and are beyond the scope of this book to unravel. They have been explored, for example, by Merchant (1980), K. Thomas (1983), Pepper (1984), R. H. Grove (1990, 1995) and Grove et al. (1998).

Global Environmentalism

The attempt to write about the history of sustainable development is made difficult by the abundance of, and the Eurocentric and Americocentric focus of, the literature on environmentalism. Accounts of the "global environmental movement" have tended to portray its history as an almost exclusively northern-hemisphere phenomenon. This ethnocentrism should make us wary of international comparisons that are in fact based on European or North American experience. Southern environmental non-governmental organizations (NGOs) began to appear from the 1970s onwards, and their number and capacity have grown rapidly. By the end of twentieth century an environmental movement was even developing in China. However, the size and influence of environmental NGOs based in industrialized countries are such that they remain the dominant force internationally. The effectiveness of developing world grass-roots organizations is often precisely in their ability to transcend locality and connect to international arenas, and this is often done through better-connected metropolitan patterns.

Concern about human relationships with environment ran deep through the medieval and modern periods in Europe. Mediterranean classical writing provided numerous forerunners of Western thinking. They influenced thinking about the destructive power of human activities in North America and Europe, particularly in the second half of the nineteenth century, most famously in George Perkins Marsh's Man and Nature (1864). Marsh observed: "man is everywhere a disturbing agent. Wherever he plants his foot, the harmonies of nature are turned to discords. The proportions and accommodations which ensured the stability of existing arrangements are overthrown". Marsh's classical education, his boyhood in Vermont and his sojourns in Europe created a truly modernist critique of industrialization and the environment demands of economic growth. Nature, he explained and demonstrated, "avenges herself upon the intruder, by letting loose on her defaced provinces destructive energies hitherto kept in check by organic forces destined to be his best auxiliaries, but which he has unwisely dispersed and driven from the field of action". Similar perceptions, if less scientifically expressed, had surfaced elsewhere in the industrializing world, most significantly perhaps in the Romantic movement and the ideas of people like the British poet William Wordsworth or John Ruskin. Concern for the conservation of nature in the U. S. A. and in Europe tapped these concerns in a direct way.

Nature Preservation and Sustainable Development

In many ways, wildlife or nature conservation has been the most deep-seated root of sustainable development thinking. Indeed, sustainable development was put forward as a concept partly as a means of promoting nature preservation and conservation. The history of nature conservation in countries of industrial area-for example in Britain or the U. S. A. —is well established. Although the intellectual roots of a concern for nature (either for its own sake, or for fear of repercussions of misuse for people) lay deeper and further back, the foundation of formal organizations to carry out and promote conservation began in the nineteenth century.

These developments were primarily aimed at promoting the protection of nature within the industrialized nations themselves. However, from an early date there was also concern about conservation on a wider geographical scale, in imperial or colonial possessions. In Africa, for example, concern about depletion of forests in the Cape Colony developed in the early nineteenth century. This, with pressure for government money for the botanic garden, led to the appointment of a Colonial Botanist in 1858. Legislation to preserve open areas close to Cape Town was passed in 1846, and further Acts for the preservation of forests (1859) and game (1886) followed (Grove 1887). Similar institutions were created in other imperial territories; in India, for example, 30 per cent of non-agricultural land in some provinces had been brought under the control of the Forest Department.

The most obvious aspect of the conservation based on this hunting ethos was complete denial of hunting to Africans. White men hunted; Africans poached. This denial was achieved through controls on firearms, and latterly by establishment of game reserves. The Cape Act for the Preservation of Game of 1886 was extended to the British South African Territories in 1891. In 1892 the Sabie Game Reserve was established (to become the Kruger National Park in 1926), and in 1899 the Ukamba Game Reserve was created in Kenya, including land in what became the Amboseli National Park. In 1900 the Kenyan Game Ordinance was passed, effectively banning all hunting except by license (Graham 1973; MacKenzie 1988; Adams 2004)

Ecology and the Balance of Nature

The science of ecology developed at the end of nineteenth century in Europe and the U. S. A. There were close links from an early date between the new science and the preservation or conservation movement, particularly in the U. K. , especially through the work of plant ecologist Arthur Tansley. Perhaps it was not inappropriate that the word "ecology" was used in popular discussion of the rise of environmentalism from the 1960s, even where ideas owed little or nothing to scientific ideas or method, although it was true that many prominent early figures in the environmental movement were trained in ecology.

Ecology has contributed to thinking about sustainability in a series of related ways. First, ecological theory has underpinned much broader thinking about the environment and human impacts upon it. This relationship, and particularly the idea that there is some kind of "balance of nature", is reviewed in this section. Second, ecology has been particularly important in thinking about tropical environments and therefore the development of colonial territories and developing countries. This is discussed in the next section. Third, there is close resonance between the acquisition of scientific understanding and the application of that knowledge to both environmental management and development. In some ways ecology was a "science of empire", and there were strategic links between science and politics, ecology and empire (Robin 1997). This ecological "managerialism", which was particularly attractive in places such as Africa at the end of the Second World War, formed an important strand in the growth of sustainable development thinking.

Ecology's most obvious contribution to sustainable development has been its scientific description and analysis of the living environment. Within ecology, a whole series of concepts had been developed to describe patterns of change in natural systems, and these came to provide a powerful conceptual basis for sustainable development. Chief among them was the concept of the ecosystem and the idea of balance between predator and prey species. These concepts underpin the close links between science of ecology and the development of conservation discussed earlier in this chapter.

Selected from "W. M. Adams, *Green Development*, *Environment and sustainability in developing world*, third edition Published by Routledge 2009".

Unit 40

Text: Sustainability

"If we do not change direction, we are likely to end up where we are headed[①]," (old Chinese proverb).

"If we make the effort to learn its language, the Earth will speak to us and tell us what we must do to survive[②]."

Sustainability

The old Chinese proverb certainly applies to modern civilization and its relationship to world resources that support it. Evidence abounds that humans are degrading the Earth life support system upon which they depend for their existence[③]. The emission to the atmosphere of carbon dioxide and other greenhouse gases is almost certainly causing global warming. Discharge of pollutants has degraded the atmosphere, the hydrosphere, and the geosphere in industrialized areas. Natural resources including minerals, fossil fuels, freshwater, and biomass have become stressed and depleted. The productivity of agricultural land has been diminished by water and soil erosion, deforestation, desertification, contamination, and conversion to non-agricultural uses. Wildlife habitats including woodlands, grasslands, estuaries, and wetlands have been destroyed or damaged. About 3 billion people (half of the world's population) live in dire poverty on less than the equivalent of U. S. $2 per day. The majority of these people lack access to sanitary sewers and the conditions under which they live give rise to debilitating viral, bacterial, and protozoal diseases. At the other end of the standard of living scale, a relatively small fraction of the world's population consumes an inordinate amount of resources with lifestyles that involve living too far from where they work in energy-wasting houses that are far larger than they need, commuting long distances in large "sport-utility vehicles" that consume far too much fuel, and overeating to the point of unhealthy obesity with accompanying problems of heart disease, diabetes, and other obesity-related maladies.

As We Enter the Anthropocene

Humans have gained an enormous capacity to alter Earth and its support systems. Their influence is so great that we are now entering a new epoch, the anthropocene, in which human activities have effects that largely determine conditions on the planet. The major effects of human activities on Earth have taken place within a miniscule period of time relative to the time that life has been present on the planet or, indeed, relative to the time that modern humans have existed. These effects are largely unpredictable, but it is essential for humans to be aware of the enormous power in their hands—and of their limitations if they get it wrong and ruin Earth and its climate as life-support systems.

Achieving Sustainability

Although the condition of the world and its human stewards outlined in the preceding paragraphs sounds rather grim and pessimistic, this is not a grim and pessimistic book. That is because the will and ingenuity of humans that have given rise to conditions leading to deterioration of Planet Earth can be—indeed, are being—harnessed to preserve the planet, its resources, and its characteristics that are conducive to healthy and productive human life. The key is sustainability or sustainable development defined by the Bruntland Commission in 1987 as industrial progress that meets the needs of the present without compromising the ability of future generations to meet their own needs. A key aspect of sustainability is the maintenance of Earth's carrying capacity, that is, its ability to maintain an acceptable level of human activity and consumption over a sustained period of time. Although change is a normal characteristic of nature, sudden and dramatic change can cause devastating damage to Earth support systems. Change that occurs faster than such systems can adjust can cause irreversible damage to them. The purpose of this book is to serve as an overview of the science and technology of sustainability—green science and green technology. This is an introduction to green science and technology and their relationship to sustainability.

 Table 1 illustrates the evolution leading from early attempts to control pollution to the current emphasis upon sustainability. Until approximately 1980, pollution control was almost exclusively driven by regulations. Pollutants were produced, but efforts were concentrated

Table 1 Evolution from regulation-driven pollution control to current systems emphasizing sustainable development

Current	**Sustainable Development** Individual and corporate responsibility Economic environmental social resource
1990s	**Design for Environment** Proactive and beyond compliance Extended product responsibility Life cycle analysis eco-efficiency
1980s	**Pollution Prevention** Reduce amounts of pollutants produced Reduce amounts of materials used by recycle and reuse
Before 1980s	**Regulation-Driven Pollution Control** Reactive with reliance on abatement Little consideration of resource consumption End-of-pipe pollution control

on so-called end-of-pipe measures to prevent their release to water, air, or land. As it became more difficult to meet increasingly stringent regulations, it was realized that a better approach was pollution prevention, reducing the amounts of pollutants and wastes at the source and employing recycle and reuse to lower levels of release while using less materials. Pollution prevention led to design for environment that went beyond simple compliance and was proactive in reducing pollutants, waste, and material consumption. Design for environment recognized that responsibility for products extended beyond the point of sale and made use of life-cycle analysis and eco-efficiency in reducing adverse environmental and resource impacts. Since the 1990s, sustainable development has come into vogue. Although the concept has taken until recently to become widely accepted as the best means of doing business, it dates back to the previously mentioned 1987 Bruntland Commission report entitled "Our Common Future" resulting from a United Nations commission chaired by the Prime Minister of Norway, Gro Harlem Brundtland. Sustainable development makes use of the concepts of green science and green technology. It emphasizes individual and corporate responsibility and considers economic, environmental, social, and resource impacts.

The Economics of Sustainability

Humans obtain food, shelter, health, security, mobility, and other necessities through economic activities carried out by individuals, businesses, and government entities. By their nature, all economic systems utilize resources (renewable and nonrenewable) and all tend to produce wastes. With these characteristics in mind, it is possible to define three key characteristics of a sustainable economic system operating within Earth's carrying capacity.

(1) The usage of renewable resources is not greater than the rates at which these resources are regenerated.

(2) The rate of use of nonrenewable resources do not exceed the rates at which renew-

able substitutes are developed.

(3) The rates of pollution emission or waste production do not exceed the capacity of the environment to assimilate these materials.

Although they are useful guidelines, these rules cannot be followed exactly. Certainly, it should be possible to keep usage of renewable resources at levels that are sustainable, and there are many cases of economic systems that have suffered grievously when such resources are not renewed at a sufficient rate. For example, the consumption of firewood in Haiti has greatly exceeded the rates at which the wood resource is replenished, and the population has suffered grievously as a result. With regard to the second point, it is not always possible to use substitutes for nonrenewable resources, such as essential metals in some applications, although greatly reduced levels of usage can often be achieved, and recycling can reduce consumption of some resources extracted from the Earth almost to the point of renewability. The third point above suffers from uncertainty regarding the capacity of the environment to assimilate wastes and pollutants. For example, until the early 1970s, there was no concern regarding known emissions of chlorofluorocarbons (freon gases) to the atmosphere because the quantities were small, the substances among the least toxic known, and their reaction in the lower atmosphere were negligible. Then it was found that they caused destruction of the essential protective stratospheric ozone layer and, as a result, the issue of their discharge into the atmosphere became very important.

The challenge of attaining global sustainability is enormous. The total burden on Earth's carrying capacity is a product of population times demand per person[④]. This leads to the conclusion that most of the increase in the burden on Earth's carrying capacity will come form the populations of developing countries. This fact also provides an opportunity, however, in that sustainable systems are easier to introduce into developing regions in which the infrastructure and economic systems are less developed and therefore more amenable to development along lines of greater sustainability.

Selected from "*Stanley E Manahan, Environmental Science and Technology, A Sustainable Approach to Green Science and Technology, Taylor & Franicis Group, USA, 2007*"

Words and Expressions

abound　　*v.* 充分说明

hydrosphere　　*n.* 水界圈（地球水面）

geosphere　　*n.* 陆界圈

estuary　　*n.* 江口，河口

dire　　*a.* 极端的，可怕的

debilitating　　*a./vt.* 虚弱的，使虚弱

obesity　　*n.* 过度肥胖

diabetes　　*n.* 糖尿病

malady　　*n.* 疾病

anthropocene *a.* 人类为宇宙中心的

epoch *n.* 纪元

stewards *n.* 管理员

grim *a.* 严酷的

pessimistic *a.* 悲观的

end-of-pipe 末端

stringent *a.* 严格的

compliance *n.* 服从，顺从

proactive *a.* 正面活动的

vogue *n./a.* 流行，时髦

assimilate *vt./vi.* 消化

grievously *ad.* 剧烈地，严重地

infrastructure *n.* 基础

amenable *a.* 服从的，顺从的

Notes

① If we do not change direction, we are likely to end up where we are headed. 可译为：如果不改变方向，我们可能就无路可走（到头了）。

② If we make the effort to learn its language, the Earth will speak to us and tell us what we must do to survive. 可译为：如果我们努力学习地球语言，地球就会跟我们说，告诉我们为了生存必须做什么。

③ Evidence abounds that humans are degrading the Earth life support system upon which they depend for their existence. 可译为：大量证据表明，人类为了生存正在破坏他们赖以生存地球生物支持系统。

④ The total burden on Earth's carrying capacity is a product of population times demand per person. 可译为：地球容量总负荷是人口乘以每个人需求。

Exercises

1. Put the following into Chinese.

    Anthropocene, grievously, stringent, grim, abound, obesity, amenable, renewability, nonrenewable resource, freon gases, Earth's carrying capacity

2. Put the following into English.

    时髦　末端处理　　人类为宇宙中心　过度肥胖　新纪元　实现可持续发展
    水界圈与陆界圈　　生活方式　　　　挑战　　　基础　　严格的　严酷的　悲观的

3. Answer the following questions.

    (1) What are three key characteristics of a sustainable economic system operating within Earth's carrying capacity based on the text?

    (2) How many people of the world's population live in dire poverty according to the text?

Reading Material: Newer Synthetic Methods

Waste prevention and environmental protection are major requirements in an overcrowded world of increasing demands. Synthetic chemistry continues to develop various techniques for obtaining better products with less damaging environmental impacts. The control of reactivity and selectivity is always the central subject in the development of a new methodology of organic synthesis. Novel, highly selective reagents appear every month. New reactions or modifications of old reactions have been devised to meet the ever-increasing demands of selectivity in modern synthesis. Periodic review articles and books appear in the literature on these newer reagents. The scope of this part is to focus on newer techniques (experimental) for improving the yield and reducing the duration of the reactions and also to discuss the need for a good synthetic design. In other words, newer methods of kinetic activation, which minimize the energy input by optimizing reaction conditions, will be discussed along with the need for an elegant synthetic design.

In most reactions, the reaction vessel provides three components (as shown in Fig. 1)
(1) solvent,
(2) reagent/catalyst,
(3) energy input.

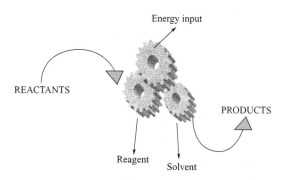

Fig. 1 Components of Chemical Reaction

Hence, efforts to green chemical reactions focus predominantly on "greening" these three components. By "greening", we mean to

(1) Use benign solvents or completely dispense with the solvent.

(2) Use alternate, more efficient and effective reagents/catalysts.

(3) Optimize the reaction conditions by using cost-effective, eco-friendly alternative processes.

The role of alternate reagents, solvents, and catalysts in greening chemical reactions is discussed in other parts.

In this part we shall see newer methods of kinetic activation of molecules in chemical reactions. Pressure and temperature are important parameters in reaction processes in chemical systems. However, it is a less well-known fact that other than thermally initiated reactions can also lead to sustainable results. The basic requirement is to capture the energy required by a reaction. The energy required for synthesis as well as that required for cooling are of interest here.

In order to minimize energy and control reactions with a view to green chemistry, attempts are being made to make the energy input in chemical systems as efficient as possible. Approaches are being taken and possibilities investigated to use until now scarcely used forms of energy, so-called non-classical energy forms, in order to optimize the duration and product yield and avoid undesired side products. Teams working in the area are also interested in the energetic aspects of the preparation of starting substances and products and the conditioning of reaction systems (e. g. , surface activation, emulsification, homogenization, degassing, etc.).

We now have six well-documented methods of activating molecules in chemical reactions, which can be grouped as follows: the classical methods, including thermal, photochemical, electrochemical, and the non-classical methods, which include sonication, mechanical, microwave.

Each of these methods has its advantages and niche areas of applications, alongside its inherent limitations. A comparative study of the techniques is given later.

What do we mean by classical and non-classical energy forms? In classical processes, energy is added to the system by heat transfer; by electromagnetic radiation in the ultraviolet (UV), visible, or infrared (IR) range; or in the form of electrical energy. On the other hand, microwave radiation, ultrasound, and the direct application of mechanical energy are among the non-classical forms.

Sonochemical Processes

Ultrasound, an efficient and virtually innocuous means of activation in synthetic chemistry, has been employed for decades with varied success. Not only can this high-energy input enhance mechanical effects in heterogeneous processes, but it is also known to induce new reactions, leading to the formation of unexpected chemical species. What makes sonochemistry unique is the remarkable phenomenon of cavitation, currently the subject of intense reaction, which has already yielded thought-provoking results.

The majority of today's practitioners accept a rationale based on "hotspot" interpretation, provided this expression is not taken literally, but rather as "a high-energy state in a small volume". One should also recall that only a small part (10^{-3}) of the acoustic energy absorbed by the system is used to produce a chemical activity. High-power, low-frequency (16-100 Hz) waves are often associated with better mechanical treatment and less importantly with chemical effects. With high frequency ultrasound, the chemistry produced displays characteristics similar to high-energy radiation (more radicals are created). One of the most striking features in sonochemistry is that there is often an optimum value for the reaction temperature. In contrast to classical chemistry, most of the time it is not necessary to go to higher temperatures to accelerate a process. Each solvent has a unique fingerprint.

Sonochemistry in heterogeneous systems is the result of a combination of chemical and mechanical effects of cavitation, and it is very difficult to ascribe sonochemistry to any single global origin, other than the overriding source of activity, namely, cavitation.

The real benefit of using ultrasound lies in its unique selectivity and reactivity enhancement. The heterogeneity of the reaction phase would be particularly significant. In fact, heterogeneous reactions are those in which ultrasound is likely to play the most important role by selective accelerations between potentially competitive pathways.

Apart from the use of ultrasound in enhancing the reactivity in organic reactions, ultrasound has varied uses in industry, such as welding, cutting, emulsification, solvent degassing, powder dispersion, cell disruption, and atomization. It was reported that the sonochemical decomposition of volatile organometallic precursors was shown to produce nanostructured materials in various forms with high catalytic activities. This has proved extremely useful in the synthesis of a wide range of nanostructured inorganic materials, including high surface area transition metals, alloys, carbides, oxides, and sulfides, as well as colloids of nanometer cluster.

Ultrasound is known to enhance the reaction rate, thus minimizing the duration of a reaction. A large number of published examples, which highlight this observation, are shown in appendix. Apart from this, it is known to induce specific reactivity, known as "sonochemical switching". Ando's research group (1984) reported that benzyl bromide, on treatment with alumina impregnated with potassium cyanide, yielded benzyl cyanide on sonication, while, without sonication, on heating the reaction mixture yielded diphenylmethanes. This work was the first experimental evidence that ultrasonic irradiation induces a particular reactivity. Further studies on sonochemical induction indicated that

1. Reactions activated by sonication are those that proceed via a radical or radical ion intermediate (electron transfer).

2. Ionic reactions (polar) mostly remain unaffected.

Use of Microwave for Synthesis

In synthetic chemistry, 1986 was an important year for the use of microwave devices. Since that year, countless syntheses initiated by microwaves have been carried out on a laboratory scale. The result is often a drastic reduction in the reaction time with comparable product yields, if microwaves are used instead of classical methods of energy input. Unwanted side reactions can often be suppressed and solvents dispensed with.

Numerous reactions, such as esterification, Diels Alder reactions, hydrolyses, or the production of inorganic pigments, have been investigated in recent years. Reactions listed in Appendix illustrate nicely the advantages of this non-classical means of energy input. Apart from the obvious advantages of the use of microwaves in chemical syntheses, microwave technologies are being tested as energy- and cost-saving alternatives. Hopes are high, for example, in the field of green extraction of pollutants from contaminated soil, or for the improvement of the breakdown of biomass waste by fermentation as part of green biorefractive.

Electro-Organic Methods

Over the past 25 to 30 years, the use of electrochemistry as a synthetic tool in organic chemistry has increased remarkably. According to Pletcher and Walsh (1993), more than 100 electro-organic synthetic processes have been piloted at levels ranging from a few tons up to 10^5 tons. Such examples include reductive dimenrization of acrylonitrile, hydrogenation of heterocycles, pinacolization, reduction of nitro aromatics, the Kolbe reaction, Simons fluorination, methoxylation, epoxidation of olefins, oxidation of aromatic hydrocarbons, etc. Many excellent reviews and publications highlight the synthetic utility of electro-organic methods. These cover a broad spectrum of applications of electrochemical methods in organic synthesis, including their use in the pharmaceutical industry. Mild reaction conditions, ease of control of solvent and counter-ions, high yields, high selectiveties, as well as the use of readily available equipment, simply designed cells, and regular organic glassware make the electrochemical synthesis very competitive to the conventional methods in organic synthesis. The use of sacrificial anodes is an effective way for the preparation of metallo-organic compounds by cathodic generation of organic anions and anodic generation of metal cations. This approach was very successful for synthesis of the organosilicon compounds. A large variety of fluorinated organosilicon compounds can be synthesized using a sacrificial Al anode and a stainless steel cathode under very mild conditions and in good yields.

Discoveries of new types of electro-organic reactions based on coupling and substitution reactions, cyclization and elimination reactions, electrochemically promoted rearrangements, recent advances in selective electrochemical fluorination, electrochemical versions of the classical synthetic reactions and successful use of these reactions in multi-step targeted synthesis allow the synthetic chemist to consider electrochemical methods as one of the powerful tools of organic synthesis.

Selected from *"Mukesh Doble, Anil Kumar Kruthiventi, Green Chemistry and Processes, Elsevier Inc. USA, 2007"*

Unit 41

Text: Green Science and Technology and Development

Green Science

Although **science** is a widely used word having somewhat different meanings in different contexts, it can generally be regarded as a body of knowledge or system of study dealing with an organized body of facts verifiable by experimentation that are consistent with a number of general laws. In its purest sense, science avoids value judgments; it involves a constant quest for truths whether they be good, such as the biochemical basis of a cure for some debilitating disease, or bad, such as the nuclear physics behind the development of nuclear bombs. However, in defining green science, it is necessary to modify somewhat the view of "pure" science. **Green science** is science that is oriented strongly toward the maintenance of environmental quality, the reduction of hazards, the minimization of consumption of nonrenewable resources and overall sustainability.

When the public thinks of environmental pollution, exposure to hazardous substances, consumption of resources such as petroleum feedstocks, and other unpleasant aspects of modern industrialized societies, chemical science (the science of matter) often comes to mind[①]. So, it is fitting that to date the most fully developed green science is *green chemistry* defined as the practice of chemistry in a manner that maximizes its benefits while eliminating or at least greatly reducing its adverse impacts[②]. Green chemistry is based upon "twelve principles of green chemistry" and, since the mid-1990s, has been the subject of a number of books, journal articles, and symposia. In addition, centers and societies of green chemistry and a green chemistry journal have been established.

Green Technology

Technology refers to the ways in which humans do and make things with materials and energy directed toward practical ends. In the modern era, technology is to a large extent the product of engineering based on scientific principles. Science deals with the discovery, explanation, and development of theories pertaining to interrelated natural phenomena of energy, matter, time, and space. Based on the fundamental knowledge of science, engineering provides the plans and means to achieve specific practical objectives. Technology uses these plans to carry out the desired objectives. Technology obviously has enormous importance in determining how human activities affect Earth and its life support systems.

Technology has been very much involved in determining levels of human population on Earth, which has seen three great growth spurts since modern humans first appeared. The first of these, lasting until about 10000 years ago, was enabled by the primitive, but remarkably effective tools that early humans developed, resulting in a global human population of perhaps 2 or 3 million. For example, the bow and arrow enabled early hunters to kill potentially dangerous game for food at some (safer) distance without having to get very close to an animal and stab it with a spear or club it into submission. Then, roughly 10000 years ago, humans who had existed as hunter/gatherers learned to cultivate plants and raise domesticated animals, an effort that was aided by the further development of tools for cultivation and food production. This development ensured a relatively dependable food supply in smaller areas. As a result, humans were able to gather food from relatively small agricultural fields rather than having to scout large expanses of forest or grasslands for game to kill or berries to gather. This development had the side effect of allowing humans to remain in one place in settlements and gave them more free time in which humans freed from the necessity of having to constantly seek food from their natural surroundings could apply their ingenuity in areas such as developing more sophisticated tools. The agricultural revolution allowed a second large increase in numbers of humans and enabled a human population of around 100 million, 1000 years ago. Then came the industrial revolution, the most prominent characteristic of which was the ability to harness energy other than that provided by human labor and animal power[3]. Wind and water power enabled mills and factories to use energy in production of goods. After about 1800, this power potential was multiplied many fold with the steam engine and later the internal combustion engine, turbines, nuclear energy, and electricity, enabling current world population of around 6 billion to grow (though not as fast as some of the more pessimistic projections from past years).

There is ample evidence that new technologies can give rise to unforeseen problems. According to the law of unintended consequences, whereas new technologies can often yield predicted benefits, they can also cause substantial unforeseen problems. For example, in the early 1900s, visionaries accurately predicted the individual freedom of movement and huge economic boost to be expected from the infant automobile industry. It is less likely that they would have predicted millions of deaths from automobile accidents, unhealthy polluted air in urban areas, urban sprawl, and depletion of petroleum resources that occurred in the following century. The tremendous educational effects of personal computers were visualized when the first such devices came on the market. Less predictable were the mind-numbing hours that students would waste playing senseless computer games. Such unintended negative consequences have been called revenge effects. Such effects occur because of the unforeseen ways in which new technologies interact with people.

Avoiding revenge effects is a major goal of **green technology** defined as technology applied in a manner that minimizes environmental impact and resource consumption and maximizes economic output relative to materials and energy input[4]. During the development phase, people who develop green technologies, now greatly aided by sophisticated computer methodologies, attempt to predict undesirable consequences of new technologies and put in place preventative measures before revenge effects have a chance to develop and cause major problems.

A key component of green technology **is industrial ecology**, which integrates the principles of science, engineering, and ecology in industrial systems through which goods and services are provided, in a way that minimizes environmental impact and optimizes utilization of resources, energy, and capital. In so doing, industrial ecology considers every aspect of the provision of goods and services from concept, through production, and to the final fate of products remaining after they have been used. It is above all a sustainable means of providing goods and services[5]. It is most successful in its application when it mimics natural ecosystems, which are inherently sustainable by nature. Industrial ecology works through groups of industrial concerns, distributors, and other enterprises functioning to mutual advantage, using each others' products, recycling each others' potential waste materials, and utilizing energy as efficiently as possible. By analogy with natural ecosystems, such a system comprises an **industrial ecosystem**.

Green Development

The concept of sustainability lies at the core of the challenge of environment and development, and the way governments, business and environmental groups respond to it. Green Development provides a clear and coherent analysis of sustainable development in both theory and practice.

Green Development offers clear insights into the challenges of environmental sustainability, and social and economic development. It is unique in offering a synthesis of theoretical ideas on sustainability and in its coverage of the extensive literature on environment and development around the world. Green Development has proved its value to generations of students as an authoritative, thought-provoking and readable guide to the field of sustainable development.

Selected from "*Stanley E Manahan, Environmental Science and Technology, A Sustainable Approach to Green Science and Technology, Taylor & Franicis Group, USA, 2007*"

Words and Expressions

verifiable *n.* 可核实的
feedstocks *n.* 原料，进料
symposia *n.* 专题报告会，讨论会，专题文集
pertaining *a.* 附属于……的，与……有关的，为……所固有的
spurt *vt./vi.* 喷射出，急剧上升，生长，发芽
stab *vt./vi.* 刺（穿，伤），企图，努力，尝试
scout *vt./vi./n.* 侦察，搜索，监视，发现，排斥
harness *vt.* 控制，治理，利用，开发
unintended consequence 缺乏研究（知识，经验，外行）的结果（结论）
visionary *a./n.* 幻想的，空想，非实际
mind-numbing 失去感觉的心情（头脑），迟钝的感觉神智（头脑）
revenge effect 报复效应
mimic *a./vt./n.* 模仿的，假装的，仿造物（品）

Notes

① When the public thinks of environmental pollution, exposure to hazardous substances, consumption of resources such as petroleum feedstocks, and other unpleasant aspects of modern industrialized societies, chemical science (the science of matter) often comes to mind. 可译为：当公众思考环境污染时，如有害物质的暴露，石油原料的消耗，现代工业社会产生其他令人不愉快方面，常常会想到化学学科（物质科学）。

② So, it is fitting that to date the most fully developed green science is *green chemistry* defined as the practice of chemistry in a manner that maximizes its benefits while eliminating or at least greatly reducing its adverse impacts. 可译为：因此，迄今为止，最成熟的

绿色科学是绿色化学，它定义为以最大限度排除或削减其不利影响时获得最大利益方式的化学实践。

③ Then came the industrial revolution, the most prominent characteristic of which was the ability to harness energy other than that provided by human labor and animal power. 可译为：接着，工业革命来临，工业革命重要的特征是能够利用能量，而不是由人力或动物提供能量（该句为倒装句，因为主语太长，把动词 came 放在主语前面）。

④ Avoiding revenge effects is a major goal of **green technology** defined as technology applied in a manner that minimizes environmental impact and resource consumption and maximizes economic output relative to materials and energy input. 可译为：避免环境报复效应是绿色技术的主要目的，绿色技术定义为使得环境影响和资源消耗最小，相对于物质与能量输入而言，经济输出最大的技术。

⑤ It is above all a sustainable means of providing goods and services. 可译为：最重要的是提供物质与服务的可持续手段。

Exercises

1. Put the following into Chinese.

   symposia, harness, pertaining, feedstocks, prominent, automobile accidents, hunter/gatherers, unintended consequence, mind-numbing.

2. Put the following into English.

   专题报告会　缺乏经验的结果　原料　　幻想　　报复效应　模仿的　利用能量
   无关的自然现象　工业生态学　资源消耗　预防措施　环境影响

3. Answer the following questions.

   （1）What are the definition of Science and Technology based on the text?
   （2）What are Green Technology and Industrial Ecology according to the text?
   （3）What is the definition of Green Chemistry?

Reading Material: Green Chemistry

Of all the "green science" green chemistry is arguably the most well developed. **Green chemistry** is the practice of chemical science and manufacturing within a framework of industrial ecology in a manner that is sustainable, safe, and nonpolluting and that consumes minimum amounts of materials and energy while producing little or no waste material. Fig. 1 illustrates this definition. There are certain basic principles of green chemistry. Some publications recognize "The Twelve Principles of Green Chemistry." This section addresses the main ones of these.

As anyone who has ever spilled the contents of a food container onto the floor well knows, it is better to not make a mess than to clean it up once made. As applied to green chemistry, this basic rule means that waste prevention is much better than waste cleanup. Failure to follow this simple rule has resulted in most of the troublesome hazardous waste sites that are causing problems throughout the world today.

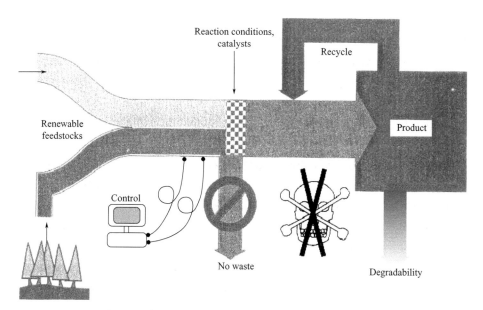

Fig. 1 Illustration of the definition of green chemistry

One of the most effective ways to prevent generation of wastes is to make sure that insofar as possible all materials involved in making a product should be incorporated into the final product. Therefore, the practice of green chemistry is largely about incorporation of all raw materials into the product, if at all possible. We would not likely favor a food recipe that generated a lot of inedible by-product. The same idea applies to chemical processes. In that respect, the concept of atom economy, expressed as the percentage of reagents that get into the final product, is a key component of green chemistry.

The pose or generation of substances that use hazards to humans and the environment should be avoided. Such substances include toxic chemicals that pose health hazards to workers. They include substances that likely to become air or water pollutants and harm the environment or organisms in the environment. Here the connection between green chemistry and environmental chemistry is especially strong.

Chemical products should be as effective as possible for their designated purpose, but with minimum toxicity. The practice of green chemistry is making substantial progress in designing chemicals and new approaches to the use of chemicals such that efficiency is retained and even enhanced while toxicity is reduced.

Chemical syntheses as well as many manufacturing operations make use of auxiliary substances that are not part of the final product. In chemical synthesis, such a substance consists of solvents in which chemical reactions are carried out. Another example consists of separating agents that enable separation of product from other materials because these kinds of materials may end up as wastes or (in the case of some toxic solvents) pose health hazards, the use of auxiliary substances should be minimized and preferably totally avoided.

Energy consumption poses economic and environmental costs in virtually all synthesis and manufacturing processes. In a broader sense, the extraction of energy, such as fossil fuels pumped from or dug out of the ground, has significant potential to damage the environment. Therefore, energy requirements should be minimized. One way in which this can be done is through the use of processes that occur near ambient conditions, rather than at elevated temperature or pressure. One successful approach to this has been the use of biological processes, which, because of the conditions under which organisms grow, must occur at moderate temperatures and in the absence of toxic substances. Such processes are discussed further later.

Raw materials extracted from Earth are depleting in that there is a finite supply that cannot be replenished after they are used. So, wherever possible, renewable raw materials should be used instead of depletive feedstocks. Biomass feedstocks are highly favored in those applications for which they work. For depleting feedstocks, recycling should be practiced to the maximum extent possible.

In the synthesis of an organic compound, it is often necessary to modify or protect groups on the organic molecule during the course of the synthesis. This often results in the generation of by-products not incorporated into the final product, such as occurs when a protecting group is bonded to a specific location on a molecule, then removed when protection of the group is no longer needed. Because these processes generate by-products that may require disposal, the use of protecting groups in synthesizing chemicals should be avoided insofar as possible.

Reagents should be as selective as possible for their specific function. In chemical language, this is sometimes expressed as a preference for selective catalytic reagents over nonselective stoichiometric reagents.

Products that must be dispersed into the environment should be designed to break down rapidly into innocuous products. One of the oldest, but still one of the best, examples of this is the modification of the surfactant in household detergents 15 or 20 years after they were introduced for widespread consumption to yield a product that is biodegradable. The poorly biodegradable surfactant initially used caused severe problems of foaming in wastewater treatment plants and contamination of water supplies. Chemical modification to yield a biodegradable substitute solved the problem.

Exacting "real-time" control of chemical processes is essential for efficient, safe operation with minimum production of wastes. This goal has been made much more attainable by modern computerized controls. However, it requires accurate knowledge of the concentrations of materials in the system measured on a continuous basis. Therefore, the successful practice of green chemistry requires real-time, in-process monitoring techniques coupled with process control.

Accidents, such as spills, explosions, and fires, are a major hazard in the chemical industry. Not only are these incidents potentially dangerous in their own right, but they also tend to spread toxic substances into the environment and increase exposure of humans and other organisms to these substances. For this reason, it is best to avoid the use or generation of substances that are likely to react violently, burn, build up excessive pressures, or otherwise cause unforeseen incidents in the manufacturing process.

The principles outlined above are developed to a greater degree in the remainder of the book. They should be kept in mind in covering later sections.

Selected from "*Stanley E Manahan, Environmental Science and Technology, A Sustainable Approach to Green Science and Technology, Taylor & Francis Group, USA, 2007*"

Unit 42

Text: Clean Technologies through Microbial Processes for Economic Benefits and Sustainability

All technological developments are aimed at improving the quality of life of a community of people. Developing countries are looking for programs reducing the risk to health and achieving sustainable, economical growth conducive to a higher per capita income. The introduction of clean technologies using microbial processes requires therefore a complete change from the presently existing industrialized economic to a more socio-economic approach. Waste management must become an integrated part of our new clean technology systems, which are often referred to as *integrated biosystems*[①] or *socio-economic biotechnology systems*.

The Sugarcane and Sugar Processing Industry as well as Clean Technology

The processing of sugar from sugarcane is practiced in many sub-tropical and tropical countries such as Australia, Brazil, Cuba, Fiji, India, Malawi, Mexico and Zimbabwe. In most cases, sugarcane farmers and sugarcane processors are different entities; the farmer works under contract with the processor to produce a certain amount of sugarcane. In some countries, however, sugarcane farmer cooperatives may have their own sugarcane-processing facilities. The sugarcane industry, from the farmer to the processor, has suffered—and still is suffering—from the instability of income due to fluctuations on the overseas market demand and prices as well as weather fluctuations affecting the yield of sugarcane. The instability of the income causes in many cases a reduction in waste management. Pollution is therefore a major concern because of the enormous variation in harvesting and sugar-processing procedures and their efficiency. There is no doubt that the introduction of microbial process systems into the sugarcane industry would not only offer a more stable income through diversifying product formation but also would help to eliminate pollution and transform the sugarcane industry into a clean-technology industry.

1. Possibilities for Cleaner Technologies through Introduction of Microbial Processes

In order to overcome the income fluctuations and pollution problems, the sugarcane industry must become more flexible, allowing other products to be produced for the local market to reduce the overseas market dependence, to compensate for the losses and return to the farmer a more stable and predictable income. A socio-economic cooperative system, combining farming, milling, processing and waste reuse could solve these problems. Microbial process industries combine very well with chemical and agricultural engineering systems.

2. Microbial Processes on the Farm

On the farm itself, tops and trash[2] from the sugarcane harvest should be collected for composting. In many areas of Australia, for example, it is prohibited to burn garden trash. Such a law should also be extended to the farms. Composting is an excellent way to produce biofertilizer with a relatively high mineral salt content. As will be outlined below, this compost can be converted into rich humus using supplementation with animal waste or anaerobic digester sludge and earthworm cultivation. The rich humus compost can be sold to the local community or spread back onto the fields. This microbial process not only would eliminate air pollution but would also reduce the costs of expensive chemical fertilizer used by the farmer, which only harm the soil microbial mineralization process.

3. Microbial Processes Attached to the Sugarmill Operation

The energy requirement for the mill should be supplemented with biogas, which has a much higher heat value compared to the present use of bagasse. This would free the bagasse, which is an excellent resource for mushroom production. Lignocellulosic fibres supplemented with solids from the anaerobic digester would give an ideal carbon: nitrogen ratio. Mushrooms are the only microorganisms capable of excreting the enzymes needed to separate the lignin economically from cellulose and hemicellulose. The residue from mushroom production will enrich the composting of the tops and trash or can be used as biofertilizer.

Some countries have already started with the diversification of their sugarcane industry, allowing better health standards in the community. For example, the National Alcohol Program in Brazil helped the national economy of Brazil significantly through the production of ethanol from sugarcane juice and molasses. The ethanol produced by the microbial yeast process was used to substitute petrol (gasoline) in the car fleet, reducing the need for importation of oil. It is possibly true to say that Brazil was one of the first, or the first country to demonstrate that the economy of a country can substantially benefit by a reduction of imports. Most of our industrialized economic thinking is orientated solely towards the export of commodities. It was very unfortunate that the program had to be halted due to lack of microbial waste management. Other countries, such as Malawi and, Zimbabwe, followed the way of the Brazilians, but with better results regarding the distillery waste In Malawi (and in India) biomethanation experiments with distillery waste have shown excellent results. In India, mesophilic biomethanation[3] of distillery spentwash[4] was very successful in a diphasic anaerobic process. The first acid phase with an organic loading of 30 kg COD/(m^3 · day) was followed by a methane phase operation enabling a 65% reduction in COD and 0.3 m^3

biogas produced per kg COD using an organic loading rate of 3.25 kg COD/(m^3 · day). With a methane phase at 5 kg COD/(m^3 · day) and a mean retention time of 27 days, a 70% reduction in COD was achieved with the same mount of biogas formed.

As anaerobic digestion does not reduce the chemical oxygen demand (COD) to the required minimum level for discharge into the environment, the effluent should be guided into a high rate oxidation pond⑤ as a polishing step. In this case the production of the protein-rich microalgae spirulina or other algae may be a key factor for the economic viability of the whole system. This would be even more the case if fish breeding or any aquaculture ponds could be added to the system. Aquaculture could, of course, be of various types.

If the microbial yeast technology for ethanol production is replaced by a bacterial Zymomonas technology, cleaner ethanol production can significantly reduce biochemical oxygen demand (BOD) and COD in the effluent. A lower cell biomass, negligible by-product formation and faster time of fermentation would further contribute to economic efficiency. In addition, special strains of Zymomonas are capable of producing not only fructose and ethanol but also other sugar products. The important factor in this consideration for flexibility is that low-, medium- and high-yielding sugarcane can be used for fermentation, composting and mushroom production.

Microbial processes can help in introducing a clean technology if appropriately introduced and maintained. They could convert the sugarcane industry into a clean industry with much greater viability and sustainability. It is the flexibility and diversification of product formation, together with the mixing of animal and human waste with stillage or other residues during anaerobic digestion or composting and mushroom production which secure not only the ecological environment but also the sustainability and lifting of living standards. This introduction of clean technologies may also attract joint venture capital investment⑥. However, it requires additional infrastructure training, which could be provided with the help of the global network of Microbiological Resource Centers.

Selected from "*Kojima Hiroyuki, Lee Yuan Kun, Photosynthetic microorganisms in Environmental Biotechnology Published in UC Berkeley, USA, 2002*"

Words and Expressions

socio-economic biotechnology systems　社会经济生物工艺学系统
sub-tropical　*a.* 亚热带的
biofertilizer　*n.* 生物化肥
earthworm　*n.* 蚯蚓
biogas　*n.* 生物气，沼气
lignocellulosic　*a.* 木素纤维的
lignin　*n.* [生化] 木质素
hemicellulose　*n.* [化] 半纤维素
molasses　*n.* [地质] 磨砾层（相），软砂岩沉积，磨拉石

export of commodity　商品出口
mesophilic biomethanation　嗜热生物甲烷化过程
protein-rich　*a.* 高蛋白的
aquaculture　*n.* 水产业
yeast　*n.* 酵母，发酵粉
zymomonas　单酵母菌
fermentation　*n.* 发酵
stillage　*n.* 滑板输送器架

Notes

① integrated biosystems　集成型生物系统
② tops and trash　枯干及废叶
③ mesophilic biomethanation　嗜热生物甲烷化过程
④ distillery spentwash　酿酒厂的废浸泡液
⑤ oxidation pond　氧化塘
⑥ This introduction of clean technologies may also attract joint venture capital investment.
可译为：引入清洁生产还可以吸引合资企业来投资。

Exercises

1. Put the following into Chinese.

(1) The sugarcane industry, from the farmer to the processor, has suffered-and still is suffering-from the instability of income due to fluctuations on the overseas market demand and prices as well as weather fluctuations affecting the yield of sugarcane.

(2) this compost can be converted into rich humus using supplementation with animal waste or anaerobic digester sludge and earthworm cultivation.

(3) The first acid phase with an organic loading of 30 kg COD/(m^3 · day) was followed by a methane phase operation enabling a 65% reduction in COD and 0.3 m^3 biogas produced per kg COD using an organic loading rate of 3.25 kg COD/(m^3 · day).

(4) It is the flexibility and diversification of product formation, together with the mixing of animal and human waste with stillage or other residues during anaerobic digestion or composting and mushroom production which secure not only the ecological environment but also the sustainability and lifting of living standards.

2. Put the following into English.

清洁技术　　微生物处理系统　　微生物矿化过程　　厌氧消化器　　微生物发酵工艺
生化需氧量　　细胞生物质　　　产品多样化　　　　蘑菇生产　　　海外市场需求

3. Answer the following questions.

　　(1) Why is the sugarcane industry suffering from the instability income?
　　(2) Why is composting an excellent way to produce biofertilizer?
　　(3) How to reduce the chemical oxygen demand of the effluent from anaerobic digestion?

Reading Material: Green Development: Reformism or Radicalism?

There is no magic formula for sustainable development. Despite the enthusiastic rhetoric, the technical guidelines and the celebrated greening of development agencies, corporations and governments, there is no easy reformist solution to the dilemmas and tragedies of poverty and environmental degradation, whether at the local or global scale. There is no "magic bullet" to defeat these threats to human well-being. Behind the slogans about environment and development lies the hard process of development itself, wherein choices "are indeed cruel". One early message from twenty-first century must be that the state of future conditions cannot be assured, even for the wealthy.

Development ought to be what human communities do to themselves. In practice, it is usually what is done to them by others, whether governments or their bankers or "expert" agents, in the name of modernity, national integration, economic growth or a thousand other slogans. Fundamentally, it is this reality of development—imposed, centralizing and often unwelcome—that the greening of development challenges. It throws attention back on the ethical questions that underlie the idea of development itself. It recognizes that societies are "developing" whether or not they are the targets of some specific government "development" scheme. In practice, in the developing world (as elsewhere) ideas, culture and the nature of society are in flux. Farming practice, production system, economy, are all sucked into the whirlpool of the world economy to some extent, moving in response to the pull of capital. There is no "real development" to be reached for that escapes this pull, and sustainable development is no magic bridge by which it can be attained.

Part of the limitation of the sustainable development thinking and the reformist technical approaches discussed in this book is their failure to address the politics of environment and development. Without a theory of how the world economy works, and without theories about the relations between people, capital and state power, sustainable development thinking—and most conservation action—is profoundly limited. In practice, plans for both development planning and environmental conservation tend to be formed by technocratic elites and imposed, although both kinds of planners seek to involve (and co-opt) local interests and both believe they are operating in the interests of some notional wider constituency. Development initiatives and the context of air-giving and project formulation have to be understood in terms of the way the world economy functions. Pollution and environmental degradation reflect economic and political structures, and have to be understood in terms of their relations to the urban, industrial cores of the world economy.

"Development" is not necessarily good; it depends on who you are, it depends on how the structures in society expose you to its hazards or open to you its fruits. It depends on how you value the changes created around you by others, and whether your own voice can gain purchase on the behemoths of state planning and business profit-seeking. Development planning involves choices, and tough decisions. Very often in the past those decisions have been taken by "expert" trained to see the world through clever but reductionist lenses, and insulated by wealth, culture and place of residence from the consequences of their decisions. However, even where planning is brought down to earth, dragged out of the tangle of government bureaucracy and politics, extracted from the spread-sheets of experts and freed from the stranglehold of consultancy contracts, the hard decisions do not go away. Sometimes improved development planning is sufficient to move towards sustainability, and "win-win" solutions are possible. At others the hard choices inherent in development still have to be made, and, when they are made, the decision comes down against the poor, the marginal, the uneducated and the powerless.

We know the limitations and failures of development, of course. They have been key elements in the litany of sustainable development since the Stockholm Conference in 1972. Indeed, sustainable development has been one of the ideas through which we have sought to recapture a sense of moral trajectory, and a means of measuring our success in driving economies and societies forwards. Since the 1980s, more and more have been added to the concept, until it groans under the weight of ideas not only about the environment, but also about equity, democracy, openness and freedom. As the economic system has become increasingly globalized, with power leaking from nation states towards transnational corporations linked in a highly interconnected global order, the attraction of the moral agenda apparently offered by sustainable development has grown. However, "sustainable development" offers no escape from the dilemmas of development. The huge achievement of debate about sustainability has been that it has expanded the horizons of development thinking to embrace the environment. Yet, it offers no resolution of the moral ambiguities inherent in development. It offers no route around development's hard choice.

The green challenge in development is not therefore simply about reforming environmental policy; it also issues a challenge to the very structures and assumptions of development. It is, first and foremost, about poverty and human need, about sustainable livelihood security. It is about the state of the environment, and the rights of people to enjoy its benefit. Debates about the mechanisms and dynamics of development have tended to obscure its ethical basis, but the concept of sustainable development is inherently and inevitably ethical. There are strong moral as well as practical reasons for putting poor people first in development planning. Goulet (1971) suggests that the "stock of underdevelopment" can be overcome only by creating "conditions favorable to reciprocity", in which "stronger patterns…offset the structural vulnerability of weaker interlocutors by being themselves rendered politically, economically and culturally vulnerable". The feasibility of such a vision as a political project can be debated, but the extent of its challenge to reformist thinking with environmental aspects of development policy is clear.

Green development focuses on the rights of the individual to choose and control his or her own course for change, rather than having it imposed. The green agenda is therefore necessarily radical, but it is also open-ended, flexible, and diverse. Green development is almost a contradiction in terms, not something for which blueprints can be drawn, not something easily absorbed into structures of financial planning, or readily co-opted by the state. It shares the very real tensions between techno-centric and eco-centric environmentalism. It requires the state of nature and the state of society to be considered together. It demands an interdisciplinary approach to analysis, training and policy. Green development is something that very often emerges in spite of, rather than as a direct result of, the actions of development bureaucracies. Green development programs must start from the needs, understanding and aspirations of individual people, and must work to build and enhance their capacity to help themselves. As Robert Chambers comments: "The poor are not the problem, they are the solution".

Green development is not about the way the environment is managed, but about who has the power to decide how it is managed. Its focus must be the capacity of the poor to exist on their own terms. At its heart, therefore, greening development involves not just a pursuit of new forms of economic accounting or ecological guidelines or new planning structures, but an attempt to redirect environmental and developmental change so as to maintain or enhance people's capacity to sustain their livelihoods and to direct their own engagements with nature. Escobar (2004) calls for "dissenting imaginations" that can think beyond modernity and the regimes of the globalized economy and the exploitation of marginalized people and nature.

"Sustainable development" is a way of talking about the future shape of the world. To conceive of the future in these terms marks the beginning of a process of political reflection and action, not the end. To call for sustainable development is not to set out a blueprint for the future but to issue a statement of intent and a challenge to action.

Selected from "*W. M. Adams, Green Development, Environment and sustainability in developing world, third edition Published by Routledge 2009*".

GLOSSARY

a blast of cyanide　氰化物爆炸
abiotic　*a*. 无生命的，无生物的
abound　*v*. 充分说明
abrade　*v*. 磨损，擦伤
access to　使用或接近的权力，机会
acclimate　*v*. 服水土，适应环境
acclimation　*n*. 服水土，顺应，适应环境
accommodate　*v*. 使适应，调节
Acropolis in Athens　希腊雅典卫城
activated sludge　活性污泥
adhesives　*n*. 黏合剂
adsorbate　*n*. 吸收质，被吸附物
aeration　*n*. 曝气
aerobic　*a*. 需氧的、有氧的
aesthetic　*a*. 美学的，审美的，有审美感的
aesthetics　*n*. 美学，美丽好坏
agglomerate　*v*. 使聚集，成团
agrarian　*a*. 土地的，农民的，农业的
aide　*n*. 助手，副官，侍从武官
airborne　*a*. 空气中的，空降的，空运的
airborne tritium　空气中的氚
air-quality monitoring　空气质量监测
aldehyde　*n*. 乙醛
aldehyde　*n*. 醛，乙醛
algae　*n*. alga 的复数，海藻，藻类
algal　*a*. 海藻的
alkane　*n*. 链烷，烷烃
alkene　*n*. 烯烃，链烯
alkysulfate　*n*. 烷基硫酸盐
allergy　*n*. 过敏症，变态反应
all that therein　里面的一切都是
alum　*n*. 铝
aluminum sulfate　硫酸铝
alveoli　*n*. 齿槽炎，肺泡炎
ambient　*a*. 周围的，包围着
amenable　*a*. 服从的，顺从的
amphibian　*n*. 两栖动物
amend　*v*. 修正，改进，改正
anaerobic　*a*. [微] 没有空气而能生活的，厌氧性的
analog-to-digital converter　模拟数字转换器
anecdotal　*a*. 趣闻的，轶事的

anion　*n*. 阴离子
annoying　*a*. 烦人的
anthropocene　*a*. 人类为宇宙中心的
anthropogenically　*ad*. 人为地
anticipate　*v*. 预料，期望
apocalypse　*n*. 启示，启示录
aquaculture　*n*. 水产业
aquatic　*a*. 水生的
aquatic life　水生物
aqueduct　*n*. 渠，水管，
arable　*a*./*n*. 可耕的，耕地
arbitrary　*a*. 独裁的，专制的，专横的，任意的
arctic regions　北极区，寒冷区
ardent　*a*. 热心的，热情洋溢的，激烈的，燃烧般的
arguably　*ad*. 可论证地
arid　*a*. 干旱，缺水
aromatic　*n*. 芳香烃
arsenic　*n*. 砷
arsenite　*n*. 亚砷酸盐
asbestos　*n*. 石棉
aspartame　*n*. 天门冬酰苯丙氨酸甲酯
asphaltic residue　沥青残留物
asphyxiate　*v*. 使（人）窒息，闷死
assimilate　*v*. 消化
asthma　*n*. 哮喘
at the tap　自来水，饮水
attributes　*n*. 特质，属性
augment　*v*. 增大，加大
austere　*a*. 苛刻的，恶劣的
authenticity　*n*. 可靠性
autotrophic　*a*. 自造营养物质的，自给营养的
backdraft = backdraught　*n*. 倒转，回程
backrubber　摩擦施药器
bacteria　*n*. 细菌（复数）；bacterium（单数）
banket　*n*. 护坡堤，填土
bar rack　格栅池
batch　*a*. 一次的分量的，一批的
batch-fed　分批投料
bedrock　*n*. 基岩
beguilingly　*ad*. 欺骗地
benthic　*a*. 底栖生物的，水底植物的

beryllium　*n*. 铍
bicarbonate　*n*. 重碳酸盐
bioassay　*n*. 生物鉴定，活体鉴定
biochemical　*a*. 生物化学的
biodegradation　*n*. 生物降解
biodiversity　*n*. 生物多样性
biofertilizer　*n*. 生物化肥
biogas　*n*. 生物气，沼气
biogenic　*a*. 生物的，生命所需的
biogeophysical　*a*. 生物地球物理的
biomass　*n*.（单位面积或体积内）生物质
biorecalcitrant　*a*. 顽强的，反抗的，对抗的
bioremediation　*n*. 生物治理
bioscrubber　*n*. 生化洗涤器
biosphere　*n*. 生物圈
biotic　*a*. 生命的，生物的
biphenyl　*n*. 联（二）苯，联苯基
bloodstream　*n*. 血流
Boltzmann equilibrium　玻耳兹曼平衡
bridge the gap　填补差距
brownish　*a*. 呈褐色的
buffer　*n*. 缓冲物（液）
bulk　*n*. 大小，尺寸
bulldozer　*n*. 推土机
bung　*v*./*n*. 堵；桶等的塞子，桶孔
burrowing mammal　穴居哺乳动物
butane　*n*. 丁烷
butter flavor　牛油香精
cacti　*n*. 仙人掌
cadmium　*n*. 镉
cafeteria　*n*. 自助食堂
calcium and magnesium　钙和镁盐
canderel　*n*. 阿斯巴甜
carbon dioxide　二氧化碳
carbon monoxide　一氧化碳
carbonate　*n*. 碳酸盐
carboxyhemoglobin　*n*. 碳氧血红蛋白
carboxylics　*n*. 羧酸
carcinogen　*n*. 致癌物，致癌因素
care giver　关怀者，照顾者，护理者
carnivore　*n*. 食肉动物
catastrophic　*adj*. 大事故的，毁灭的，灾祸的
catch-basin　雨水井，沉泥井
categorize　*vt*. 把……分类
cedar　*n*. 红松杉
centipede　*n*. 蜈蚣

chador　*n*. 黑布
centrifuge　*n*. 离心分离机
characterization　*n*. 表征，性能描写
characterize　*vt*. 表征
cheluviation　*n*. 螯合淋溶作用
chelating agent　螯合剂
china clay　瓷土
chloramine　*n*. 氯胺
chlorella　*n*. 小球藻
chlorides and sulphates　氯化物和硫化物
chlorine　*n*. 氯气
chlorine dioxide　二氧化氯（面粉漂白剂，又称Dyox）
chlorine dioxide　二氧化氯
chlorofluorocarbon（Freon）　*n*. 氯氟烃（氟里昂）
cholera　*n*. 霍乱
churn　*n*. 搅乳器 *v*. 搅拌，搅动
cinder　*n*. 煤渣，焦渣
citric acid　柠檬酸
clams　*n*. 蚌
clarifier　*n*. 澄清器，澄清池
clay-baking　烧瓷，烤瓷
climatology　*n*. 气候学，风土学
clinker　*n*.（煤在火炉、熔炉等中燃烧后所留下的）渣滓，熔渣，熔块
closed-loop　闭合回路，闭环
cluster　*v*./*n*. 集结，成团
coagulation　*n*. 凝固，絮凝
coastal　*a*. 近海的，海岸的
cob　*n*. 玉米棒子
coining the phrase　解释词组，编造词组
colloid　*n*. 胶体
colloidal　*a*. 胶体的，乳化的
combustible　*a*. 容易着火燃烧的
commandeer　*v*. 征用，强占
comminutor　*n*. 粉碎机
communities　*n*. 社区
complex symptomology　综合征
compliance　*n*. 符合，顺从，一致
confidence interval　可靠的区间，可靠的间隔
coniferous　*a*. 针叶树的，松柏类植物的
conservation　*n*. 保护，保存
conservative　*a*. 守恒的
consolidation　*n*. 加强，协同，合并，凝固
coordinate　*a*/*vt*. 同等的，并列的 调整，整理
corer　*n*.（苹果、梨等）去心器，挖核器
cornerstones　*n*. 基石

cottager　n.住在农舍（别墅）者，佃农
crayfish　n.龙虾
creosote bush　杂酚油灌木
criteria　n.判据
criterion　（pl. criteria）n.标准，规范
crockery　n.陶（瓦）器、瓦罐
cropland　n.农田，植作物之农地
crucial　a.至关紧要的
currency　n.货币，传媒
cyanide　n.氰
cyanide fishing　氰化物捕集
cyanide　n.氰化物
cycloalkene　n.环烯烃
data acquisition system　数据获取系统
date back to　追溯到
debilitating　a./vt.虚弱的，使虚弱
debris　n.有机残渣，腐质，残骸
debris　n.碎片，残骸
decibel　n.分贝
deciduous　a.落叶性的
decontamination　n.净化、消除……的污染
decoupled　a.去偶的，分离的
defacing　n.磨损，损伤外观
deforestation　n.砍伐树木，除去森林
degrade　v.降解
degreasers　n.去（油）污剂，脱脂剂
demolition　n.拆除，推翻
derelict　a.被弃的，被遗弃的
desalination　n.（海水）脱盐
detoxification　n.消毒，去毒，戒毒
detritivore　n.食碎屑者，食腐质者，食腐动物
desertification　n.（土壤）荒漠化，沙漠化
detonation　n.爆炸，起爆
detritus　n.碎岩，泥河
diacetyl　n.二乙酰
dilemma　n.进退两难，困境，二难推理
diplomat　n.外交官，有手腕者
diabetes　n.糖尿病
dire　a.极端的，可怕的
discernible　a.鉴别的，识别的
discharge　v./n.排出，排出物
discharge A from B　从B上卸A
disintegrated rock　崩解性岩石
discrete　a.分散的，独立的
disseminate　vt.传播，宣传，扩散
disequilibrium　n.不均衡，不安定

dispose of　处理，处置
disruptive　a.使破裂的，分裂性的
divalent　n.二价的
diverse　a.种类不同的
dividend　n.红利，额外津贴
dizziness　n.头晕，混乱
d-limonene　n.d-柠檬油精
domestic　a.生活的，家庭的
dose-response　用量（剂量）响应
downstream　ad./a.下游地，下游的
dump　v.n.倾销，倾倒
dragout　n.废酸洗液
dwindle　v.缩小，减小，变坏
Dutch elm　荷兰榆
dysentery　n.痢疾
earthworm　n.蚯蚓
EDTA　乙二胺四乙酸
effusions　n.流出物，出口流体
elaborate　a.精巧的，精细的
electorate　n.选民，选区，有选举权者
electroplating facility　n.电镀设备
emission　n.散发，排放物
empirically　ad.以经验为根据地，经验地
enact　v.制定法律，颁布，扮演
encompass　v.围绕，包围，包含，包括
endangered species　濒临危险物种
end-of-pipe　末端
endure　v.忍受，持久
entrap　v.收集，诱捕
environmental disturbance　环境破坏
enzyme　n.酶
epidemic　n./a.流行（的），传染（的）
episode　n.事件，插曲
episode　n.一个事件，插曲
epoch　n.纪元
equalization　n.平衡，平衡化，同等化
equalization basin　n.平衡洗涤槽
esthetic depreciation　感觉下降
estuary　n.河口，江口
ethanol　n.乙醇
eutrophic　a.发育正常的，营养良好的
evacuation　n.消除，除清，撤离
everglades　n.湿地，沼泽地
evergreen forests　常青林
exacerbate　v.激怒，使恶化
ex ante　事前

excavate　$v.$ 挖掘，开凿，挖出，挖空
excreta　$n.$ 排泄物，粪，便
excreted material　排泄物
expeditiously　$ad.$ 迅速地
explicitly　$ad.$ 明晰地，清楚地
export of commodity　商品出口
expose the considerable gaps　揭示大的差别
exotic　$n.$ 恶习（性）
extrude　$v.$ 挤压出，挤压成，突出，伸出，逐出
fabrication　$n.$ 生产，加工
facility　$n.$ 环境，设备
facultative　$a.$ 特许的
faintest　$a.$ 微弱的，轻微的
feast　$n.$ 宴席；$v.$ 享受，使……愉快
feedlot　$n.$ 牧场
feedstock　$n.$（送入机器或加工厂的）原料
fermentation　$n.$ 发酵，激动
ferocious rate　极大速率，惊人速率
ferrolysis　$n.$ 铁解作用
ferrocyanide　$n.$ 氰亚铁酸盐，亚铁氰化物
ferrous　$a.$ 含有铁的
fibrous　$a.$ 含纤维的，纤维性的
filamentous　$a.$ 细丝状的，纤维所成的，如丝的
filter　$v.$ 过滤
filtration　$n.$ 过滤
firecrackers　$n.$ 爆竹
fishery　$n.$ 渔场，渔业
fixture　$n.$ 装置
flammable　$a.$ 易燃的
flash flooding　暴洪，猝发洪水
flocculant　$n.$ 凝聚剂
flocculation　$n.$ 絮凝（沉淀法）
flowering seasons　开花季节，花粉季节
flowrate　$n.$ 流速
fluffy　$a.$ 松散的，易碎的
fluorescence　$n.$ 荧光发射
fluorine　$n.$ 氟
food web　食物网
foothill　$n.$ 山麓小丘，丘陵地带
formaldehyde　$n.$ 甲醛
foundry　$n.$ 铸工厂；玻璃（制造）厂
fragment　$n.$ 碎片
framework　$n.$ 结构，框架，构架
full-scale　工业规模
fungi　$n.$（fungus的复数形式）真菌类
fungicides　$n.$ 杀真菌剂

garbage　$n./v.$（丢弃或喂猪等之）剩饭残羹，垃圾
gene transfer　基因转移
genetic　$a.$ 遗传的，基因的
geogenic　$a.$ 断裂地貌的，破裂带地貌的
geosphere　$n.$ 陆界圈
geostatistics　$n.$ 地球统计学
global change　全球（地球）变化
globule　$n.$ 小球，小珠，滴
glutaric acid　戊二酸
grab sample　攫取样品
granite　$n.$ 花岗岩
grapefruit　$n.$ 朱栾，葡萄柚（一种植物）
greenhouse gas　$n.$ 导致温室效应的气体，如二氧化碳、甲烷等
grievously　$ad.$ 剧烈地，严重地
grim　$a.$ 严酷的
grit　$n.$ 粗砂
ground-state　基态
guideline　$n.$ 准则，指导路线
habitats　$n.$ 动植物
halide　$n.$ 卤化物
halogen　$n.$ 卤素如 Br、I、F、Cl
halogenated　$a.$ 卤化的，卤代的
harmony　$n.$ 协调，一致
harness　$vt.$ 控制，治理，利用，开发
hatchery pond　鱼塘
the haves and the have nots　富人与穷人
havoc　$n.$ 浩劫，大破坏
hazards　$n.$ 有害物
heap　$v.$ 堆积
heat-trapping　$a.$ 捕热的
hemicellulose　$n.$［化］半纤维素
hemlock　$n.$ 铁杉
hepatitis　$n.$ 肝炎
herbicide　$n.$ 除草剂，阻碍植物生长的化学剂
herbivore　$n.$ 食草动物（植物）
heterogeneity　$n.$ 不均匀性，多相性，异种，异质
heterogeneous　$a.$ 非均相的，多相的
heterotrophic　$a.$ 非自养的，异养的
hexachlorobenzene　$n.$ 六氯苯
hexavalent　$a.$ 六价的
hierarchy　$n.$ 层次，层级
highly turbid water　高度（非常）浑浊的水
Hispanic　$a.$ 西班牙的
homeostasis　$n.$ 动态平衡
human excrement　人类排泄物

human welfare   人类的福利
humus   n. 腐殖质
hydraulic   a. 水力的，水压的
hydrogen peroxide   过氧化氢
hydrolysis   n. 水解
hydrometallurgical   a. 湿法冶金学［术］的
hydroperoxyl   n. 氢过氧化
hydrophilic   a. 亲水的，吸水的
hydrosphere   n. 水界圈（地球水面）
hydrostatic   a. 静水力学的，流体静力学的
hydroxyl radical   氢氧根自由基
hygroscopic   a. 吸水的，吸湿的
hypochlorite   n. 次氯酸盐
hypothesis   n. 假想，假设
illustrative   a. 用作说明的，解说性的
impact analysis   作用分析
implementation   n. 补充
igneous rock   火山岩，岩浆岩
in contrast to   与……相反
in question   上述的，所讨论的
inadequate   a. 不充足的，不适当的
incinerator   n. 焚化炉
inclusive   a. 包括的、包含的，包括许多或一切的
incubate   v. 孵卵
incubation   n. 孵卵，抱蛋，［医学］潜伏期
indicator   n. 指示物
indiscriminate   a. 不分皂白的，不加选择的
inedible   a. 不适于食用的，不可食的
inert   a. 无活动的，惰性的，迟钝的
inevitable   a. 不可避免的
influent   a./v. 流入的，支流
infrared   n./a. 红外线，红外线的
infrastructure   n. 基础
inorganic   a. 无机的，人造的
insofar as   在……范围，到……程度
institution   n. 公共机构设施，机关
institutional   a. 惯例的，制度的，慈善机构的
instrument-guided   导航
insult   vt./n. 损害，刺激
intact   a. 完整无缺的
integral   a. 完整的，组成的
integrity   n. 完整，完整性，完全
intestine   a./n. 内部的，肠
intriguing   a. 引人兴趣的
invariably   ad. 不变地，总是
inventory   n. 目录，报表

invertebrate   n. 无脊椎动物
ion exchange   n. 离子交换
isoalkane   n. 异烷烃
isotope   n. 同位素
It is anticipated that   可以预料
jackhammers   n. 汽锤
jeopardize   v. 危及，使受危害
jurisdiction   n. 权限
kinetics   n. 动力学
lactic acid   乳酸
Lake Erie   伊利湖
landfill   n. 垃圾掩埋场
landfilling   n. 土地掩埋
lax   a. 松懈的，不严格的，松弛的
leftovers   剩余物
legislation   n. 立法，法律的制定（或通过）
legislative   a. 立法的
let alone   更不用说
lignin   n. ［生化］木质素
lignocellulosic   a. 木素纤维的
limestone   n. 石灰石
lithosphere   n. 岩石圈，陆界
litter   n. 杂乱的废物
Lockheed   n.（公司名）洛克希德
lodge   n./v. 容纳，寄存，存放，（向有关当局）提出（声明）
log   n. 原木，圆形木材，圆木
logic   n. 逻辑学
longevity   n. 长寿，耐久性
lot   n. 一块地、一块地皮
low-lying   低（洼，标高）的、位置很低的
lymph   n. 淋巴，黏液，血清
magpie collection   胡乱收集，混杂收集
malady   n. 疾病
malformed larvae   畸形幼虫
mammals   n. 哺乳动物
manatee   n. 海牛科
manifest   v./a. 表明，显示
manure   n. 粪肥
maple   n. 枫，枫木
marble   n. 大理石
masonry   n. 砖石建筑
maze   n. 迷宫，混乱，错综复杂（事物）
membrane   n. 膜
mesh grid   细网，编织网
mesopause   n. 中间层顶

mesophilic biomethanation	嗜热生物甲烷化过程
mesosphere	n.中间层,散逸层(同温层上部最低温度区,高度在50~90m处)
metabolic rates	新陈代谢速度
metabolism	n.新陈代谢
methanogen	n.[微]产甲烷生物
methodology	n.方法论
methyl chloroform	三氯乙烷
methyl radical	甲基自由基
metric ton	吨,即1000kg
microalgae	[复]n.[植]微藻类(指肉眼看不见的藻类)
microbe	n.微生物,细菌
microbial	a.微生物的,由细菌引起的
microbial disinfectants	微生物消毒剂
microcosms	n.微观世界
microorganism	n.微生物,微小动植物
milkweed	n.牧草
millipede	n.千足虫
mimic	a./vt./n.模仿的,假装的,仿造物(品)
mind-numbing	失去感觉的心情(头脑),迟钝的感觉神智(头脑)
miscellaneous	a.(混)杂的,杂项的,各种各样的,多方面的
mite	n.螨,蚤
mitigate	v.减缓,使缓和
moderately turbid water	中等浑浊的水
molasses	n.[地质]磨砾层(相),软砂岩沉积,磨拉石
monarch butterfly	君主蝴蝶
mortality rate	死亡率
motel	n.汽车旅馆
muffler	n.消声器
multiple observations	多重观察
multiple scales of space	多尺度空间
municipal	a.城市的,市政的
municipal sewage treatment plant	市政污水处理厂
municipality	n.市政当局,自治市
mussel	n.河蚌
nausea	n.恶心,反感
nematode	n.线虫类
nesting environment	居住环境
neutralization reaction	中和反应
nitrate	n.硝酸盐,硝酸钾
nitrogen oxide	氧化氮
nonmetallic	a.非金属的
nonnative =exotic	a.外来的,非本土的
not vice versa	反过来就不一样
NTA	亚氨基三乙酸
nutrasweet	n.天冬甜素
null hypothesis	虚假设,解消;假设无效[无价值]假说;零假说
obesity	n.过度肥胖
objective interpretations	客观解释,客观说明
obligate	v./a.使负义务,有责任的
obstacle	n.障碍物
ocean liner	海轮
offensive	a.令人不快的,讨厌的
off-set	v.补偿
omnivore	n.杂食动物
on file	存卷归档
onslaught	n.冲击
ooze	v./n.(慢慢)渗出(物),徐徐流出(物)
oratory	n.讲演术
orchard	n.果园
organic	a.有机的,机体的
organic vanadium	有机钒
Oriental	n.东方人(尤指中国人和日本人);a.东方诸国的,亚洲的,东方的,(珍珠等)最优质的
outbound	a.驶往国外的,出境的
outfall	n.出口,排出
outgoing	a.外逸的,外出的
out there	向那边,到战场,现在的情况下
overarching	a.首要的
overdevelopment	n.过度开发
overfishing	n.过度捕鱼
overflow	v.溢出,泛滥
overharvesting	n.过度砍伐
oxidation/reduction	氧化/还原
oxidizer	n.氧化剂
oxidizing agents	氧化剂
oxygenase	n.(加)氧酶
oyster	n.蚝
ozonation	n.臭氧氧化
ozone	n.臭氧
packed column	n.填料柱(塔)
padding	n.填充剂
palatable	a/ad.可口的(地),受欢迎的(地)
parasitic	a.寄生的
panacea	n.万能药,治百病的灵药
participant	n.参加者,参与者

particulates n. 颗粒物
passive sampler n. 手动取样器
pathogen n. 病菌，病原体
pathogenic a. 致病的，病原的，发病的
pathological a. 病理学的，与疾病有关的，有病的
pebble n. 小圆石，小鹅卵石
pedosphere n.（地球）表土层，土壤圈，土界
peer-reviewed 同行评议（审）的
pelagic a. 深海的，大洋的
penetrate v. 穿透，渗透，看穿，洞察；刺入
perception n. 感觉，感知
perchloroethylene n. 全氯乙烯
perennial a. 四季不断的，常年的
peroxyacylnitrate n. 过氧酰基硝酸酯
perspex n. 塑胶玻璃，透明塑胶［化］聚合的2-甲基丙烯酸甲酯
pertaining a. 附属……的，与……有关的，为……所固有的
pessimistic a. 悲观的
pesticide n. 杀虫剂，农药
phenol n. 苯酚，石炭酸
philosophical a. 哲学的，理性的，自然科学研究的
phosphate n. 磷酸盐
phosphorescence n. 磷光发射
photodissociation n. 光解
photoexcite v. 光激发
photoionization n. 光电离
physiological a. 生理学的，生理学上的
phytoplankton n. 浮游植物群落
pile driver 打桩机
pilot-scale 小规模试验，中试
pine n. 松树
pitch n. 音调
plain sedimentation 普通沉淀法
planetary adj. 行星的，漂泊不定的
plasma n. 血浆
plumbing n. 管道
pneumonia n. 肺炎
poaching n. 偷捕鱼
pollen n. 花粉
pollster n. 民意测验专家，整理民意测验结果的人
polychlorinated biphenyl 多氯代联苯
polymer n. 聚合物
polymeric bead n. 聚合物单体
polysaccharide n. 多醣，聚糖，多聚糖

polystyrene 聚苯乙烯
popcorn n. 爆米花
post－audit n. 后检查，后审查
potassium permanganate 高锰酸钾
precipitation n. 沉淀作用，沉淀物，析出
precursor n. 先驱
predatory a. 食肉（捕食）的
predefined intervals 预定的间隔
predominantly ad. 主要地
preemptive a. 有先买权的，有强制收购权的，抢先的
premise n. 前提
primary air pollutant n. 一次大气污染物
priori n. 先验，预知
prioritize v. 优先处理
proactive a. 正面活动的
proliferate v. 细胞繁殖，激增
prolonged a. 长时间的，长期的
promulgate v. 颁布，公布
propane n. 丙烷
proponent n. 建议者，提议者
proportionality n. 比例（性），相称
protein-rich a. 高蛋白的
protozoa n. 原生动物，原形动物
province of Ontario 安大略省
pulp and paper 纸浆
pulverization n. 磨碎，粉化
pulverize v. 磨碎，研末
purchase v. 购买，采购
purification n. 净化，纯化
putrefaction n. 腐烂，腐败
putrefactive a.（容易）腐败（烂、朽）的
putrescible a. 易腐烂的
pyrite n. 黄铁矿，硫铁矿（FeS_2）
pyrolysis n. 高温分解
qualitative a. 定性的
qualitative logic 定性逻辑学
quantitative a. 定量的
quantitative logic 定量逻辑学
quantum n. 量子
quarry n. 采石场
quibble n.v. 诡辩，支吾，双关语
quaternary a. 由四元素（或四基）构成的
quiescent a. 宁静的，平静的
raceway paddle wheel 套管式桨轮
radionuclide n. 放射性核素

radon *n.*氡
rag *n.*抹布，碎屑，石板瓦
random sampling 随机采样
rangland *n.*牧场
randomize *v.*使随机化，完全打乱，（使）作任意排列
rarefied *a.*稀薄的
rationalization *n.*合理，有理化
raze *v.*铲平，拆毁
reclaim *a.*开拓的
reclamation *n.*（废料的）回收，改造
refractory *a.*难控制的，难熔的
refuse *n.*弃物，垃圾，废物
remainder *n.*残余物
rendering plants 炼油厂
reoccurring question 再发生问题
replenish *vt.*添满，充足
replicas *n.*复制品
representative *a.*（有）代表性的
reproducible *a.*重复的
reptile *n.*爬行动物
residential *a.*住宅的，居住的
respiration *n.*呼吸
respired gas 呼吸气体
resurgent *a.*复活的
retrieval *n.*（可）重新获得，（可）收回
retrospective *a.*追溯的，回顾的
revenge effect 报复效应
reverse osmosis 反向渗透
rhetoric *a.*花言巧语的
rhinoceros *n.*犀牛
rhythmic *a.*有节奏的
rodenticides *n.*杀鼠剂
roosting ground 栖息地，居住地
rotary saw 转锯，带锯
rudimentary *a.*基本的，初步的，早期的
runoff *n.*排水，流放口
saccharin *n.*糖精
rustling *a.*沙沙响的，急促运动的
sagebrush *n.*灌木蒿丛，蒿属植物
salience *n.*显著，特征
salvage *v.*抢救；废物处理
sanitary *a.*公共卫生的，清洁的，清洁卫生的
scarring *n.*伤疤
scavenger *n.*清除剂，清洁工，食腐动物，拾荒者
scenario *n.*计划说明书；方案，剧情，设想

scout *v./n.*侦察，搜索，监视，发现，排斥
scraper *n.*刮刀
screening *n.*（用拦污）栅隔离
scrub *v.*使（气体）净化
secondary air pollutant 二次大气污染物
sedimentation *n.*沉淀，沉降
seepage *v.*渗透，渗溢
segregation *n.*隔离，分离
semi-arid shrublands 半干旱灌木地
semipermeable *a.*半渗透
sewage *n.*污水，废水
sewer *n.*下水道
shaft *n.*（矿）井，通道，轴
shearing *n.*切应力
shingle *n.*屋顶（墙面）板
silt *n.*淤泥，残渣，煤粉，泥沙
skewed *n/a.*偏离（的），曲解（的）
skimming *n.*浮渣
slaughter *n.*屠宰
sloppy *a.*潮湿的，溅湿的，溅污的
slurried *a.*泥浆的，浆化的
slurry *n.*泥浆，浆
smelter *n.*熔炉，冶金厂，冶炼者
socialize *v.*使社会化，使社会主义化
socio-economic biotechnology systems 社会经济生物工艺学系统
socio-political 社会政治的
sod *n.*草地
sophistication *n.*改进，复杂化
sow bug 潮虫
spatial patterns 空间模式，空间格局
spatio-and temporal-distribution 空间与时间分布
speciation *n.*物种形式
splenda *n.*三氯蔗糖
spirulina *n.*螺旋藻属
spurt *v.*喷射出，急剧上升，生长，发芽
squabble *v.*争吵，搞乱
stab *v.*刺（穿，伤），企图，努力，尝试
stack *n./v.*烟囱，排气管，堆积
stagnation *n.*停滞，迟钝；萧条
stain *v./n.*沾污，弄脏，污点
standard deviation 标准偏差
steady-state 稳态的
stewards *n.*管理员
stewardship 乘务员（服务员）的职位，工作
stillage *n.*滑板输送器架

stipulate v. 约定，规定
Stokes equation 斯托克斯方程
straightforward a. 简明的，明确的
stratopause n. 同温层顶
stratosphere n. 同温层，平流层
streamflow n. 溪流量，溪流速
stringent a. 严格的，严厉的
stunning fish 令人眩晕的鱼
stump n. 树桩，残余，烟头 v. 掘去树桩，砍成树桩，绊倒，难住，截去
sturgeon n. 鲟鱼
subdisciplines n. 分支学科
subregional scale 亚分区范围
subsonic a. 亚声速的
subtle a. 敏感的，微妙的，精细的
subtleties n. 细微的区别
sub-tropical a. 亚热带的
sulfate n. 硫酸盐
sulfonics n. 磺酸
sulfur dioxide 二氧化硫
superficial a. 肤浅的，表面的
supersonic a. 超声速的
susceptibility n. 敏感性，敏感度，易感受性
sweetener n. 甜味剂
switchgear n. 电力设施，接电装置
symbiotic a. 共生的
symphony orchestra 交响乐团
symposia n. 专题报告会，讨论会，专题文集
symptom n. 症状，征兆
synchrony n. [物] 同步（性）
syndrome n. 症状
synergistical a. 合作的，协同的
tar n. 焦油，柏油
tarry n./a. 焦油，柏油状的，煤胶物质
tartrate n. 酒石酸盐
tattoos n. 文身，刺青
temperate climates 温和气候
tenacious lichen 坚韧的青苔
terminology n. (总称) 术语，专门用语
termiticides n. 杀白蚁剂
terpine-based 可用松油烯稀释的，松油烯型
terrestrial n./a. 陆地（的），地球（的），陆生的
thermistor n. 热敏电阻器
tidal a. 潮汐的，潮水的
time-consuming 耗时的
tip n./v. 垃圾场/使倾卸、倾倒、倾翻

topography n. 地形学
torrential rainstorm 暴雨
toxicity n. 毒性
toxicological a. 毒物学的，毒理学的
toxin n. 毒素
trace levels 微量，痕量
transatlantic freighter 横过大西洋货轮
transect v. 横断，横切 n. 横断面
trash n. 垃圾，无价值的东西
tribes n. 部落，群落
trichlorotrifluoroethane n. 三氟三氯乙烷
trickling filter 滴滤池（器）
trimming n. 装饰物
tritium n. 氚，超重氢
trivalent a. 三价的
trophic a. 营养的
tropical cyclones and hurricanes 热带气旋和台风
tropopause n. 对流层顶，休止层
troposphere n. 对流层
trout n. 鳟鱼
tsunamis n. 津浪，海啸地震
tuberculosis n. 结核病，肺结核
tundra n. 冻原带，苔原，冻土带
turbidity n. 浑度，浑浊性，混乱
turbulent a. 湍流的，湍动的
typhoid n./a. 伤寒，似班伤疹伤寒
ubiquitous a. 无所不在的，普遍存在的
ultraviolet light 紫外光
underlying a. 潜在的，在下的
unduly ad. 过度地，过分地
unicellular a. 单细胞的
unintended consequence 缺乏研究（知识，经验，外行）的结果（结论）
uniqueness n. 独特性，唯一性
unlined a. 无衬里（炉衬、镶衬、衬砌）的
unpleasant odors 难闻气味
unshared a. 独享的
unsightly a. 难看的，丑的
upgrade n. 升级，上升
upstream ad. 在上游地，向上游地
urbanization n. 城市化
urine n. 尿
valences n. (化合) 价，原子价
vandalism n. 故意破坏
vascular a. 充满活力的
verifiable n. 可核实的

vermin *n.* 害兽
version *n.* 版本
via *prep.* 经由，通过
virulent *a.* 致命的，极毒的
visionary *a./n.* 幻想的，空想，非实际
vogue *n./a.* 流行（的），时髦（的）
vomiting *n.* 呕吐
vulnerability *n.* 脆弱
vulnerable *a.* 易受攻击的，易受……的攻击
waterborne *ad.* 水中的
watershed *n.* 分水岭
well-being 幸福，福利

whereby *a.* 借（由，因）此，利用它，凭（为）什么、靠什么
whilst *conj.* ＝while
with great care and caution 以极细心和谨慎的态度
World Meteorological Organization 世界气象组织
wrangle *v.* 争论，口角
wreak *vt.* 造成，报复
yeast *n.* 酵母，发酵粉
yucca *n.* 丝兰植物
zoning *n.* 分区，区域划分（商业区、住宅区、工业区等）
zymomonas *n.* 单酵母菌